Engineering Design and Optimization of Thermofluid Systems

Engineering Design and Optimization of Thermofluid Systems

DAVID S.K. TING
Turbulence and Energy Lab, University of Windsor, Ontario, Canada

This edition first published 2021

Registered Office
John Wiley & Sons, Inc., 111 River Street, Hoboken, NJ 07030, USA

Editorial Office
111 River Street, Hoboken, NJ 07030, USA

For details of our global editorial offices, customer services, and more information about Wiley products visit us at www.wiley.com.

Wiley also publishes its books in a variety of electronic formats and by print-on-demand. Some content that appears in standard print versions of this book may not be available in other formats.

Library of Congress Cataloging-in-Publication Data is Available:

ISBN 9781119701606 (hardcover)
ISBN 9781119701613 (ePDF)
ISBN 9781119701668 (ePub)

Cover Design: Wiley
Cover Image: skodonnell/ E+/Getty Images

Set in 9.5/12.5pt BKM-STIXTwoText by SPi Global, Chennai, India

SKY10024631_020221

Engineering design is the synthesis of science and art for practical applications. Engineering Design and Optimization of Thermofluid Systems is very much a subset of engineering as described by J.A.L. Waddell, "Engineering is the science and art of efficient dealing with materials and forces … it involves the most economic design and execution … assuring, when properly performed, the most advantageous combination of accuracy, safety, durability, speed, simplicity, efficiency, and economy possible for the conditions of design and service."

> *The difference between science and the arts is not that they are different sides of the same coin… or even different parts of the same continuum, but rather, they are manifestations of the same thing. The arts and sciences are avatars of human creativity.*
>
> – *Mae Jemison*

> *After a certain high level of technical skill is achieved, science and art tend to coalesce in esthetics, plasticity, and form. The greatest scientists are always artists as well.*
>
> – *Albert Einstein*

This book is dedicated to the everyday artistic engineers who unceasingly put into effect human creativity to forge a better future for the generations to come.

Contents

Preface

This book is primarily designed for senior undergraduate engineering students interested in Engineering Design and Optimization of Thermofluid Systems. It invokes basic undergraduate mathematics, thermodynamics, fluid mechanics, and heat transfer concepts. The book aims at stimulating every keen mind to appreciate design and optimization of engineering thermofluid systems.

David S-K. Ting
June 20, 2020

Acknowledgments

This book would have remained an ambitious dream, if not for the overflowing help from above and around the author. From above, timely and sufficient grace inspirited the author all the way through, from onset to finish. The earlier than anticipated completion is nothing short of a miracle. Each and every individual mentioned below played a pivotal yet unique role in this miracle.

The striving and fulfillment of this dream starts at home. The Turbulence and Energy (T&E) Laboratory (http://www.turbulenceandenergylab.org/) is the home for forging enthusiasts into experts, including many artistic engineering experts. These skillful T&E experts are recognized in the captions of their creative figures. Yang Yang and Xi Wang also assisted in overcoming a few technical hurdles that severely tangled the decrepit author. Gnanesh Nagesh deserves a special mention for fashioning the many top-notch figures. Dr. Mehdi Ebrahimi sympathetically furnished practical details associated with the compressed air energy storage project.

The eagle eyes of Dr. Jacqueline A. Stagner captured every tilde littered throughout the manuscript, from Preface to Appendix. Dr. JAS' contribution far exceeds proofreading; she also inspired Project A.7 Desert Expedition in the Appendix. It would be a huge loss if the reader misses the opportunity to time travel with her in Starship T&E.

The recessive gene inherited from mom and dad has made possible the creation of idiosyncratic humor fastidiously placed throughout the book. The author is convinced that environmental factors, i.e., his three sisters, brother, and the enchanting rainforest of Borneo, are to blame for these puns that may be above the appreciation threshold of some readers. The author is aware that he has to work on designing better jokes and optimizing the placement of these buffooneries to maximize students' learning. Another thought-to-be childish dream, the Allinterest Research Institute (https://allinterestresearchinstitute.ca/), has bridged daydreaming with reality. Thanks to many naive fantasies, some dreams do become reality.

Mother Teresa is right, "If you want to change the world, go home and love your family." The endeavor was fueled, from the beginning to the end, by love from Naomi, Yoniana, Tachelle, and Zarek Ting. As with all engineering systems, there are many constraints to overcome, so also, true love is woven with many constructive criticisms and sarcasms for the visionary to exercise his faith muscles. The quality of the book was thus substantially enhanced. Surely there is still much room for further improvement. As with optimization, this book has been optimized within the constraints of life. The future is yet filled with hope of eventual perfection, with progressive betterment to tread.

This book is accompanied by a book companion site: www.wiley.com/go/ting

1

Introduction

To develop a complete mind: Study the science of art; Study the art of science.
– Leonardo da Vinci

Chapter Objectives

- Understand what design and optimization of thermofluid systems mean.
- Differentiate engineering from science.
- Discern development, design, and analysis.
- Become familiar with the design process.
- Be aware of the existing books on thermofluid system design and/or optimization.
- Appreciate the organization and contents of the book.

Nomenclature

HVAC	heating, ventilation, and air conditioning
I_{dir}	direct radiation on a horizontal surface
KISS	keep it simple, stupid
LED	light-emitting diode
PV	photovoltaic
UWCAES	underwater compressed air energy storage
X, x	(design) variables or influencing parameters
Y	a variable, the objective function

1.1 What Are Design and Optimization of Thermofluid Systems?

Design and optimization of thermofluid systems are

> the design and, subsequently, optimization of the design of engineering systems involving significant fluid flow, thermodynamics, and/or heat transfer.

To more fully understand Design and Optimization of Thermofluid Systems, we need to clearly comprehend the four main terms:

1) design
2) optimization
3) thermofluid[1]
4) systems.[2]

Within this context,

1) *design* is the creation of an engineering system which will provide the desired result, and
2) *optimization* is taking the workable design one step further, attaining not just a better but the best design.

There usually exist a few unavoidable constraints, putting practical limits within which the optimal design is bounded. The optimal car may be the one performing the best in terms of mileage. For a typical middle-class engineer with four mouths to feed, however, the price of the car may be the deciding factor, limiting the selection to within a low-budget ceiling.

Example 1.1 ***Design a residential solar thermal energy storage system***

Given
An engineering student living in a temperate climate region wishes to store the thermal energy harnessed from the sun when it shines during the day, for residential use during the night.

Find
An appropriate storage system.

Solution
A workable design is running a glycol-water line from the solar thermal collector into an adequately large insulated water tank. Glycol-water is appropriately employed to prevent freezing. The temperature of the stored fluid has to be sufficiently high for the intended usage. Reasonable drops in the temperature from the solar collector to the storage tank and to the delivery end use must be accounted for, as some losses are inevitable.

The initial workable design, however, is probably not the best design as it may occupy the entire basement. The use of phase-change material will probably keep the size in check. Molten salt is also worth exploring, especially when dealing with larger utilization, such as a multiple-housing residence. Comparing different existing options, such as off-the-shelf tank sizes and storage media to achieve the best option is called optimization. Since the budget, as well as the available space for the storage tank, are likely limited, the optimization of the residential solar thermal energy storage system is thus subjected to budget, space, and other constraints.

Example 1.1 hints that a *workable design* does not necessarily need to be the best design. In fact, it typically is not. When the project is adequately large and there are (financial) backings for it, optimization is invoked to deduce the best design. Furthermore, for a company to compete in mass-selling of such systems, progressively better designs which are cheaper to manufacture are

1 The term thermofluid encompasses thermodynamics, fluid mechanics, and heat transfer.
2 A system is an orderly collection of integrated parts forming a unitary whole. An internal combustion engine is a familiar everyday engineering system.

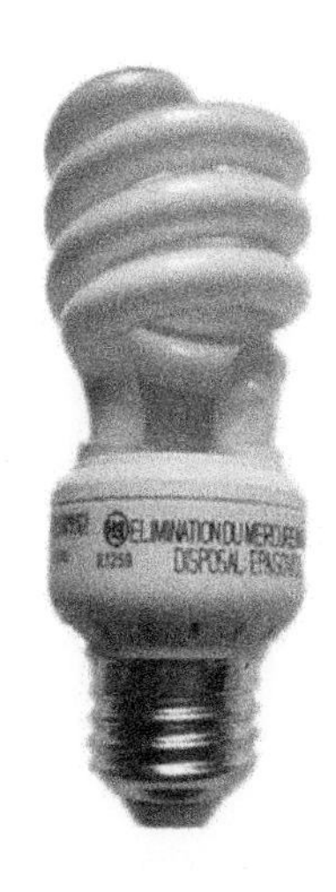

Figure 1.1 Workable versus optimal design of electricity-driven household light bulbs. Source: Photos taken by X. Wang and Y. Yang.

necessary. By and large, there will be budgetary, space, and other constraints. Other constraints for a thermal storage tank can be a maximum workable storage temperature, particular charging and discharging rates, etc. In some sense, moving from a feasible design to an optimum design is like progressing from an "ad hoc art and/or experience" to a "systematic scientific artistic endeavor."

A familiar design versus optimization exemplification is the three types of light bulb for everyday usage, see Figure 1.1. The incandescent light bulb is a workable design, and it has been satisfying our need since Thomas Edison invented it in 1879. Much later, the fluorescent light bulb is optimized in terms of energy usage and cost. For this reason, the compact fluorescent light bulb has finally squeezed out its archetype after being in the market for a couple of decades, the duration for the price to drop to a competitive level. Over the long run, the LED (light-emitting diode) light bulb is the best, because the money saved due to its low wattage and very long life span far exceed the high initial cost. In short, the incandescent light bulb, with a typical life span of 1,000–2,000 hours, is a workable design. The compact fluorescent light bulb, which lasts on the ballpark of 10,000 hours and uses around 75% less energy, is currently the optimum design. The LED light bulb, which outlasts the fluorescent by up to 50,000 hours while using 90% less energy, is the fruit of the latest design and optimization endeavor, and it is expected to be the new optimum design in a few years, as its manufacturing cost drops.

1.2 Differentiating Engineering from Science

The challenging tasks associated with thermofluid systems' design and optimization are only to be executed by individuals well educated and trained in engineering, i.e. competent engineers. But what is engineering? How does it differ from science? Science may be defined as the systematic knowledge of the physical world that is testable, repeatable, and predictable. Concisely,

> *Science* is the systematic knowledge of the physical world.

Simply put,

> *Engineering* is putting science into practice.

Figure 1.2 The millennia-old spoked wheel for horse chariots (created by S. Akhand). Shown are four-spoke chariot wheels resembling those found in the Red Sea, which are attributed to the powerful Egyptian army, as recorded in Exodus, Chapter 14.

By and large, engineering was initiated for, and still is, the exploitation of science to create practical systems to make life easier for society. In relation to the context of the material covered in this book,

> Engineering is the science and art of efficient dealing with materials and forces … it involves the most economic design and execution … assuring, when properly performed, the most advantageous combination of accuracy, safety, durability, speed, simplicity, efficiency, and economy possible for the conditions of design and service.
>
> J.A.L. Waddell

Let us look briefly at the millennia-old wheels, sketched in Figure 1.2. Horse chariots date further back than the Old Testament, where the Pharaohs were largely feared because of their vast number of powerful horse chariots. Durable wood was the material adopted, and the forces at play included the load on the chariot and the required torque. As per "economic design and execution," the wood has to be readily available locally, or relatively accessible and affordable to acquire from a not-too-distant land, or from subject nations as tributes under one's dominance. Accuracy may be viewed as the wood that does not expand or contract excessively with moisture and/or changes in the weather. Safety and durability may be perceived as keeping the soldiers from falling off as they charge the chariots forward into partially-rocky or muddy fields[3] at great speeds. Note that speed, to a large extent, decides the fate of the riding warriors. Simplicity and efficiency can easily be inferred from the spoke design, including the number of spokes. This becomes particularly obvious when contrasted with the predecessor of the spoked wheels, the clumsy, spoke-less, solid wood wheels; see Figure 1.3. For war chariots, securing sharp weapons on the outer side the (spoked) wheel further illustrates ingenious, effective design for the intention.

Further to the differentiation between science and engineering, a scientist is an expert in science, whereas an engineer creatively converts the scientific findings into useful applications. A good scientist indiscriminately strives to improve all kinds of knowledge, irrespective of any potential

3 It is worth mentioning that these powerful chariots can become handicapped on muddy and/or hilly ground. For this reason, foot soldiers prevail in mountainous battlefields.

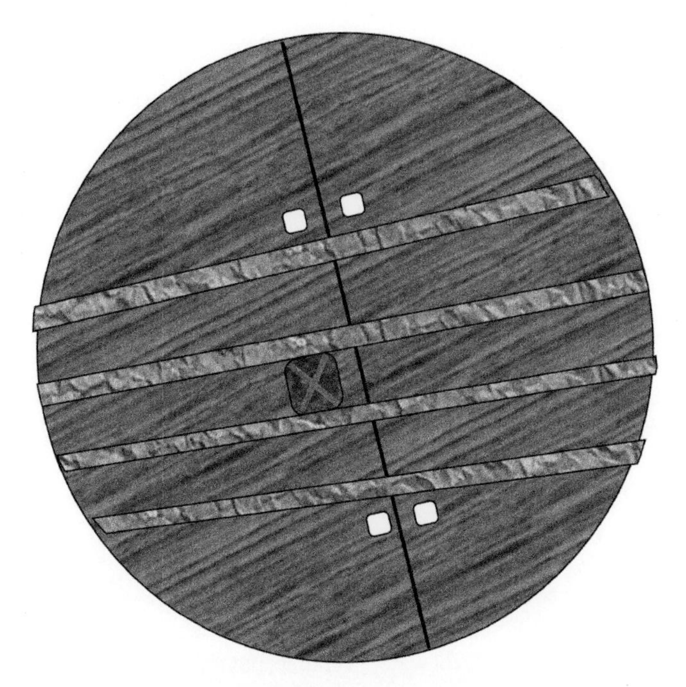

Figure 1.3 A sketch of, presumably, the world's oldest wheel, the Ljubljana Marshes Wheel, found in the Ljubljana Marshes in 2002 (Gasser, 2003). It has been radiocarbon-dated to before 3100 BCE. Source: Y. Yang.

usage, of the physical world. An applied scientist undertakes only applications-oriented scientific endeavors. This includes an engineering researcher who develops ideas that advance the frontiers of knowledge but may not be applied for a number of years. In other words, good engineers are not short-sighted; the prospective applications need not be immediately cognizable. Engineers may be regarded as professionals who design and develop, creatively converting theoretical concepts into useful applications on a daily basic. What exactly do engineers do? They link theory with practical applications. Bona fide engineers possess an extensive theoretical knowledge, the ability to think creatively, and a knack for obtaining practical results. The materials covered in this book aim at fostering the forging of amateur engineering students into fully-fledged creative engineers. While thermofluids is the subject of coverage, much of the knowledge delineated in this book, especially the core element, optimization, can equally be employed to improve solid mechanics and also process and production line processes. The aforementioned wheels for horse chariots clearly fall under the solid mechanics, not thermofluids, stream. Designing sound wheels for muddy thoroughfares, however, would encompass solid mechanics, thermofluids, and dynamics.

1.3 Development, Design, and Analysis

Development is the initial phase of an intended venture, where different methods through which a project may be realized are explored, analyzed, compared, and tested. Once the basic method is decided, design takes place. *Design* is the establishment of the exact way the various relating parts are to be put together so that the entire system functions properly. It generally involves the employment of concepts from engineering science coupled with a creative touch to make it work, and work elegantly. As such, to design is to create, devise, and/or forge artistically. To design is to invoke fundamental principles to an open-ended problem to procure one or more possible solutions. This is different from *analysis*, which is the application of fundamental principles to a well-defined problem to attain the solution. Note that there is a lack of openness and creativity in analysis.

1.4 The Design Process

In general, the design process may be roughly divided into the following steps.

1) *Identify the need.* What is the problem we have to solve? Keep in mind that engineers are problem-solvers.
   Let us assume the need is the stabilization of an intermittent renewable energy grid.
2) *Conception.* Establish an insight into the desirable end result and define the project.
   The end result is a stable grid which balances supply with demand.
   Energy storage can be utilized to mitigate the intermittence of the existing renewable energy grid.
3) *Synthesis.* Synthesize one or more possible ways to accomplish the end result.
   Pumped hydro, compressed air, underwater compressed air, flywheel, battery, hydrogen, etc., or a combination of some of these are all possible solutions. Assume that we wish to explore the underwater compressed air option as there is a body of water available, but not the elevation to permit pumped hydroelectric storage. Put the major pieces such as the motor, the compressor, the piping network, the underwater accumulator, the expander, and the generator together. Note that to realize this, the engineer should have acquired the required knowledge essential to the pieces of the engineering system puzzle involved.
4) *Operation conditions and limits.* Outline the operating conditions and spell out the constraints.
   Estimate the values of the major parameters. Storage capacity: how much energy storage is needed? What are the required storing and discharging rates? Storage duration: how long should the energy be stored? How much money can be spent? How deep is the water?
5) *Analysis.* Analyze the conceptual solution to deduce its feasibility. If it is infeasible, evaluate alternative plan(s). Analytical, numerical, or experimental analyses may be invoked.
   Perform the basic thermodynamics and fluid mechanics analyses. For example, are the available depth and volume of water adequate?
   Construct and test a pilot-scale system. Perform a parametric study, if appropriate and feasible.

Example 1.2 ***Design a daily routine for maximizing life span***

Given

A large pool of data correlating exercise and diet with life span of a typical human being is available.

Find

The healthiest (longest) life span of an average human being, based on the available data.

Solution

1) The objective function, Y = life span.
2) $Y = f(X_1, X_2)$, where X_1 is the number of hours of exercise per week, and X_2 is the amount of food in kilograms consumed every week.
3) In the considered case, $X_i = x_i$, and thus, $Y = f(x_1, x_2)$.

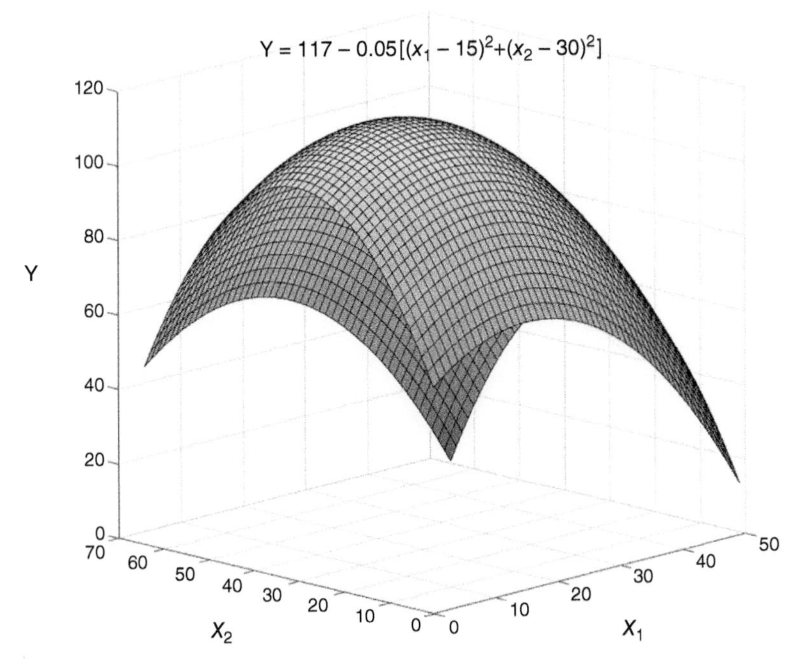

Figure 1.4 Life span as a function of exercise and diet. Source: Y. Yang.

4) $0 < x_1 < 63$ hours, $0 < x_2 < 77$ kg.
   The premise is that only someone who is dead is doing absolutely no exercise. Also, no one can work out for more than 63 hours (9 hours per day) every week. Furthermore, an average human being has to eat at least some food every day to stay alive, and no one can consistently consume more than 77 kg (11 kg per day) of food every week. In other words, an individual is expected to expire outside of these limits.

After compiling the available data, curve (surface) fitting leads to

$$Y = 117 - 0.05\left[(x_1 - 15)^2 + (x_2 - 30)^2\right]$$

This is plotted in Figure 1.4.

We see that the longest life span is 117 years, i.e., $Y = 117$ years at $x_1 = 15$ hours and $x_2 = 30$ kg.

It is noted that in this design, which happens to be the optimization of human life span, type and intensity of the exercise and contents of the food have not been considered. Other design parameters, such as quantity and quality of sleep, also demand our attention. The mental, psychological, or spiritual aspects also come into play, not to mention our typical complaint regarding our many problems associated with the inheritance of some bad genes.

One could protest that 117 years is simply too old to be considered as a good optimum, and dyeing of hair, along with cosmetic surgery etc., can only uplift the façade. This is very much in line with the view of Steve Jobs, who argued that "Design is not just what it looks like and feels like. Design is how it works." The sound functioning of the life span optimization stratagem is built into this groove. In other words, if a 100+year-old person can perform a couple of hours of sound exercise, along with savoring a couple of kilograms of victuals every day, the individual must also have no problem conducting the regular washroom business, and thus, the living being is not old but lively. Pointedly, the design is working!

In real life, it is not always feasible to optimize the design, especially under the pressure of money and time, both are essential for the survival of the company. Nevertheless, engineers should adopt the habit of striving for quality in their designs. Steve Jobs correctly said, "Be a yardstick of quality. Some people aren't used to an environment where excellence is expected."

Example 1.3 ***Stabilize the intermittent renewable energy grid***

Given

The increase in renewable energy into the power grid causes a heightened grid stability challenge. To rephrase it, the sporadic supply of natural energy, such as that harnessed by a wind turbine, mismatching with irregular demand introduces unprecedented challenges.

Find

A solution to mitigate the grid intermittence.

Solution

Possible steps in the design and optimization process are:

1) Identify the need.
   To stabilize the intermittent renewable energy grid.
2) Develop a conceptual solution.
   Battery, compressed air, hydrogen, flywheel?
3) Estimate values of major parameters.
   How much energy/power? For how long (storing, releasing)? How much will it cost?
4) Construct, test, and modify.
   Try out a scaled model.
5) Management and financial review.
   Financially feasible? Environmentally acceptable?
   The cost may depend on the size and the volume (numbers of units to be sold), and the length of the payback period.
6) Refine and optimize parameters.
   Efficiency is a major concern, how and how much can we improve; availability/cost of materials used, size, number of accumulators, etc.?
7) Field test for meeting performance, reliability, and safety goals.
   UWCAES (underwater compressed air energy storage) prototype, identifying shortcomings, opportunities for improvement.
8) Manufacture and market.
   Sampling/monitoring users' experiences – failure rates, failure modes, life expectancy, etc.

Iterate as needed.

A good engineer should never forget common sense. Sketching the disposition of physical parts and possible relative positions is generally a valuable aid. The KISS (Keep It Simple, Stupid) design philosophy is always a helpful guide. As the case may be, it seems like some advanced thinking is required to keep it simple. Richie Norton stated that, "Simplicity is complex. It's never simple to keep things simple. Simple solutions require the most advanced thinking." There is an upper limit, beyond which we would be overdoing it. This is nicely put by Albert Einstein, "Make everything as simple as possible, but not simpler."

1.5 Existing Books on Thermofluid System Design and/or Optimization

Some existing books which expound, with varying degree of details, on particular aspects of the materials discussed in this book include:

Arora, J.S. (1989). *Introduction to Optimum Design*. New York: McGraw-Hill.
Balaji, C. (2011). *Essentials of Thermal System Design and Optimization*. New York: CRC Press.
Bejan, A., Tsatsaronis, G., and Moran, M. (1996). *Thermal Design and Optimization*. New York: John Wiley & Sons, Inc.
Boehm, R.F. (1987). *Design Analysis of Thermal Systems*. New York, John Wiley & Sons, Inc.
Burmeister, L.C. (1998). *Elements of Thermal-Fluid System Design*. Upper Saddle River, NJ: Prentice-Hall.
Dhar, P.L. (2016). *Thermal System Design and Simulation*. Cambridge, Academic Press.
Edgar, T.F. and Himmelblau, D.M. (1988). *Optimization of Chemical Processes*. Part II. Boston, McGraw-Hill.
Jaluria, Y. (1998). *Design and Optimization of Thermal Systems*. New York McGraw-Hill.
Penoncello, S.G. (2015). *Thermal Energy Systems: Design and Analysis*. Boca Raton, FL: CRC Press.
Stoecker, W.F. (1989). *Design of Thermal Systems*, 3rd ed. New York: McGraw-Hill.
Suryanarayana, N.V. and Arici, Ö. (2003). *Design and Simulation of Thermal Systems*. New York: McGraw-Hill.

Arora (1989) and its later versions are very comprehensive and far-reaching in scope, covering a wide range of engineering designs, including many non-thermofluid systems. The optimization methods covered in Balaji (2011) are somewhat unconventional. Bejan et al. (1996) offer a more advanced book which is not quite suited for an undergraduate engineering curriculum. While the title of Boehm's (1987) book does not spell out optimization, it is briefly covered as the final chapter. Edgar and Hummelblau's book (1988) is a specialized monograph geared for chemical engineering students. Burmeister (1998), Dhar (2016), Jaluria (1998) and their newer editions, and Stoecker (1989) are probably the four textbooks which are closest to covering the subject matter on design and optimization of thermofluid systems, with Stoeker (1989), or, more correctly, its first edition published by McGraw-Hill in 1971, as the classic text. Penoncello (2015) deals extensively with thermal systems analysis and design. Optimization is only briefly mentioned in the last chapter. It is worth noting that the second edition of this book is also available (Penoncello, 2018). Suryanarayana and Arici (2003) comprehensively review thermodynamics, fluid mechanics, and heat transfer, but optimization is not included in their textbook. There are other books on design and/or simulations of thermofluid systems. They, however, do not typically cover optimization. This book aims at balancing the coverage of the essential elements; engineering economics, the prevailing thermofluid systems and devices and their modeling, and generic and selected advanced optimization methods.

The following books may also be of interest:

Fox, R. (1971). *Optimization Methods for Engineering Design*. Reading, MA: Addison-Wesley.
Goldberg, D. (1989). *Genetic Algorithms in Search, Optimization, and Machine Learning*. Boston: Addison-Wesley.
Rao, S. (1996). *Engineering Optimization*. New York: Wiley-Interscience.

1.6 Organization of the Book

Money talks and, thus, basic engineering economics is conveyed in Chapter 2. Common thermofluid devices such as valves, ducts, pipes, and fittings are reviewed in Chapter 3. Chapter 4 presents the fundamentals of heat exchangers. The focus is on the most prevailing indirect-contact heat exchangers. To enable modeling, the system under consideration must be accurately described by mathematical equations. Therefore, equations are covered in Chapter 5, where pertinent curve fittings are highlighted. Once the mathematical model is established, we move on to thermofluid system simulation in Chapter 6. Most prevalent, robust sequential, and simultaneous solution methods are expounded. With the functioning of the system simulated, the problem can be formulated for optimization. Chapter 7 delineates the formulation of the concerned system, clearly defining the objective function and relevant constraints. For differentiable objective functions, the calculus approach discussed in Chapter 8 can nail the optimum via rigorous differentiation. For constrained problems, the Lagrange Multiplier can convey the sensitivity of the solution with respect to modest relaxation of, or changes in, the constraint. The beef of the curriculum is Chapter 9, where the standard search approaches are detailed. With the bullet-proof Exhaustive Search as the base, versatile single-variable elimination methods, Dichotomous Search, the Fibonacci search and the Golden Section search are explained. For multi-variable problems, Lattice Search, Univariate Search, Steepest Ascent/Descent methods are viable for unconstrained problems. Penalty-function and Search-along-a-constraint methods can be resorted to for constrained multi-variable problems. For thermofluid systems, the objective function and the constraints can often be expressed as sums of polynomials. As elaborated in Chapter 10, geometric programming is especially suited for solving this kind of problems. A few large-scale, real-world, as well as some envisioned, projects are included in the Appendix.

Problems

1.1 Hot water storage

Assume that the necessary storage energy for Example 1.1 is 2 kJ and water is the medium. How big is the required storage tank?

1.2 Water temperature leaving a solar thermal collector

The solar radiation is 1500 W/m^2. Water ($c_p \approx 4.2$ kJ/kg·K) at 12°C enters a solar thermal collector at 0.1 kg/s. The available surface area for collecting the solar radiation is 20 m^2. What is the (ideal, maximum) temperature of the water leaving the solar collector?

1.3 Sizing a solar thermal water tank

Assume the daily solar irradiance (direct radiation on a horizontal surface), I_{dir}, for Los Angeles is equal to that of May 1, 1990 as summarized in Table 1.1.

Table 1.1 Solar irradiance for Los Angeles on May 1, 1990.

Time	8:00	9:00	10:00	11:00	12:00	13:00	14:00	15:00	16:00	17:00
I_{dir} [kJ/h·m^2]	109	915	1332	2108	2480	2889	2521	1821	837	128

Design a solar thermal system (collector) to supply 0.003 m^3/s of 50°C hot water between 9:00 a.m. and 3:00 p.m., where the makeup water is at 20°C. What is the required collector area? If the maximum available collector area is 200 m^2, what is the size of the required hot water storage tank?

1.4 Battery storage for a solar photovoltaic system

Assume the daily solar irradiance (direct radiation on a horizontal surface), I_{dir}, for Los Angeles is equal to that in Table 1.1. Design a solar photovoltaic (PV) system using common commercial PV panels to supply a total of 10 kJ of electricity in a day. How many panels are needed? What is the required area? If the maximum available collector area is 200 m^2, and a minimum of 1 kW of power is needed between 9:00 a.m. and 3:00 p.m., what is the size of the required battery storage?

1.5 Improve the horse chariot wheels

You are asked to improve the design of a horse chariot such as that for the powerful biblical pharaoh. How would the performance of the horse chariot vary with the number of spokes, say, from 2 to 200 spokes? What is the proper sequence concerning development, design, optimization, and research? What do you call the process of varying the number of spokes in an effort to improve performance?

1.6 Thermoelectric wristwatch

Design a wristwatch that runs on thermoelectric power based on the temperature difference between the human body and the ambient temperature. Follow the steps delineated in this chapter. Estimate the needed power and size the thermoelectric power generator accordingly. See Synder (2019) for an overview of this sexy technology.

1.7 Sleeping Beauty's life span

Suppose additional data on sleeping hours and life span is available, modify the equation in Example 1.2 to include x_3, the number of hours of sleep per month. Create a new equation so that the same maximum life span is achieved at $x_3 = 240$ hours.

1.8 Water transport network

Water from a reservoir is to be transported at 0.07 m^3/s via a 15-cm diameter commercial steel pipe piping network with three flanged elbows and to be discharged into the open atmosphere at 25 m above the free surface of the water in the reservoir. A pump is to be located at 5 m below the free water surface. What is the required head which needs to be supplied by the pump?

Hint: Invoke the conservation of energy for pipe flow. The equations can be found in Chapter 3 on Thermofluid Devices, where fluid transport in a piping network is recapped. For example, from the inlet of the pump to the discharge outlet into the open atmosphere at 25 m above the free surface of the water in the reservoir, conservation of energy can be written as

$$P_1/\rho g + \tfrac{1}{2}U_1^{\,2}/g + z_1 + h_{pump} = P_2/\rho g + \tfrac{1}{2}U_2^{\,2}/g + z_2 + h \qquad (1.1)$$

where P is pressure, ρ is density, g is gravity, U is average velocity, z is elevation, h_{pump} is head supplied by the pump, h_L is head loss associated with the piping network. The head loss consists of the major head loss in the straight pipe sections and the minor head loss associated with the fittings. The major head loss can be deduced from

$$h_L = f\ (L_{pipe}/D)\ (\tfrac{1}{2}U^2/g) \tag{1.2}$$

where f is the friction factor, and D is the diameter of the pipe. The minor losses can be estimated from

$$h_{L,minor} = K_L\ \tfrac{1}{2}U^2/g \tag{1.3}$$

where K_L is the loss coefficient.

1.9 Storing energy underwater

An underwater air accumulator is needed to store 70 kJ of energy during the low-demand hours when there is plenty of wind to harness energy from a wind turbine. Provide two workable options in terms of the size of the accumulator and the depth at which it is placed underwater. See Wang et al. (2016) to appreciate the background of this promising technology.

1.10 Cool a solar photovoltaic panel to boost efficiency

The energy conversion efficiency of a solar photovoltaic (PV) panel is known to decrease as the PV cell temperature increases (Wu et al., 2018; Fouladi et al., 2019; Yang et al., 2019). Devise a passive turbulence generator which can effectively lower the cell temperature by 2°C, at solar noon on July 1, in Windsor, Ontario, Canada, or the location where you reside, assuming that the wind is prevailing at 7 m/s over the PV panel.

1.11 Renewable water desalination system

Design a workable renewable energy system for remote water desalination for a typical family of four in Bathurst Inlet, Nunavut, Canada. Soni et al. (2017) estimated a typical case for India with approximately 100 liters of water consumption per capita per day. They presented a simple four-stage still for the water desalination process based on reduced vapor pressure. Their wind and solar driven system for a solar flux of 850 W/m^2 for six hours a day and 1–5 m/s wind can meet the fresh water needs of rural and urban communities.

1.12 Seasonal thermal storage for a heating greenhouse

Design a seasonal thermal energy storage system for heating a one-acre greenhouse in South Western Ontario, or another temperate location. For the South Western Ontario setting, see Semple et al. (2017).

References

Arora, J.S. (1989). *Introduction to Optimum Design*. New York: McGraw-Hill.

Balaji, C. (2011). *Essentials of Thermal System Design and Optimization*. New York: CRC Press.

Bejan, A., Tsatsaronis, G., and Moran, M. (1996). *Thermal Design and Optimization*. New York: John Wiley & Sons, Inc.

Boehm, R.F. (1987). *Design Analysis of Thermal Systems*. New York, John Wiley & Sons, Inc.

Burmeister, L.C. (1998). *Elements of Thermal-Fluid System Design*. Upper Saddle River, NJ: Prentice-Hall.

Dhar, P.L. (2016). *Thermal System Design and Simulation*. Cambridge, Academic Press.

Edgar, T.F. and Himmelblau, D.M. (1988). *Optimization of Chemical Processes*. Part II. Boston, McGraw-Hill.

Fouladi, F., Henshaw, P., Ting D.S-K., and Ray, S. (2019). Wind turbulence impact on solar energy harvesting. *Heat Transfer Engineering* 41(7): 407–417.

Fox, R. (1971). *Optimization Methods for Engineering Design. Reading*. MA: Addison-Wesley.

Gasser, A. (2003). *World's oldest wheel found in Slovenia*. Republic of Slovenia, Government Communication Office, March.

Goldberg, D. (1989). *Genetic Algorithms in Search, Optimization, and Machine Learning*. Boston: Addison-Wesley.

Jaluria, Y. (1998). *Design and Optimization of Thermal Systems*. New York: McGraw-Hill.

Penoncello, S.G. (2015). *Thermal Energy Systems: Design and Analysis*. Boca Raton, FL: CRC Press.

Rao, S. (1996). *Engineering Optimization*. New York: Wiley-Interscience.

Semple, L., Carriveau, R., and Ting, D.S-K. (2017). A techno-economic analysis of seasonal thermal energy storage for greenhouse applications. *Energy and Buildings* 154: 175–187.

Snyder, J. (2019). http://thermoelectrics.matsci.northwestern.edu/, accessed November 20. 2019.

Soni, A., Stagner. J.A., and Ting, D.S-K. (2017). Adaptable wind/solar powered hybrid system for household wastewater treatment. *Sustainable Energy Technologies and Assessments* 24: 8–18.

Stoecker, W.F. (1989). *Design of Thermal Systems*, 3rd ed. New York: McGraw-Hill.

Suryanarayana, N.V. and Arici, Ö. (2003). *Design and Simulation of Thermal Systems*. New York: McGraw-Hill.

Wang, Z., Ting, D.S-K., Carriveau, R., Xiong, W., and Wang, Z. (2016). Design and thermodynamic analysis of a multi-level underwater compressed air energy storage system. *Journal of Energy Storage* 5: 203–211.

Wu, H., Ting, D.S-K., and Ray, S. (2018). Flow over a flat surface behind delta winglets of varying aspect ratios. *Experimental Thermal and Fluid Science* 94: 99–108.

Yang, Y., Ting, D. S-K., and Ray, S. (2019). Convective heat transfer enhancement downstream of a flexible strip normal to the freestream. *International Journal of Thermal Sciences* 145: 106059.

2

Engineering Economics

No society can surely be flourishing and happy, of which the far greater part of the members are poor and miserable.

– Adam Smith

Chapter Objectives

- Appreciate what engineering economics is and why it is important.
- Comprehend the worth of money with respect to time.
- Understand interest rates, and the influence of frequency of interest compounding.
- Become familiar with the money flow series and the cash flow diagram.

Nomenclature

A	payment amount for a uniform series of payments
C_{RF}	cost recovery factor
G	the gradient of a constant gradient series of payments
HVAC	heating, ventilation, and air conditioning
i	interest; i_{eff} is the effective interest rate, the annual interest rate, or the annual interest rate after inflation
j	annual inflation rate
m	annual frequency, i.e. number of times the interest is calculated every year
n	time period
P	purchasing power (money); P_0 is the purchasing power at time zero, P_n is the purchasing power at the end of the nth period.
x	a variable

2.1 Introduction

There is no dispute concerning the truthfulness of the quote by Adam Smith, "No society can surely be flourishing and happy, of which the far greater part of the members are poor and miserable." Equally true is the saying by Winston Churchill, "You don't make the poor richer by making the rich poorer." Bringing these two wise statements together, we see that industrialization which mobilizes many breadwinners, including engineers, is not necessarily a bad business. It goes without saying that quality designs not only can bring in the needed profit to sustain the engineering industry, they also tend to solve the poverty issue by eliminating the tedious laborious jobs which pay little.

What is Engineering Economics? To answer that, we first need to understand what economics is. *Economics* is the body of knowledge dealing with the production, distribution, and consumption of goods and services. *Engineering economics* is the subset of economics that applies economic principles in evaluating engineering decisions. For example, if there are multiple engineering designs which can perform the required task, a good engineer who is vested with sound engineering economics would favor the cheapest design or, more befittingly, the one that grants the most profit. To be environmentally responsible, a good balance between profitability and system efficiency and environmental friendliness is strived for. The best solutions are the ones which are simultaneously most sustainable and profitable.

This chapter is compiled based on materials from numerous sources. Stoecker (1989), Burmeister (1998), Jaluria (1998), and Whitman and Terry (2012) are particularly germane.

2.2 Worth of Money with Respect to Time

Money is a measure of worth. It is the dominant medium of exchange used for paying for goods, services, etc. Commodity money can range from precious metals, such as gold or silver, to special stones. Currency, on the other hand, is most easily understood as printed dollars. The Canadian dollar is the primary medium of exchange in Canada. In the United States, the cold, hard cash of Canada, the loonie, is materially depreciated to a couple of American quarters, if it is accepted. The power, stability, and influence of a nation on the international stage have a serious impact on the worth of its currency. This is the reason why most residents in less-stable nations prefer to hang on to the strong, stable, and internationally-envied American dollar. Let us look at two age-old questions, that is, is time money? Is knowledge money?

Example 2.1 ***Time, knowledge, power, and money***

Given

Knowledge, power, time, and work are four elements which we constantly strive for and try to balance. These four factors are said to be related.

Find

If working hard or gaining more knowledge increases the amount of money one makes.

Solution

$$\text{Knowledge} = \text{Power} \tag{E2.1.1}$$

$$\text{Time} = \text{Money} \tag{E2.1.2}$$

$$\text{Power} = \text{Work/Time} \tag{E2.1.3}$$

Substitute Eqs. (E2.1.1) and (E2.1.2) into Eq (E2.1.3)

$$\left(\text{Knowledge}\right) = \text{Work/(Money)} \tag{E2.1.4}$$

or,

$$\text{Money} = \text{Work/Knowledge.} \tag{E2.1.5}$$

We see that with increasing hard work, the amount of money we earn increases accordingly. On the other hand, money decreases with increasing knowledge!

Are you wasting your time trying to gain more knowledge?

While the debate concerning whether more knowledge amounts to more or less money continues, there is time-tested consensus that "time is money." The dispute over "time is money" rests only on the conversion. Whatever your view may be, rest assured that no amount of money can bring back an inch of time. Although not as certain as money cannot buy back time, or every human being has to pay taxes and face death, as pointedly expressed by Benjamin Franklin, "In this world nothing can be said to be certain, except death and taxes," money or, more correctly, currency slides in the opposite direction as the upward-flying sparks: "Yet man is born unto trouble, as the sparks fly upward" (Job 5:7). In short, the worth of a currency diminishes with time. This depreciation is quantified by the economic term called inflation. This will be expounded when we discuss the effective interest rate.

As money cannot buy back time, one has to pay a "levy" for spending the yet-to-be-earned money, e.g. by borrowing it from someone who has the money, such as the bank. Thereupon, interest is introduced to account for the rental fee for borrowing money. As such, interest gives money a time value in the positive sense, i.e. you should keep some of your income in the bank to earn some interest, so that you may survive "rainy days" in the future.

A common time period for counting interest is annually, or yearly. The *purchasing power* at the end of the first time period,

$$P_1 = (1 + i)\ P_0 \tag{2.1}$$

Here, purchasing power, P, signifies the amount of money, with subscript '0' denoting that it is at the beginning, i.e. time zero, and 'i' is the interest. The purchasing power at time zero, P_0 is called the *principal*, and P_1 is called the *future worth*, at the end of the first time period. The interest i in Eq (2.1) is the *simple interest*, simple because it does not consider the change (gain) in investment due to accumulation of interest with time but merely applies it on the original amount, the principal, at the end of the period (year). At the end of the second period, we have

$$P_2 = (1 + i)\ P_1 = (1 + i)^2\ P_0 \tag{2.2}$$

It is clear that the interest of the first period has been compounded by the end of the second period. At the end of n interest periods,

$$P_n = (1 + i)^n\ P_0. \tag{2.3}$$

If the utilized period is one year, then, it is compounding annually, i.e. the annual interest is added to the principal at the end of each year.

Table 2.1 Money value with time periods.

Period (Year)	Interest earned or paid during the period	Money at the end of the period
1	$P_0 i$	$P_0 + P_0 i$
2	$P_0 (1+i) i$	$P_0 (1+i) + P_0 (1+i) i$
3	$P_0 (1+i)^2 i$	$P_0 (1+i)^2 + P_0 (1+i)^2 i$
...		
n	$P_0 (1+i)^{n-1} i$	$P_0 (1+i)^{n-1} + P_0 (1+i)^{n-1} i$

Table 2.1 summarizes the change in the money with the number of time periods. The amount of money at the end of the nth period, the last column in Table 2.1, is as expressed by Eq (2.3). It is clear that the purchasing power increases with the number of periods over which the interest is compounded. Imagine you have \$1,000,000, the interest, i, is 5%, and the period is one year. At the end of the first year, you will have \$1,000,000 + \$1,000,000×5% = \$1,050,000. If n is 7, then, at the end of the seventh year, you will have $\$1{,}000{,}000(1+0.05)^{7-1} + \$1{,}000{,}000(1+0.05)^{7-1} (0.05) =$ \$1,340,095.64 + \$67,004.78 = \$1,407,100.42. Note that the interest earned in the seventh year alone is \$67,004.78. In other words, you make the most money during the last period, where the capital for that period is the highest.

2.2.1 Compound Interest and Effective Interest

Table 2.1 shows the formulae for change in purchasing power with respect to time. The initial purchasing power is the principal, P_0. Consider the base case that the time period for counting the interest, i, is annual. At the end of the nth year, the purchasing power, or, the amount of money

$$P_n = P_0 (1 + i)^{n-1} + P_0 (1 + i)^{n-1} i \tag{2.4}$$

This can be simplified into

$$P_n = P_0 (1 + i)^n \tag{2.4a}$$

which is Eq (2.3). Note that only at the end of each year is the interest, i, realized. For this single annual compounding interest, the amount of money at the end of the nth year is given by Eq (2.3). Anyone who has money to lend would quickly be dawned with the notion of maximizing the interest, namely, one should calculate the interest more frequently than merely once a year. If the interest is compounded m times a year, the interest rate during each period would be i/m. In this case, the total number of periods in n years is nm and, thus,

$$P_n = (1 + i/m)^{nm} P_0 \tag{2.5}$$

For quarterly compounding, m = 4, and for monthly compounding, m = 12.

Once one is on the train of thought about charging the interest more frequently, greed becomes the train track. This will speed up to the extreme of continuous compounding, where m approaches infinity. In plain English, the increasing purchasing power can be calculated every instant. For this continuous compounding ultimate, the amount of money at the end of the nth year,

$$P_n = e^{ni} P_0 = P_0 \exp(ni) \tag{2.6}$$

Example 2.2 ***The effect of the frequency of interest compounding***

Given

$100 is set aside to earn interest over the next 12 years. The interest rate i = 5%.

Find

The amount of money at the end of the 12th year, if the interest is compounded annually, quarterly, monthly, daily, and continuously.

Solution

Compounded annually, Eq (2.4a):

$$P_n = P_0 \, (1 + i)^n = 100 \, (1 + 0.05)^{12} = \$179.59$$

Compounded quarterly, Eq (2.5):

$$P_n = (1 + i/m)^{nm} \, P_0 = (1 + 0.05/4)^{12\times4} \, 100 = \$181.54$$

i.e. $1.95 more than simple annual compounding.
Compounded monthly, Eq (2.5):

$$P_n = (1 + i/m)^{nm} \, P_0 = (1 + 0.05/12)^{12\times12} \, 100 = \$181.98$$

i.e. $2.40 more than simple annual compounding.
Compounded daily, Eq (2.5):

$$P_n = (1 + i/m)^{nm} \, P_0 = (1 + 0.05/365)^{12\times365} \, 100 = \$182.20$$

i.e. $2.62 more than simple annual compounding.
Compounded continuously, Eq (2.6):

$$P_n = e^{ni} \, P_0 = P_0 \, \exp(ni) = 100 \, \exp(12 \times 0.05) = \$182.21$$

i.e. $2.63 more than simple annual compounding.

It is obvious that more frequent compounding gives a better return; however, the additional gain due to further increases in compounding frequency subsides rapidly with the interest compounding frequency. For example, only one extra penny is gained by increasing the compounding frequency from 365 times per year to infinity. It is noticeable that the interest rate itself has a much larger impact than the compounding frequency. As an illustration, raising the interest to 6% would lead to P_{12} of $201.22 to $205.44, with increasing compounding frequency from once to infinite times per year, respectively, instead of from $179.59 to $182.21 with the 5% annual interest rate.

Supposedly, greed abounds in the government. As the government is the body which prints the currency, it is not difficult for them to print more to pay for services and goods. The price for this printing is the devaluation of the currency. As long as there is no significant overprinting, or the governing officials are not in conflict with superpowers of the world, the citizens only have to deal with (well-accepted) inflation, i.e., depreciation of the dollar's value. Economically speaking, inflation causes the purchasing power to decrease with respect to time. This can be expressed mathematically as

$$\left(1 + j\right) \, P_1 = P_0 \tag{2.7}$$

where j = annual inflation rate. It is noted that inflation is typically deduced annually. Equation (2.7) says that the purchasing power at the end of the year, times (1+j), is equal to the purchasing power at the beginning. After n years, the future purchasing power,

$$P_n = P_0/\left(1 + j\right)^n \tag{2.8}$$

While j decreases the value of the dollar, i counters it. The outcome of this battle is indicated by the effective interest rate,

$$i_{eff} = (i - j) / (1 + j) \tag{2.9}$$

Since the inflation rate is an annual rate, we need to bring the interest rate, compounded more frequently than once a year, to the equivalent annual interest rate, i.e. the effective interest rate. This equivalent (annual) interest rate, i_{eff}, gives the same future worth as the compounding interest rate, i.e.

$$P_n = (1 + i_{eff})^n P_0 = (1 + i/m)^{nm} P_0. \tag{2.10}$$

Rearranging this, we have

$$i_{eff} = (1 + i/m)^m - 1 \tag{2.11}$$

For instance, i = 10% compounded monthly gives i_{eff} = 10.47%, when the inflation j = 0. Suppose further to this 10.47% annual interest rate there is a 7% inflation, then, invoking Eq (2.9) with the annual interest rate i = 10.47%, we have

$$i_{eff} = (i - j) / (1 + j) = (10.47\% - 7\%) / (100\% + 7\%) = 3.24\%$$

To put it another way, while the i = 10% monthly compounding interest rate inspires one to put money into savings, the 7% inflation rate devalues the investment. It is money-wise to note that it is quite possible to have a negative effective interest rate. For example, if the simple annual interest rate is 5%, instead of 10.47%,

$$i_{eff} = (i - j) / (1 + j) = (5\% - 7\%) / (100\% + 7\%) = -1.87\%.$$

Therefore, the safest savings by trusting your money to a big financial institute in a guaranteed account may entail a negative investment. Sometimes, it is better to take some calculated risks.

2.2.2 Present Worth Factor

Instead of projecting the present worth, P_0, into the future, as Eqs. (2.3–2.6) do, we can also express a future amount in terms of today's value. For the simple annual interest case, the *present worth* of a future amount at the end of the nth year,

$$P_0 = P_n/(1 + i)^n \tag{2.12}$$

For compounded interest, we have

$$P_0 = P_n/(1 + i/m)^{mn} \tag{2.13}$$

The *compound-amount factor* is simply

$$P_n/P_0 = (1 + i/m)^{mn} \tag{2.14}$$

The inverse of this is called the *present-worth factor*,

$$P_0/P_n = 1/(1 + i/m)^{mn} \tag{2.15}$$

2.3 Money Flow Series

Whether money makes the world go around or not, other than the very frugal few, the rest of us have money flowing in and out of our pockets like we consume water and pay washroom visits. The frequent in- and out-flowing of money is generally inevitable, especially in the business setting. Therefore, tools such as cash flow diagrams can be invoked to ease the management of money.

2.3.1 Cash Flow Diagram

A cash flow diagram, Figure 2.1, is a diagram depicting deposits and withdrawals, and the value of money, with respect to time. For this arbitrary case, you inherited \$50k and moved out of your parents' house. By the end of the first year, you have spent \$52k. Fortunately, you found on a nice job and worked hard the second year, totaling an annual income of \$59k at the end of year two. These ups and downs continue until our days on earth expire, leaving some money, or debts, for our children to savor.

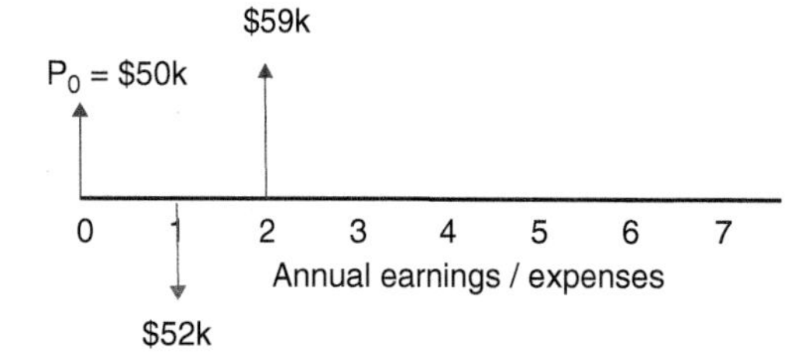

Figure 2.1 Cash flow diagram. Source: D. Ting.

Uniform Series of Payments/Savings

In business, a loan is frequently borrowed in a lump sum, and then it is paid in fixed payments over the duration of the loan. A mortgage is one such example; so is a car loan. Equally common is our income, which is typically received in regular monthly or biweekly payments. This simple kind of uniform series of annual payments, A, is depicted in the cash flow diagram in Figure 2.2. If the simple annual interest is applicable, then the future worth,

$$P_n = A\left[(1 + i)^{n-1} + (1 + i)^{n-2} + \ldots + 1\right] = A\left[(1 + i)^n - 1\right]/i \tag{2.16}$$

We can rewrite this as

$$A = P_n\, i/\left[(1 + i)^n - 1\right] \tag{2.16a}$$

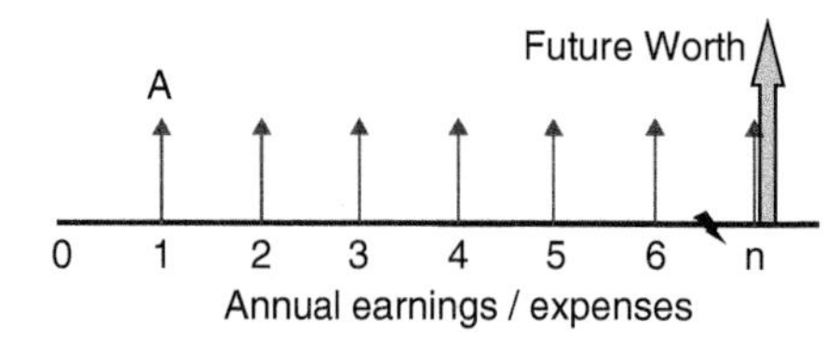

Figure 2.2 Future worth of a uniform series of payments. Source: D. Ting.

Example 2.3 ***Uniform savings for a future worth (expense)***

Given

A graduate who has secured a stable position in an engineering firm plans to set aside equal amounts of investments annually for a planned purchase of \$16,000 in 10 years. The bank provides an annual interest of 8%.

Find

How much must be saved each year?

Solution

The problem is depicted in Figure 2.3.

From Eq (2.16a),

$$A = P_n i / \left[(1+i)^n - 1\right]$$

$$= \$16000 \times 0.08 / \left[(1+0.08)^{10} - 1\right]$$

$$= \$1104.47$$

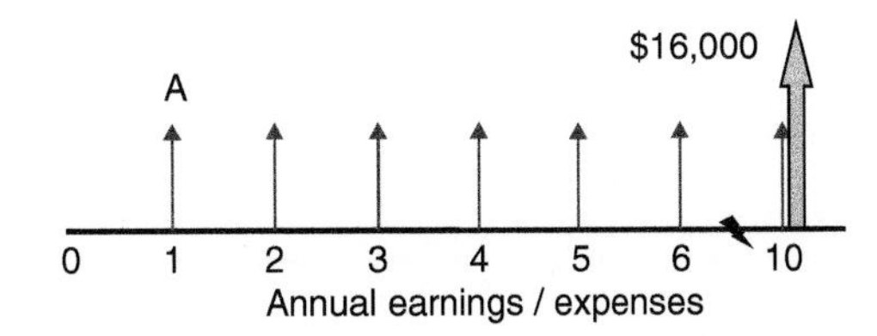

Figure 2.3 Save a uniform series of investments for a future purchase. Source: D. Ting.

If you are buying a car, for example, you probably would like to compare the price you pay via the loan option with that of paying it off up-front. In this case, you need to know the present worth. As portrayed in Figure 2.4, the present worth,

$$P_0 = A/(1+i)^1 + A/(1+i)^2 + \ldots + A/(1+i)^n \tag{2.17}$$

Multiplying both sides by $(1 + i)^n$ gives

$$P_0\,(1+i)^n = A\,\left[(1+i)^n/(1+i)^1 + (1+i)^n/(1+i)^2 + \ldots + (1+i)^n/(1+i)^n\right] \tag{2.17a}$$

This can be simplified into

$$P_0\,(1+i)^n = A\,\left[(1+i)^{n-1} + (1+i)^{n-2} + \ldots + 1\right] \tag{2.17b}$$

The right-hand side can be reduced according to Eq (2.16) and, thus, we can write

$$P_0\,(1+i)^n = A\,\left[(1+i)^n - 1\right]/i \tag{2.17c}$$

or

$$P_0 = A\,\left[(1+i)^n - 1\right] / \left[i\,(1+i)^n\right] \tag{2.17d}$$

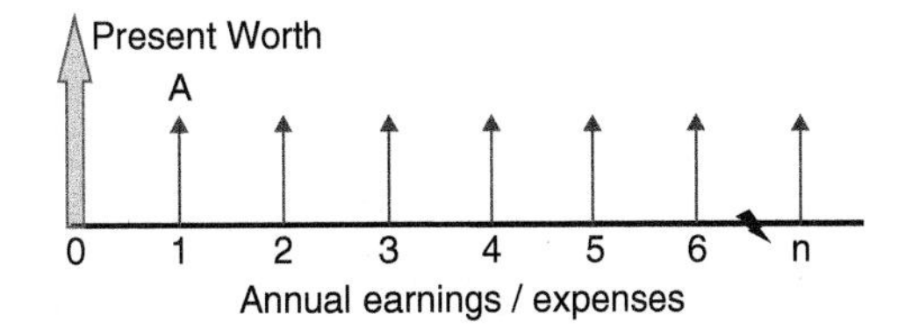

Figure 2.4 Present worth of a uniform series of payments. Source: D. Ting.

Example 2.4 ***Uniform monthly payments of a loan with monthly-compounding interest***

Given

You borrow \$1,000 from a loan company that charges 15% nominal annual interest compounded monthly. You can afford to pay \$38 per month on the loan.

Find

How many months will it take to repay the loan?

Solution

Figure 2.5 shows the flows of money. The present worth can be expressed in terms of the uniform monthly payments. Specifically, present worth,

$$P_o = A\ \{[(1+i/m)^{mn} - 1] / [(i/m)(1+i/m)^{mn}]\} \quad (E2.4.1)$$

$$\$1000 = \$38\ \{ [(1 + 0.15/12)^{12\times n} - 1] / [(0.15/12)(1 + 0.15/12)^{12\times n}] \}$$

$$N = 2.676 \text{ years} = 32.11 \text{ months}$$

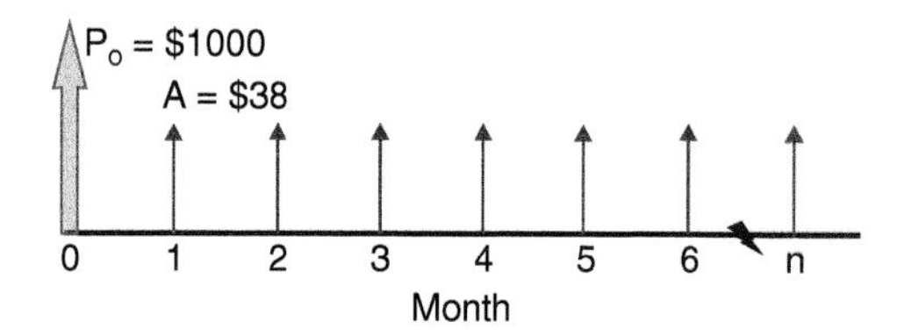

Figure 2.5 Number of uniform payments for a loan. Source: D. Ting.

What if the first payment takes place at time zero? This is quite common when financing a car. In this case, the entire series of payments is advanced by one period. Assuming a simple annual interest rate with uniform annual payment, A, as shown in Figure 2.6, the future worth is simply Eq (2.16) multiplied by an extra period of interest, i.e., (1 + i), i.e.

$$P_n = A\ (1+i)\ [(1+i)^n - 1]\ /i. \quad (2.18)$$

Similarly, if m payments are made each year, with the same compounding frequency, the future worth is

$$P_n = A\ [(1+i/m)^n - 1]\ /i/m \quad (2.19)$$

Note that the interest used is the nominal annual interest rate.

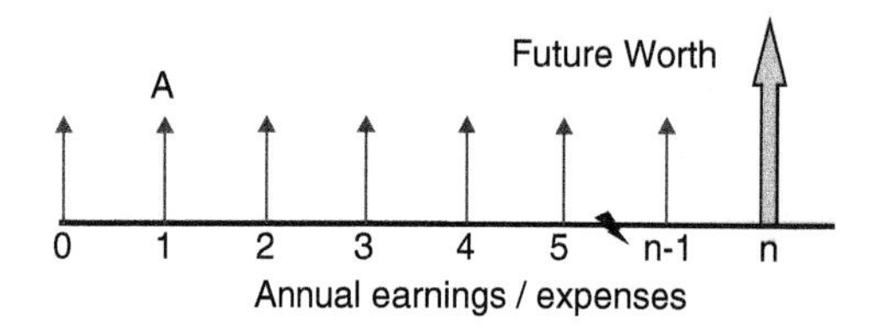

Figure 2.6 Future worth of a uniform series of payments advanced by a period. Source: D. Ting.

Constant-Gradient Series of Payments

Instead of uniform payments, we could also have incremental payments with a constant gradient. Consider a simple annual interest, A, with constant-gradient annual payment,

$$A = (n - 1)\ G \tag{2.20}$$

where G is the gradient, as illustrated in Figure 2.7. The present worth

$$P_0 = G/(1+i)^2 + 2G/(1+i)^3 + \ldots G\ (n-1)/(1+i)^n \tag{2.21}$$

This can be rewritten as

$$P_0 = G\left\{\frac{1}{i}\left[\frac{(1+i)^n - 1}{i(1+i)^n} - \frac{n}{(1+i)^n}\right]\right\} \tag{2.22}$$

An example of this is the cost of maintenance of equipment, which tends to increase progressively as the equipment ages.

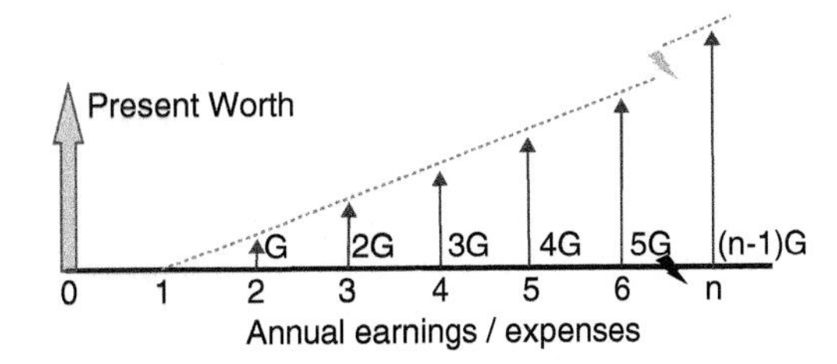

Figure 2.7 Present worth of a constant-gradient series of payments. Source: D. Ting.

Example 2.5 ***Constant-gradient payments***

Given

The annual cost of energy for a facility is $3,000, payable at the end of the first year. It increases by $300 each year thereafter. The interest rate is 9% compounded annually.

Find

The present worth of the 12-year series of energy costs.

Solution

The problem is illustrated in Figure 2.8. The payments can be considered as a series of uniform payments plus a series of constant-gradient payments. From Eqs. (2.17d) and (2.22), we can write

$$P_0 = A\left[\frac{(1+i)^n - 1}{i(1+i)^n}\right] + G\left\{\frac{1}{i}\left[\frac{(1+i)^n - 1}{i(1+i)^n} - \frac{n}{(1+i)^n}\right]\right\} \tag{E2.5.1}$$

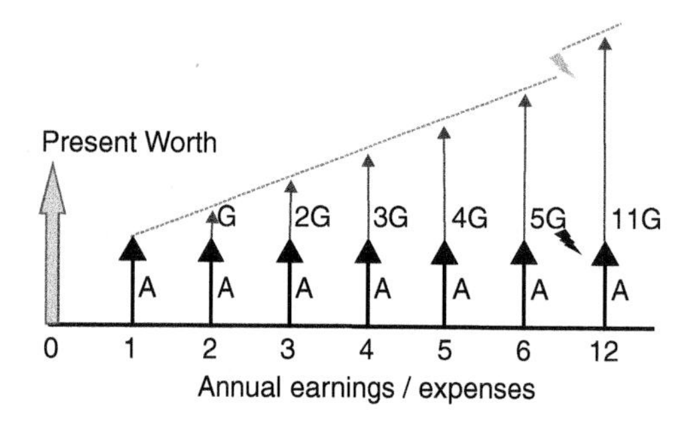

Figure 2.8 Present worth of the sum of a uniform series and a constant-gradient series of payments. Source: D. Ting.

The series is a combination of two series, a uniform series of \$3,000 and a gradient series with G=\$300. This is shown in Figure 2.8, where A = \$3,000, and G = \$300. Substituting these values into Eq (2.1) gives

$$P_o = \$3000\ \left[(1.09)^{12} - 1\right] / \left[0.09\ (1.09)^{12}\right]$$
$$+\ \$300\ \left\{(1/0.09)\ \left[\left(1.09^{12} - 1\right) / \left(0.09 \times 1.09^{12}\right) - 12/1.09^{12}\right]\right\}$$
$$= \$21482.18 + \$9647.70$$
$$= \$31130$$

The underlying elements behind various money flows, and the present and future worth of money, have thus been explained. We can move forward and apply what we have learned to practical problems.

Example 2.6 ***Select the better option to invest***

Given

After working for a big heating, ventilation, and air conditioning (HVAC) company for a number of years, two engineers decide to partner and start their own company manufacturing heat pipes, for thermal energy recovery in HVAC systems. After a long elimination process, they are down with two options.

Option A

Cost of building = \$700,000

Profit = \$500,000/yr, at end of the first year
\$400,000, at the end of the second year
\$200,000, at the end of the third year
\$150,000, at the end of the fourth year
\$100,000, at the end of the fifth year
Salvage at the end of the fifth year is \$600,000

Option B

Cost of building = \$800,000

Profit = \$100,000/yr, at end of the first year
\$200,000, at the end of the second year

$300,000, at the end of the third year
$400,000, at the end of the fourth year
$500,000, at the end of the fifth year
Salvage at the end of the fifth year is $800,000

Find
The better option if the effective interest rate, i_{eff} = 5%/yr.

Solution
To compare the options, we bring all costs and profits into the present worth. From Eq (2.3), we can write

$$P_0 = P_n/(1 + i_{eff})^n \qquad \text{(E2.6.1)}$$

For Option A,

Cost of building = $700,000, i.e., P_0 = -700,000
First year profit = $500,000, i.e., $P_0 = 500{,}000 / (1 + 0.05)^1 = 476{,}190$
Second year profit = $400,000, i.e., $P_0 = 400{,}000 / (1 + 0.05)^2 = 362{,}812$
Third year profit = $200,000, i.e., $P_0 = 200{,}000 / (1 + 0.05)^3 = 172{,}768$
Fourth year profit = $150,000, i.e., $P_0 = 150{,}000 / (1 + 0.05)^4 = 123{,}405$
Fifth year profit = $100,000, i.e., $P_0 = 100{,}000 / (1 + 0.05)^5 = 78{,}353$
Salvage at fifth year = $600,000, i.e., $P_0 = 600{,}000 / (1 + 0.05)^5 = 470{,}116$
Total present worth = $1,453,760

For Option B,

Cost of building = $800,000, i.e., P_0 = -800,000
First year profit = $100,000, i.e., $P_0 = 100{,}000 / (1 + 0.05)^1 = 95{,}238$
Second year profit = $200,000, i.e., $P_0 = 200{,}000 / (1 + 0.05)^2 = 181{,}406$
Third year profit = $300,000, i.e., $P_0 = 300{,}000 / (1 + 0.05)^3 = 259{,}151$
Fourth year profit = $400,000, i.e., $P_0 = 400{,}000 / (1 + 0.05)^4 = 329{,}081$
Fifth year profit = $500,000, i.e., $P_0 = 500{,}000 / (1 + 0.05)^5 = 391{,}763$
Salvage at fifth year = $800,000, i.e., $P_0 = 800{,}000 / (1 + 0.05)^5 = 626{,}821$
Total present worth = $1,083,460

Therefore, Option A is a better investment.

2.3.2 Rate of Return, Benefit-Cost Ratio, and Capital Recovery Factor

To make a good business decision, a business-minded engineer would evaluate an intended investment from different perspectives. Other than present worth, rate of return, benefit-cost ratio, and/or capital recovery factor can also be assessed to complement the investment assessment. These evaluation parameters are discussed here. Also gaining acceptance is life cycle analysis and/or assessment. This cradle-to-grave analysis is very involved and extensive. Interested readers can refer to Sørensen (2011) and Klöpffer and Grahl (2014) for more information.

Rate of Return
The rate of return quantifies the effective interest rate that corresponds to a zero present worth. Two familiar terms for rate of return are the IRR (Internal Rate of Return) and the ROI (Return on

Investment). As this is an internally-derived indicator, it does not require an (external) alternate, or option, to compare. Let us deduce the rate of return for the two options in Example 2.6.

Example 2.7 *Worthiness of a project based on the rate of return*

Given

A facility for manufacturing heat pipes for thermal energy recovery in HVAC systems has the following financial details. The prevailing interest rate, $i_{eff} = 5\%/yr$.

$$\text{Cost of building} = \$700{,}000 \tag{E2.7.1}$$

Profit = \$500,000 first year, \$400,000 second year, \$200,000 third year, \$150,000 fourth year, \$100,000 fifth year.

Salvage at the end of the fifth year is \$600,000.

Find

If the project is worth pursuing, based on the rate of return.

Solution

To deduce the rate of return, we set the present worth of the project to zero, i.e.,

$$0 = -700{,}000/(1+i_{eff})^0 + 500{,}000/(1+i_{eff})^1 + 400{,}000/(1+i_{eff})^2 + 200{,}000/(1+i_{eff})^3$$
$$+150{,}000/(1+i_{eff})^4 + 100{,}000/(1+i_{eff})^5 + 600{,}000/(1+i_{eff})^5$$

This gives

$$i_{eff} = 49\%/yr.$$

Let us also determine the rate of return for Option B in Example 2.6 to compare and contrast.

$$0 = -800{,}000/(1+i_{eff})^0 + 100{,}000/(1+i_{eff})^1 + 200{,}000/(1+i_{eff})^2 + 300{,}000/(1+i_{eff})^3$$
$$+400{,}000/(1+i_{eff})^4 + 500{,}000/(1+i_{eff})^5 + 800{,}000/(1+i_{eff})^5$$

This leads to

$$i_{eff} = 31\%/yr.$$

With typical annual interest rate of around 5%, both investments are excellent. As the duration is the same for both cases, and are relatively short term, the one that grants a 49% effective interest rate per year is the preferred choice.

Benefit-Cost Ratio

Another criterion, as far as the merit of an investment is concerned, is the benefit-cost ratio. This can be straightforwardly quantified by bringing all returns to present worth and divide it by the present worth of the total cost. Let us illustrate this using Example 2.6 again.

Example 2.8 ***Benefit-cost ratio of two possible investments***

Given

Two options for manufacturing heat pipes for thermal energy recovery in HVAC systems are:

Option A

Cost of building = \$700,000

Profit = \$500,000 first year, \$400,000 second year, \$200,000 third year, \$150,000 fourth year, \$100,000 fifth year.

Salvage at the end of the fifth year is \$600,000.

Option B

Cost of building = \$800,000

Profit = \$100,000 first year, \$200,000 second year, \$300,000 third year, \$400,000 fourth year, \$500,000 fifth year.

Salvage at the end of the fifth year is \$800,000.

Find

The benefit-cost ratio of these options. Assume the effective interest rate, $i_{eff} = 5\%/yr$.

Solution

To compare the options, we bring all benefits to a present worth. From Eq (2.3), we can write

$$P_0 = P_n/(1 + i_{eff})^n. \tag{E2.8.1}$$

Similarly, we bring all costs to a present worth.

For Option A,

Benefits:

First year profit = \$500,000, i.e., $P_0 = 500{,}000 / (1 + 0.05)^1 = 476{,}190$
Second year profit = \$400,000, i.e., $P_0 = 400{,}000 / (1 + 0.05)^2 = 362{,}812$
Third year profit = \$200,000, i.e., $P_0 = 200{,}000 / (1 + 0.05)^3 = 172{,}768$
Fourth year profit = \$150,000, i.e., $P_0 = 150{,}000 / (1 + 0.05)^4 = 123{,}405$
Fifth year profit = \$100,000, i.e., $P_0 = 100{,}000 / (1 + 0.05)^5 = 78{,}353$
Salvage at fifth year = \$600,000, i.e., $P_0 = 600{,}000 / (1 + 0.05)^5 = 470{,}116$

Total benefit in present worth = \$1,683,644

Cost:

Cost of building = \$700,000, i.e., $P_0 = -700{,}000$

Total present worth of cost = \$700,000

Benefit-cost ratio = 2.41

For Option B,

Benefits:

First year profit = \$100,000, i.e., $P_0 = 100{,}000 / (1 + 0.05)^1 = 95{,}238$
Second year profit = \$200,000, i.e., $P_0 = 200{,}000 / (1 + 0.05)^2 = 181{,}406$
Third year profit = \$300,000, i.e., $P_0 = 300{,}000 / (1 + 0.05)^3 = 259{,}151$

Fourth year profit = \$400,000, i.e., $P_0 = 400{,}000 / (1 + 0.05)^4 = 329{,}081$
Fifth year profit = \$500,000, i.e., $P_0 = 500{,}000 / (1 + 0.05)^5 = 391{,}763$
Salvage at fifth year = \$800,000, i.e., $P_0 = 800{,}000 / (1 + 0.05)^5 = 626{,}821$
Total benefit in present worth = \$1,883,460

Cost:

Cost of building = \$800,000, i.e., $P_0 = -800{,}000$

Total present worth of cost = \$800,000

Benefit-cost ratio = 2.35

Once again, Option A is a better investment.

In practice, it is good to look at a couple different criteria when evaluating an investment and/or comparing options to invest. The benefit-cost ratio provides an indication of the amplification of the investment. It is obvious that a project with a benefit-cost ratio of one or less is not worth pursuing. Like the rate of return, the benefit-cost ratio does not reveal the size of the project. The money flow diagram and calculations, as demonstrated in Example 2.6, disclose more details; therefore, it serves as the base evaluation standard, which can be complemented by the rate of return and/or the benefit-cost ratio.

Capital Recovery Factor

Sometimes it is convenient to treat an investment in terms of a capital cost. The most obvious one is presumably a mortgage, which is typically paid off via a series of uniform payments. This is similar to a one-time investment which reaps equal (income) payment, A, at the end of a number, n, of equal intervals. According to Eq (2.17d), the present worth

$$P_0 = A \left[(1 + i)^n - 1\right] / \left[i\,(1 + i)^n\right] \tag{2.17d}$$

where A is a uniform payment. This can be alternatively written in terms of the Capital Recovery Factor, C_{RF}, where

$$A = C_{RF}\,(i, n)\;P_0 \tag{2.23}$$

Example 2.9 ***Capital recovery factor of an investment***

Given

An investment opportunity offers \$700 per month profit for every \$100,000 invested. The annual interest rate is 6%. What if the profit is \$8,400 per year?

Find

How long does it take to recover the investment, i.e., before you start making a profit?

Solution

As the payment is monthly, let us convert all cash flows in terms of month. From Eq (2.11), the annual effective rate,

$$i_{eff} = (1 + i/m)^m - 1. \tag{2.11}$$

This can be rearranged to get the monthly interest rate which is equivalent to an annual interest rate of 6%,

$$i = 12\left[\left(i_{eff}+1\right)^{1/12}-1\right] = 5.841\%.$$

From Eq (2.17d), we can write

$$A = C_{RF}(i, m)\ P_0 = (i/m)\ (1+i/m)^{mn} / \left[(1+i/m)^{mn}-1\right]\ P_0,$$

where $i = 0.05841$, and $m = 12$. Substituting, we have

$$700 = \left\{(0.05841/12)\ (1+0.05841/12)^{12n} / \left[(1+0.05841/12)^{12n}-1\right]\right\}\ 100000.$$

This gives $12n = 244.8$ months.

For the annual payment of \$8,400 at 6% annual interest rate, we have

$$A = C_{RF}\left(i_n, m\right)\ P_0 = i_n\ \left(1+i_n\right)^n / \left[\left(1+i_n\right)^n - 1\right]\ P_0,$$

or

$$8400 = \left\{0.06\ (1+0.06)^n / \left[(1+0.06)^n-1\right]\right\}\ 100000.$$

This gives $n = 21.5$ years, or, 258 months.

2.4 Thermo-economics

The profitability and productivity of an entity can be viewed as the transfer of available energy, which we call exergy. Accordingly, the economic efficiency can be perceived as thermodynamic efficiency. In other words, if the efficiency is high, not much exergy is needed and, hence, the cost is low. This led to the coining of the phrases *thermo-economics* and/or *exergo-economics*, for energy is always needed to accomplish a project. As such, the required amount of thermodynamically-available energy, exergy, has a physical basis for measuring worth. In a nutshell, exergy is cost. If we take sustainability into consideration, exergy is a truer measure of cost than money. After all, the more we deplete the available energy, exergy, the less sustainable planet Earth is. Note that the cold, hard cash is not replaced, but supplemented by exergy, in thermo-economics. In other words, both capital and operational costs, along with thermodynamic efficiencies, are included in thermo-economic analysis. Interested readers can check out Bejan et al. (1996), El-Sayed (2003), and Bryant (2012).

Problems

2.1 Effective interest rate

The nominal interest rate is 7%. You wish to know the difference in the frequency of compounding. Find the effective (annual) interest rate, if the nominal interest rate of 7% is compounded (i) quarterly, (ii) monthly, (iii) weekly, (iv) daily, and (v) continuously.

2.2 Uniform savings for a future investment

You wish to have \$60,000 to invest in 8 years at an annual interest of 7%. What is the annual savings that needs to be deposited at the end of each year?

2.3 Prepare to replace a thermofluid system
A thermofluid system has a useful life of 10 years and will cost $25,000 to replace. Find the sum that annually should be put into an account bearing 3%/yr interest so that the system can be replaced at the end of the tenth year; the first deposit occurs at the end of the first year of the machine life.

2.4 Paying off a loan via monthly payments
You wish to borrow $500,000 and the bank charges 15% interest monthly. What is the monthly payment required to pay off the loan in 30 years?

2.5 Interest needed to make annual savings into a future worth
You put away $10,000 at the end of each year, for an anticipated future worth of $74,420 in 5 years. What is the corresponding annual interest rate?

2.6 Annual deposits earning monthly interest
You deposit $10,000 each year at a monthly interest of 8%. What is the amount you have in 10 years?

2.7 Monthly payments for a building
You wish to buy a building with monthly payments of $1500 and a $20,000 down payment. The bank offers a 30-year loan at 5.75% interest compounded monthly. What is the maximum building price that you can afford?

2.8 Equating two cash flows
Your classmate puts away $1,000, $1,200, $1,400, and $1,600, three, four, five, and six years, respectively, after graduating from high school. You wish to have the same amount of worth with three equal-amount savings for the zeroth, first, and second year after high school. What is the annual amount that you have to put away? Assume an annual interest rate of 5%.

2.9 Is knowledge money?
The average high school graduate earns $48,000/yr from age 17 to 67. A typical BSc in engineering pays $24,000/yr from age 17 to 22, beyond which the engineering graduate earns, on average, $80,000/yr from age 23 to 67. Those BSc engineering graduates who continue their education with an MSc degree pay $24,000/yr for two additional years after their BSc degrees. A typical MSc engineering graduate earns an average of $85,000/yr from age 25 to 67. Draw the cash-flow diagrams of high school, BSc, and MSc graduates. What are the respective present worths of the three education levels? Is knowledge money? What does it take to make the extra degree worthwhile?

References

Bejan, A., Tsatsaronis, G., and Moran, M. (1996). *Thermal Design & Optimization*. New York: John Wiley & Sons, Inc.

Burmeister, L.C. (1998). *Elements of Thermal-Fluid System Design*. Upper Saddle River, NJ: Prentice-Hall.

Bryant, J. (2012). *Thermoeconomics: A Thermodynamic Approach to Economics*, 3rd ed. Nottingham: VOCAT International Ltd.

El-Sayed, Y.M. (2003). *The Thermoeconomics of Energy Conversions*. Amsterdam: Elsevier.

Jaluria, Y. (1998). *Design and Optimization of Thermal Systems*. New York: McGraw-Hill.

Klöpffer, W., and Grahl, B. (2014). *Life Cycle Assessment (LCA): A Guide to Best Practice*. Weinheim: Wiley-VCH.

Sørensen, B. (2011). *Life-Cycle Analysis of Energy Systems: From Methodology to Applications*. Cambridge: Royal Society of Chemistry.

Stoecker, W.F. (1989). *Design of Thermal Systems*, 3rd ed. New York: McGraw-Hill.

Whitman, D., and Terry, R.E. (2012). *Fundamentals of Engineering Economics and Decision Analysis*. San Rafael, CA: Morgan & Claypool Publishers.

3

Common Thermofluid Devices

Education then, beyond all other devices of human origin, is the greatest equalizer of the conditions of men, the balance-wheel of the social machinery.

– Horace Mann

Chapter Objectives

- Appreciate common thermofluid devices.
- Categorize thermofluid devices into major, moderate, and minor classes.
- Become familiar with typical valves.
- Recap the fluid dynamics associated with pipe flow.
- Differentiate major and minor head losses.
- Perform basic piping and/or duct network designs.

Nomenclature

A area
a acceleration
C a constant
D diameter; D_h is the hydraulic diameter
F force; F_{vis} is viscous force
f friction coefficient
g gravity
H height
h head; h_L is head loss, $h_{L,minor}$ is the minor head loss
K_L loss coefficient
L length; L_e is entrance length, L_{pipe} is pipe length
m mass; m' is mass flow rate
P pressure; ΔP is pressure difference
r radius
Re Reynolds number
U velocity; U_{avg} is the average velocity

W width
x x direction
y y direction
z z direction, vertically up, against gravity

Greek and Other Symbols

γ specific weight
μ kinematic viscosity
ρ density
τ shear; τ_w is the shear along the wall
ε roughness
$\forall$ volume; $\forall'$ is volume flow rate

3.1 Common Components of Thermofluid Systems

Nearly all mechanical, chemical, and environmental engineers routinely tap into knowledge of many thermofluid components. These individual components are called devices. Each device is an invention or contrivance created for a specific purpose. When designing a typical thermofluid system, varying pieces of these subsystems are synthesized. They are also analyzed as parts of the complete system in the design, and the analysis is repeated to optimize the design. In engineering practice, re-invention of the wheel should be avoided, unless there are no existing wheels which can adequately perform the task at hand, so to speak. By and large, the design and optimization process involves selecting from commercially available devices and synthesizing them into the desired system. Custom fabrication of these devices is not economically feasible, and thus is only resorted to when there is no readily available, off-the-shelf items that can perform the specific task. Even then, or more so, good understanding of the inner workings of these devices is essential.

Some of the most common thermofluid systems include power plants and air conditioning systems. These are extensively covered in various textbooks. The associated *major* components or devices are combustors, boilers, turbines, condensers, engines, etc. Heat exchangers, pumps, blowers, and fans are generally considered *moderate* devices, whereas compressors and motors may fall into the major or moderate class. The *minor* components include pipes, ducts, fittings, and valves, though some valves are rather complex and, thus, may be more appropriately treated as a moderate component at times. In agreement with Steve Jobs, who said, "Our goal is to make the best devices in the world, not to be the biggest," our goal should be making the best, not the clumsiest, devices. More so, according to H.G. Wells, who asserted, "There is nothing in machinery, there is nothing in embankments and railways and iron bridges and engineering devices to oblige them to be ugly. Ugliness is the measure of imperfection," functional engineering devices are meant to be beautiful. Talking about beauty, it is often captured and more-or-less perpetuated in many photos. The instrument, in this case, is the photographer, not the camera. Eve Arnold accurately contended, "It is the photographer, not the camera, that is the instrument." We can paraphrase this as, "It is the engineer, not the composition that is the design."

3.2 Valves

Relatively comprehensive references for valves include ASHRAE (2008), Nesbitt (2007), and Smith and Zappe (2004), among others. Valves are a device for regulating the flow of a fluid, and at times, other materials, through a conduit. Other than regulating the flow, valves are also utilized to prevent back flow and to regulate pressure, via relieving excess pressure, for example. A typical thermofluid system generally employs many valves. For this reason, their characteristics can substantially influence the system performance. Needless to say, they tend to contribute to a significant portion of the cost.

The following should be considered when choosing a valve.

1) *Type of transport material.* This is typically a fluid, but is it in the liquid, gaseous, vapor, and/or mixed phase? Does the phase of the transport material change along the conduit? Is it corrosive or erosive, especially when the medium contains solids, e.g. a slurry?
2) *Purpose of the valve.* Is the valve for either letting the flow through or stopping the flow completely? Is it required to control the flow rate or to direct the flow into different flow paths? Does it need to prevent back flow?
3) *Ranges of operating conditions.* What are the lowest and highest flow rates, temperatures, and pressures involved?
4) *Frequency of operation.* How often do you have to operate the valve? Must the operation be realized manually, automatically, or both?

Depending on the purpose, different types of valves are called upon. The more common types of valves, in alphabetical order, include (i) ball valves, (ii) butterfly valves, (iii) gate valves, (iv) globe valves, (v) needle valves, (vi) pinch valves, (vii) plug valves, (viii) poppet valves, and (ix) saddle valves. Each of these is briefly discussed in the following.

3.2.1 Ball Valves

In a sense, a ball valve is a passage through a ball. As can be seen in Figure 3.1, a ball valve consists of a precision ball held between two circular seals or seats. The flow passage changes from the fully closed position to the fully open one via a 90-degree turn. Ball valves are compact, they can be

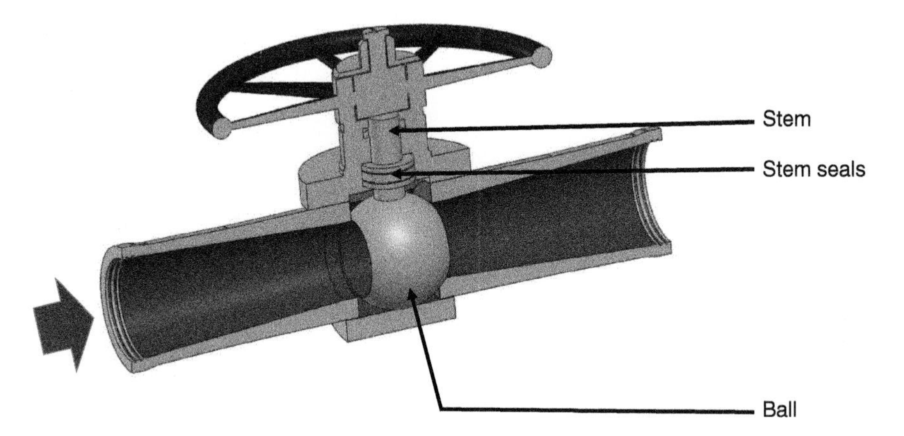

Figure 3.1 Schematic of a ball valve. Source: G. Nagesh.

actuated quickly and are relatively easy to maintain. They require a low operating torque and are largely leak-tight in shut-off.

3.2.2 Butterfly Valves

The term "butterfly" was coined for this type of valve because the rotatable disk for controlling the flow resembles butterfly wings; see Figure 3.2. Butterfly valves typically have a cylindrical, flanged-end body with an internal, rotatable disk.

Figure 3.2 Schematic of a butterfly valve. Source: S.K. Mohanakrishnan.

3.2.3 Gate Valves

As the name hints, gate valves are meant to be fully open or fully shut. In this manner, they are utilized to isolate a particular portion of a pipe system and/or directing the flow into a particular flow passage. Gate valves are one of the most common types of valves in the industry. Two common gate valves, sluice and wedge, are depicted in Figure 3.3. When fully open, there is no obstruction, as compared to a butterfly valve, in the flow path. Therefore, the gate imposes very little additional (minor) head losses. These valves can be operated via a hand-wheel, air-powered diaphragm, electric motion, piston actuator, etc.

3.2.4 Globe Valves

When the flow rate needs to be regulated over a range, rather than either zero or the maximum value, we can resort to globe valves. Unlike a ball valve, which has a ball with a hole through it, a globe valve consists of a movable, disk-type element and a stationary ring seat in a roughly spherical body that resembles a globe and, hence, the name globe valve. It is worth mentioning that the head loss caused by a globe valve is somewhat higher than that of a ball valve. Figure 3.4 shows that a globe valve controls the flow rate via positioning a plug with respect to the stationary seat.

The ports of a globe valve can be aligned with the flow or pipe, or, they may be at an angle with respect to the flow. The three primary body designs for globe valves are (i) tee-pattern or Z-body, (ii) angle-pattern, and (iii) wye-pattern or Y-body. The tee-pattern, as depicted in Figure 3.4, is the prevailing body type, with a Z-shaped diaphragm. The horizontal setting of the seat allows the stem and disk to move perpendicular to the flow. This design has the smallest flow coefficient (flow rate per unit pressure drop), that is, the largest pressure drop (per unit flow rate). By this very nature,

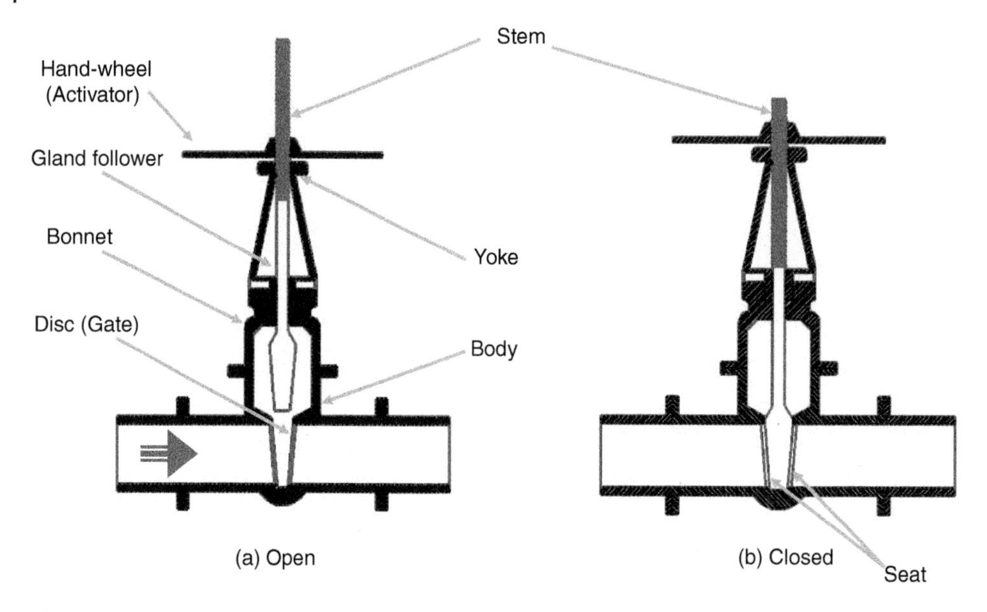

Figure 3.3 Schematics of sluice and wedge gate valves. Source: G. Nagesh.

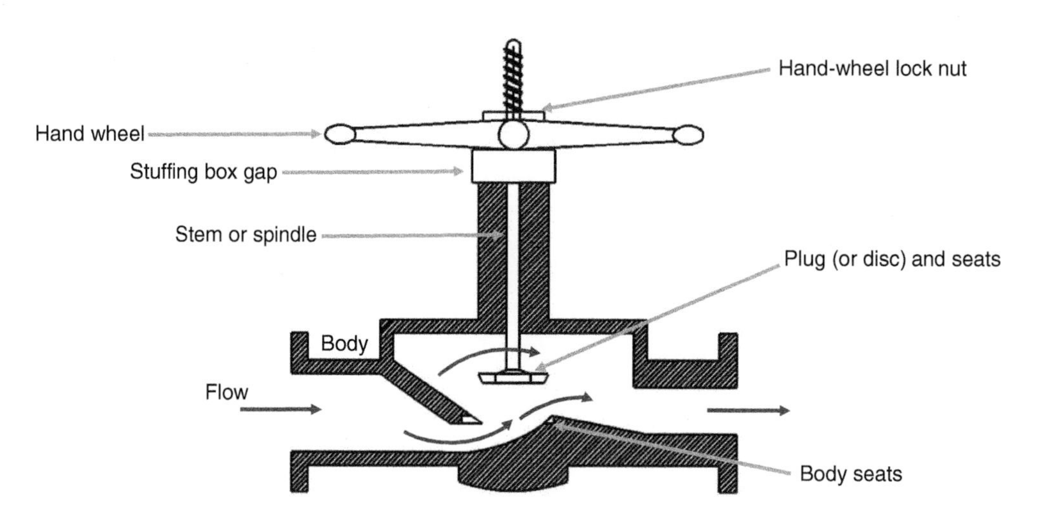

Figure 3.4 A schematic of a globe valve. Source: G. Nagesh.

they are employed in severe throttling purposes. Figure 3.5 shows an angle-pattern globe valve, in comparison to tee-pattern and wye-pattern globe valves. The incoming fluid is forced through a 90-degree turn. This kind of angled supply valve is frequently employed for corrosive or thick (viscous) fluids. They are also utilized for pulsating flow, as they are capable of handling the sluggish effect of this type of flow. The periodically required draining can be readily achieved by having an outlet that is directly downward, in the same direction as gravity. Note that regular draining can drastically reduce corrosion and clogging. These angle-pattern globe valves have slightly lower flow coefficients than wye-pattern globe valves.

To mitigate the high pressure drop associated with globe valves, a wye-pattern can be implemented. For wye-pattern globe valves, the seat and stem are angled at about 45 degrees, furnishing a straighter flow passage at the fully open position and, thus, imposing the least flow resistance. To remove debris, the valve can be removed and a rod can be used to push and drain the debris.

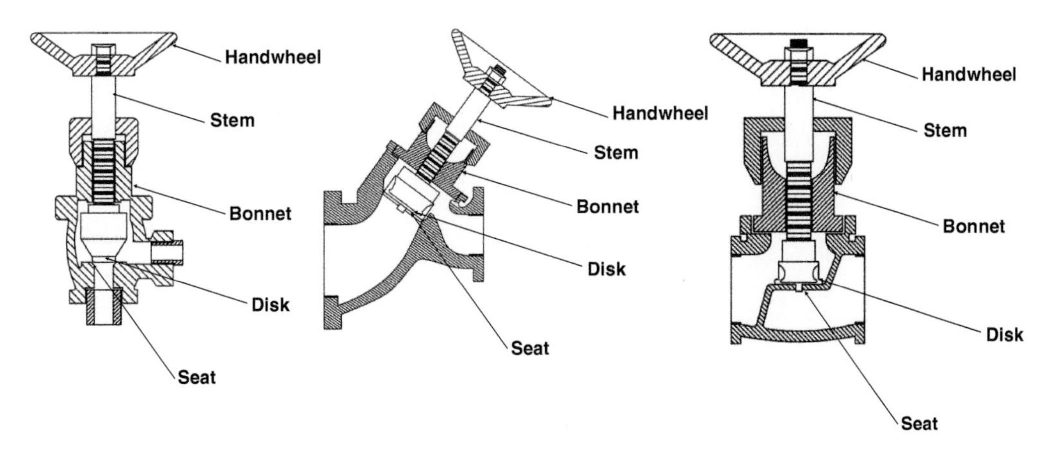

Figure 3.5 Contrasting tee-pattern (Z-body) with angle-pattern and wye-pattern (Y-body) globe valves. Source: X. Wang.

In short, globe valves are easy to shut off, moderately easy to throttle. They have a shorter stroke compared to gate valves. The three configurations, tee, wye, and angle, offer unique capabilities. It is relatively easy to machine or resurface the seats. Some of the disadvantages of globe valves include a larger pressure drop compared to gate valves. Because of the pressure under the seat, they also require a greater force to seat the valve. Some flow throttling also occur under the seat.

3.2.5 Needle Valves

For finer flow control, we can make use of needle valves. Needle valves control the flow via a spindle which adjusts a tapered pin, as shown in Figure 3.6. As the pressure involved is relatively high, only a small orifice with a long, tapered seat serves as the flow passage. To accurately and precisely control the flow, a needle-shaped plunger on the end of a screw, which perfectly fits the valve seat, is employed. A leak-tight seal is created when the plunger sits snugly in the seat.

Figure 3.6 A schematic of a needle valve. Source: G. Nagesh.

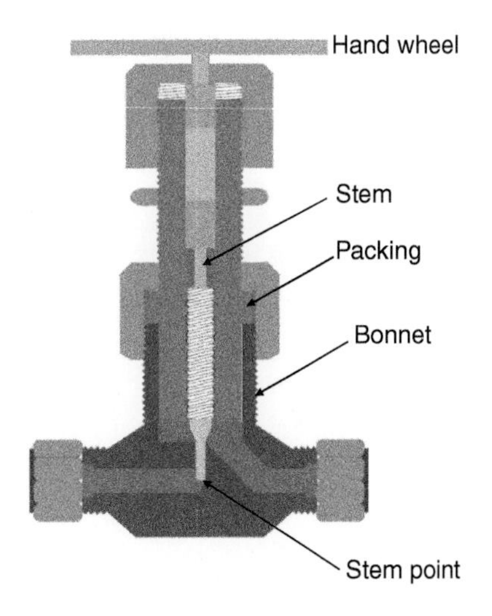

3.2.6 Pinch Valves

Pinch valves, as sketched in Figure 3.7, control the flow via a pinching effect. A pinch valve works like a tap, and it has three main parts: (i) a body or housing, (ii) an internal rubber sleeve, and (iii) end connections. To slow down the flow, the rubber sleeve is pinched and, as a result, the flow passage closes in. It is particularly suited for slurry flows, where the medium contains some amount of solid. It can also be used when the medium is completely solid.

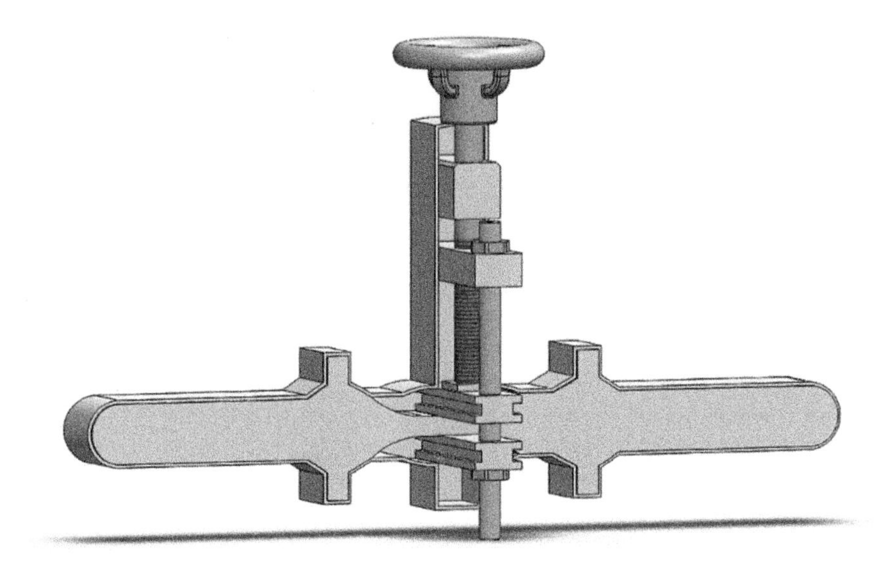

Figure 3.7 A schematic of a pinch valve. Source: S.K. Mohanakrishnan.

3.2.7 Plug Valves

Similar to a ball valve, but a plug, instead of a ball, is used in a plug valve (Figure 3.8). The plug is typically a cylindrical or cone-shaped tapered plug with a hole through it. It can be rotated 90

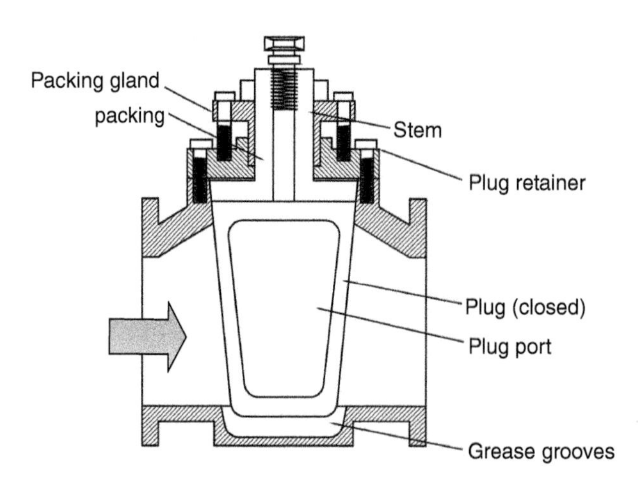

Figure 3.8 A schematic of a plug valve. Source: X. Wang, edited by D. Ting.

degrees with respect to the flow. Plug valves are simple and, therefore, economical. They are good for quick shutoff, but are not meant for accurate flow control.

3.2.8 Poppet Valves

A poppet valve consists of a tapered, round, or oval plug positioned on top of an opening via a compressing spring; see Figure 3.9. It is often used as a safety or pressure-relief valve; it releases some of the gas inside a container until the pressure is below a threshold for safe operation. For this kind of application, the poppet valve is often referred to as a relief valve. A poppet valve is also used for controlling the timing and quantity of gas into a thermofluid system, such as an engine. It can also serve as a check valve to prevent reverse flow. A safety-relief valve is a subject worthy of its own standalone merits. This is because of the high stakes involved, especially when the involved pressure is high and, more so, when in high-risk systems such as a nuclear plant. Hellemans (2009) is an excellent handbook on safety-relief valves. It follows ASME and International Codes and Standards. Air valves, which are used for removing air and wastewater gases from liquid piping systems (AWWA, 2016), are a somewhat different kind of relief valves.

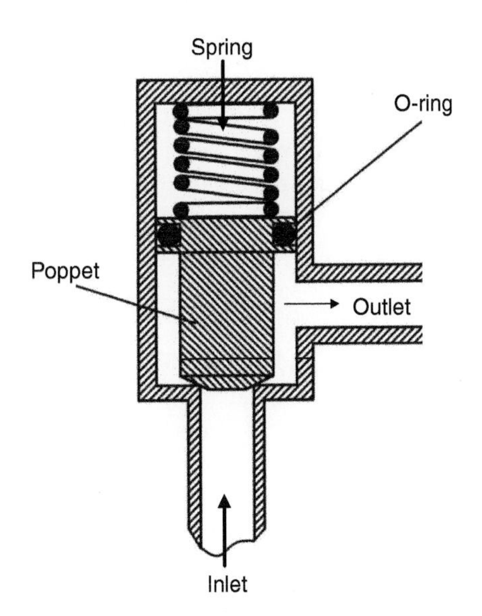

Figure 3.9 A schematic of a poppet valve. Source: X. Wang, edited by D. Ting.

3.2.9 Saddle Valves

As the name infers, a saddle valve "saddles" on a pipe, as illustrated in Figure 3.10, and directs a low volume of flow from the main, low-pressure pipe. An everyday application is for regulating a small amount of feed water to a humidifier in a residential home heating system. Saddle valves are also commonly used for supplying water to an ice-maker in a refrigerator. For these familiar applications, the saddle valves furnished with rubber seals are self-tapping. They can be installed with the main water supply running. The pin hole pierces the existing water line by turning the valve clockwise all the way. Once properly tapped, the amount of flow out and through the saddle valve is dictated by the counter-clockwise turning of the valve.

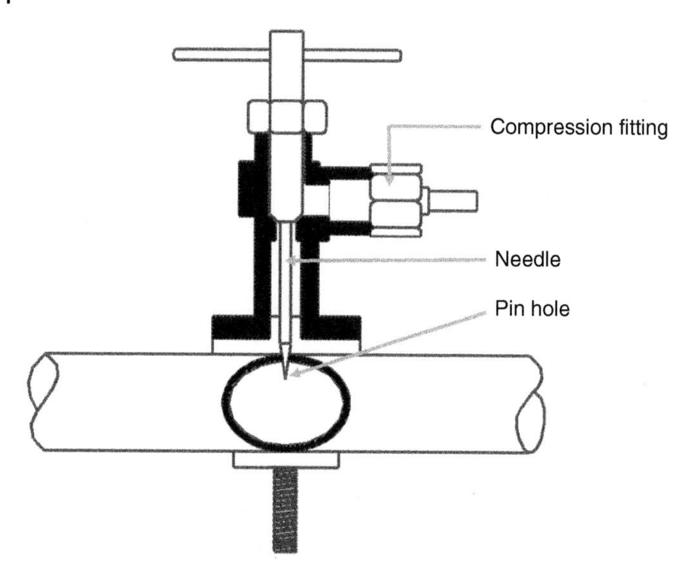

Figure 3.10 A schematic of a saddle valve. Source: G. Nagesh.

3.2.10 Some Comments on Valves

Globe, butterfly, plug, ball, cone, gate, and check valves are common in water distribution systems. The American Water Works Association, for example, is a regulating organization. They produce comprehensive documents for the industry. For example, AWWA (2006) is a manual detailing the design, selection, installation, testing, and maintenance of the mentioned distribution valves utilized in potable water systems in the United States.

3.3 Ducts, Pipes, and Fittings

Flow in a pipe is everywhere, when dealing with thermofluid systems. An obvious natural thermofluid system is the human body's blood-distribution system. According to the Franklin Institute (2019), the total length of all our blood vessels adds up to more than 100,000 km. That is a couple of loops around the earth! It is thus fitting to recap pipe flow, before we move on to review minor head losses in common fittings. We shall contain our scope to incompressible, viscous flow, such as a water delivery system with a pump, as shown in Figure 3.11. Another everyday example is a conditioned air distribution system. The key points about pipe or duct flow include the appreciation of the velocity profile, flow development, and shear stress distribution. It is important to differentiate the flow into laminar and turbulent flows. As pressure is required to move the fluid, energy consideration and head losses are some important and practical concepts to master. A selected recap of the involved fluid mechanics is given below. Readers are referred to standard fluid mechanics textbooks, such as Çengel and Cimbala (2018) and Gerhart et al. (2016) for further details.

3.3.1 Laminar and Turbulent Flow

The most eminent non-dimensional parameter in fluid mechanics is the Reynolds number. The Reynolds number denotes the flow's inertia with respect to fluid viscosity. According to Newton's

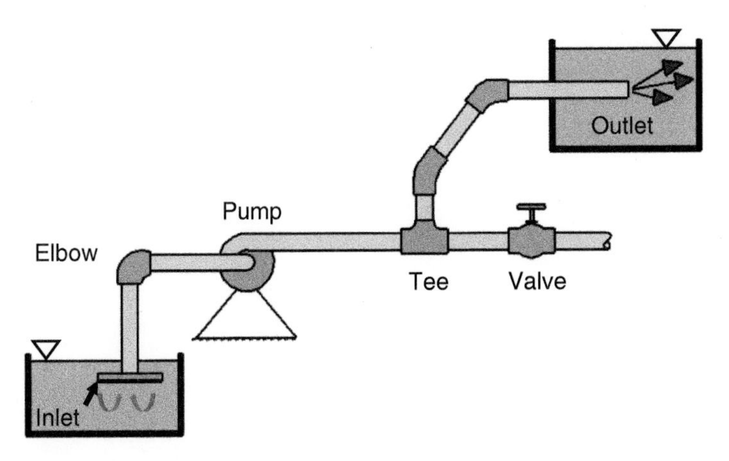

Figure 3.11 A water moving system with a pump. Source: Y. Nagaraja.

second law, force is equal to mass times acceleration. In respect to that, the inertial force of a moving fluid at velocity U is

$$F = ma = m'U = \rho \forall' U = \rho \left(D^2 U\right) U = \rho D^2 U^2 \tag{3.1}$$

where m is mass, m' is mass flow rate, a is acceleration, ρ is fluid density, ∀ is volume, ∀' is volume flow rate, and D is the diameter of the pipe. For Newtonian fluids, the shear,

$$\tau = \mu \, dU/dy \tag{3.2}$$

where μ is the fluid viscosity, and y is the direction perpendicular to the flow; in pipe flow, y is in the reverse radial direction, i.e. y = 0 at the wall, and it is equal to the radius of the pipe at the pipe center. Under the circumstances, the viscous force,

$$F_{vis} = \tau A = \mu A \, dU/dy = \mu \left(D^2\right) (U/D) = \mu DU \tag{3.3}$$

where A is area, and scaling has been applied. Dividing Eq (3.1) by Eq (3.3), we get the Reynolds number,

$$Re = \text{inertial force/viscous force} = \rho D^2 U^2 / (\mu DU) = \rho DU/\mu \tag{3.4}$$

When the Reynolds number (Re) is small, the viscous force dominates and, hence, the flow is smooth, as sketched in Figure 3.12a. This is somewhat analogous to a situation where there are many peacekeepers but only a few troublemakers. An event that takes place will proceed smoothly. We call this type of smooth-flowing phenomenon where the viscous force prevails laminar flow. The Reynolds number of the flow can be enlarged by increasing the velocity and/or the size of the pipe. With increasing Re, the inertial force amplifies and, thus, the viscosity starts to lose its ability to keep the disturbances associated with the inertia of the "rumbling" fluid in check. Consequently, the flow becomes transitional, as depicted in Figure 3.12b, which shows intermittent spots or bursts of turbulence in an otherwise calm, laminar stream. Further increases in Re lead to progressively more turbulent flow, see Figure 3.12c. In terms of the peacekeepers-troublemakers similitude, as increasingly more general people partake in causing trouble, the peacekeepers lose their ability to subdue the disturbances and, hence, disorders increase. In a nutshell, turbulent flow is unsteady, three-dimensional, and random in space and time (Ting, 2016). The critical Reynolds number for a flow stream to change from laminar to turbulent is a function of pipe roughness, vibrations, upstream fluctuations, disturbances from the ambient, etc. The rule of thumb in typical engineering practice is to assume the pipe flow is laminar when the Reynolds number is less than 2300, and fully turbulent when it is greater than about 4000.

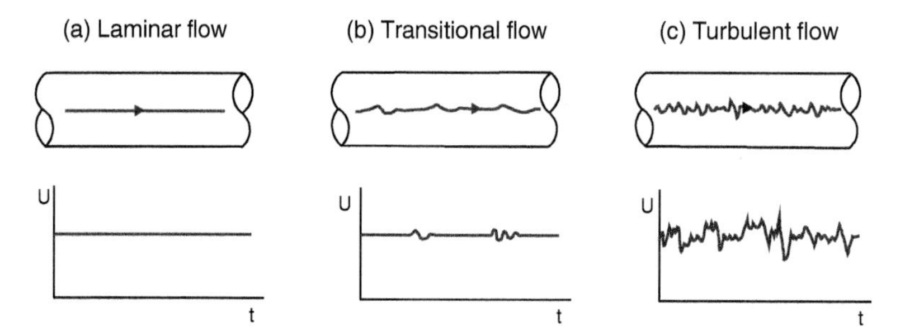

Figure 3.12 (a) Laminar, (b) transitional, and (c) turbulent pipe flow. Source: Y. Yang.

3.3.2 Entrance to Fully Developed Pipe Flow

Figure 3.13 shows the development of a flow entering a pipe with a uniform velocity profile. Note that the velocity gradient for the initially uniform velocity profile is zero. Let us board Ms. Valerie Frizzle's Magic School Bus and follow the flow into the pipe. As we are dealing with a fluid with a finite viscosity, once inside the pipe, the layer of the fluid next to the wall sticks to the wall. This next-to-the-wall layer drags the adjacent layer and slows it down. This layer subsequently slows down its neighboring layer and so on. The consequence of this depreciating dragging from the wall to the center of the pipe is a velocity gradient, dU/dy, where $y = 0$ at the wall, and $y = D/2$ at the center of the pipe, where D is the diameter of the pipe. Shortly after the entrance, the velocity gradient is large, as a significant portion of the flow in the core of the pipe is still moving with a uniform velocity. The velocity gradient lessens as we move farther into the pipe. The region with a non-zero velocity gradient is called the boundary layer. The *boundary layer* is the layer within which viscous effects are important. Outside the boundary layer, we have the inviscid core, where viscous effects are negligible and the velocity profile is uniform. The boundary layer changes the initial inertial velocity profile with respect to distance from the pipe entrance.[1] As the velocity gradient, and thus shear, decrease as we travel farther into the pipe, the boundary layer develops, i.e. thickens. With decreasing velocity in the boundary layer, continuity or conservation of mass causes the velocity at the center of the pipe to increase.

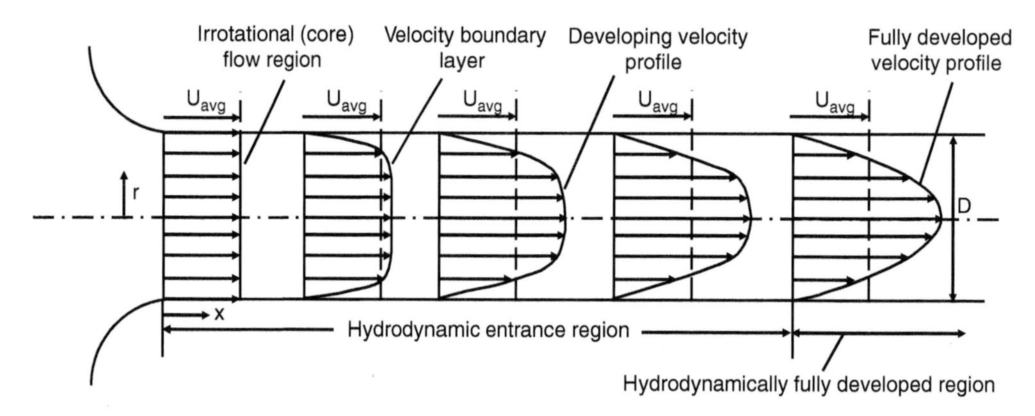

Figure 3.13 The development of flow in a straight circular pipe. Source: Y. Yang.

1 Initially, it is "inertial" velocity profile because the inertia of the moving fluid overshadows the viscosity.

When we travel sufficiently far into the pipe in the Magic School Bus, the boundary layer eventually reaches the center of the pipe. At this location, the flow becomes *fully developed* and the flow is entirely viscous. The distance from the entrance to where the flow becomes fully developed is called the (hydrodynamic) *entrance length*, L_e. Since the development is asymptotic, standard practice takes the location where the wall shear stress, the friction factor, reaches within 2% of the fully developed value as the threshold. We see that in the entrance region, where the boundary layer is developing, the effect of viscosity spreads from next to the wall into the core of the pipe. Consequently, the pressure drop per unit length of pipe, $\Delta P/\Delta x$ (or $\partial P/\partial x$), with respect to the distance from the entrance decreases, that is, the flow decelerates, as illustrated in Figure 3.14. Beyond the entrance length, the flow is fully developed, and the pressure drop per unit length of pipe with respect to distance from the entrance remains constant, unless there are other forces, such as the gravitational force, acting on it. Constant $\Delta P/\Delta x$ implies constant shear, that is, the shear remains constant once the flow is fully developed.

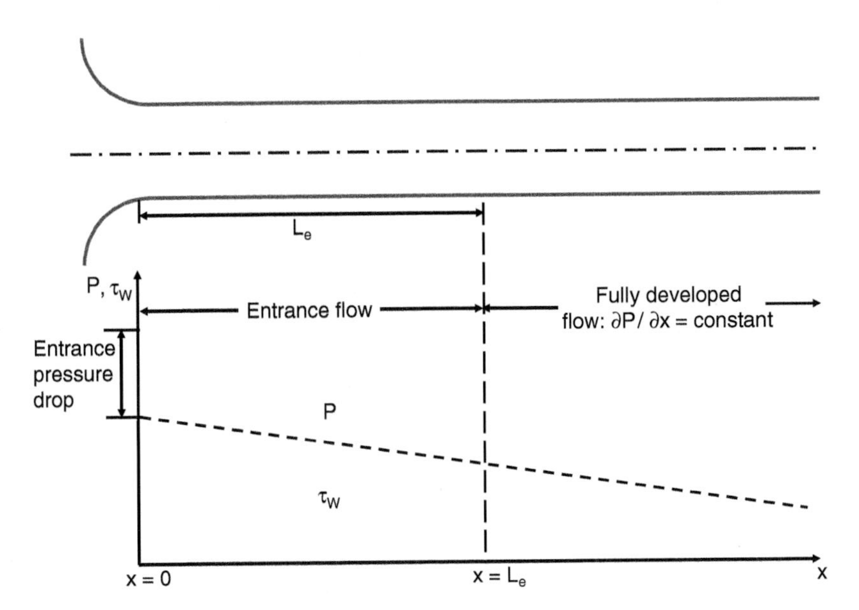

Figure 3.14 The pressure drop with respect to pipe length decreases to a constant value, as the flow becomes fully developed. Source: Y. Yang.

Laminar Entrance Length

It has be found that for laminar flows,

$$L_e/D \approx 0.06\ \rho UD/\mu \tag{3.5}$$

or

$$L_e \approx 0.06\ \text{Re}\ D \tag{3.5a}$$

for Re less than approximately 2300. We note that the entrance length increases with Re, as shown in Figure 3.15, and it can be up to 138 pipe diameters for laminar flow.

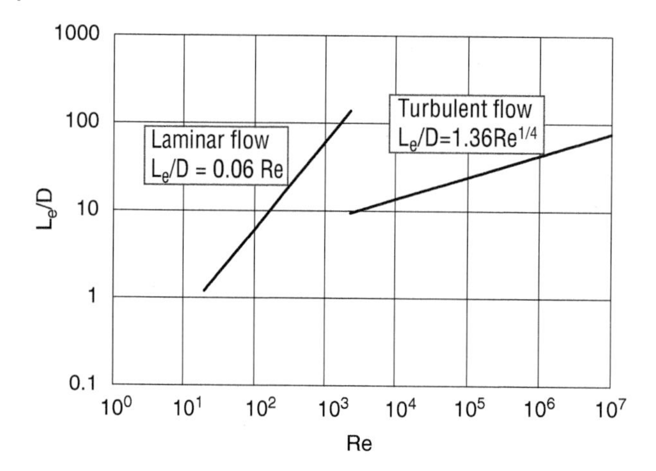

Figure 3.15 The entrance length for laminar and turbulent flow in a straight, circular pipe. Source: Y. Yang.

Turbulent Entrance Length

There is a lot more inertia in a turbulent flow and, hence, the flow develops much faster.[2] As a result, the entrance length for turbulent flow is much shorter. Moreover, as shown in Figure 3.15, the increase in L_e with respect to Re is drastically slower than that for the laminar counterpart. In short, for a turbulent flow,

$$L_e/D \approx 1.36\ Re^{1/4} \tag{3.6}$$

or

$$L_e/D \approx 4.4\ Re^{1/6} \tag{3.6a}$$

As most flows encountered in thermofluid systems are turbulent, the entrance length is reasonably short. Therefore, assuming the flow of interest is fully developed, with constant $\Delta P/\Delta x$, is acceptable in most practical applications.

3.3.3 Friction of Fully-Developed Pipe Flow

To convey the fluid through the piping system, we need to exert the appropriate amount of pressure. This is accomplished via a fan, a blower, or a pump. The required pumping power is dictated by the pressure drop, as discussed in Section 3.3.2. Dimensionally, the pressure drop for a fully-developed pipe flow in a smooth pipe is a function of the flow velocity, U, pipe length, L_{pipe}, pipe diameter, D, and the fluid viscosity, μ, that is,

$$\Delta P = f\left(U, L_{pipe}, D, \mu\right) \tag{3.7}$$

This is portrayed in Figure 3.16, where $\Delta P = P_1 - P_2$.

Invoking dimensional analysis, we see that there are five parameters, ΔP, U, L_{pipe}, D, and μ. The dimensions associated with these five parameters are mass, m, length, L, and time, t. Therefore,

2 The inertia is not restricted to the flow direction. Turbulent flow is three-dimensional, implying that the inertia is acting in all three directions. The inertia acting in the radial direction hastens the viscous effect initiated from next to the wall. Thereupon, the boundary layer development is expedited.

there are two non-dimensional parameters. Carrying out the dimensional analysis, one can find that these normalized parameters can be $D\Delta P/(\mu U)$ and L_{pipe}/D, or,

$$D\Delta P/(\mu U) = f\left(L_{pipe}/D\right) \tag{3.8}$$

This says that the pressure drop is a function of the length of pipe. Recall that for fully developed flows, the pressure drop per unit length is a constant, that is, ΔP is proportional to L_{pipe}. Thereupon, we can introduce a proportional constant, C, and make more explicit Eq (3.8), that is,

$$D\Delta P/(\mu U) = C\, L_{pipe}/D \tag{3.8a}$$

This can be rearranged into pressure drop per unit pipeline, that is,

$$\Delta P/L_{pipe} = C\mu U/D^2 \tag{3.8b}$$

For an incompressible flow in a uniform pipe, the volumetric flow rate is simply the product of the cross-sectional area and the average velocity,

$$\forall' = AU \tag{3.9}$$

Substituting for the average velocity from Eq (3.8b), this becomes,

$$\forall' = AU = \left(\pi D^2/4\right)\ \Delta P D^2/\left(C\mu L_{pipe}\right) = (\pi/4C)\,\Delta P D^4/\left(\mu L_{pipe}\right) \tag{3.9a}$$

For a pipe with a circular cross-section, C = 32 and, thus, the volume flow rate,

$$\forall' = \pi\Delta P D^4/\left(128\mu L_{pipe}\right) \tag{3.9b}$$

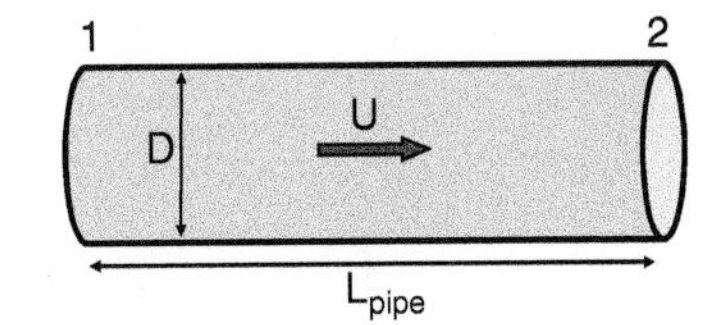

$\Delta P = P_1 - P_2 = f(U, L_{pipe}, D, \mu)$.

Figure 3.16 Parameters influencing the pressure drop of a fully-developed pipe flow. Source: D. Ting.

Substituting C = 32 for a circular pipe into Eq (3.8a) gives

$$\Delta P/L_{pipe} = 32\mu U/D^2 \tag{3.10}$$

or

$$\Delta P = 32\mu U L_{pipe}/D^2 \tag{3.10a}$$

Dividing both sides by $\frac{1}{2}\rho U^2$ leads to

$$\Delta P/\left(\tfrac{1}{2}\rho U^2\right) = 64\ \left[\mu/(\rho U D)\right]\ \left[L_{pipe}/D\right] \tag{3.10b}$$

Noting that $\mu/(\rho UD) = 1/Re$, thus, Eq (3.10b) can be rearranged into

$$\Delta P = \left[64/Re\right]\ \left[L_{pipe}/D\right]\ \left[\rho U^2/2\right] \tag{3.10c}$$

The term inside the first square brackets is a dimensionless quantity called the friction factor. Understandingly, this *friction factor* is a function of Reynolds number and, for fully developed laminar flow, it is simply

$$f = 64/Re \tag{3.11}$$

This is plotted in the Moody chart as f versus Re as depicted in Figure 3.17. Using the Darcy friction factor, we can rewrite Eq (3.10c) as

$$\Delta P = f\ (L_{pipe}/D)\ (\tfrac{1}{2}\rho U^2) \tag{3.12}$$

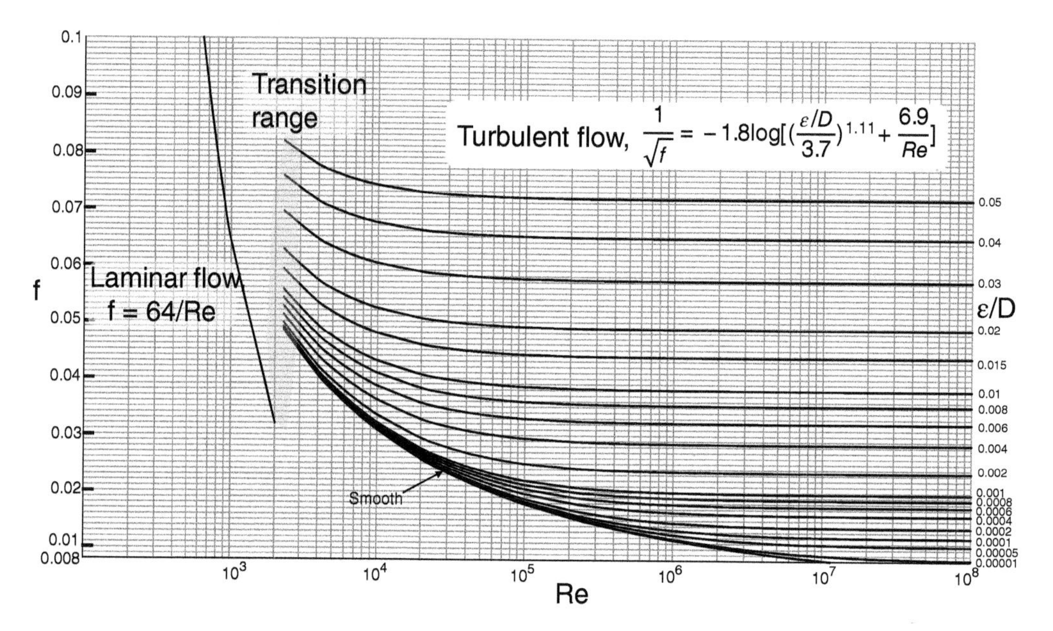

Figure 3.17 The Moody chart, friction factor versus Reynolds number. Source: Y. Yang.

It is interesting to note that, though we started with a smooth pipe with negligible roughness, the friction factor in the laminar-flow region is not influenced by the pipe roughness. This is because of the dominating viscous force keeping all potential disturbances created by the roughness in the boundary layer in check.

With an increasing Reynolds number, the viscous police started to lose control of the progressively more momentous and aggressive mob. The transitional flow region in the Moody diagram, as depicted in Figure 3.17 for Reynolds numbers between 2000 and 4000, is an uncertain regime. The friction factor escalates with Re, as denoted by the shaded, transitional region. Keep in mind that when the flow inertial force becomes more-or-less equal with respect to fluid viscosity, back and forth victory wavers; see Figure 3.12. This wavering contributes to the band of uncertain friction factors.

The pipe roughness has a direct impact on the transition to turbulent flow. In other words, roughness creates inertia in the boundary layer, which can amplify and propagate into the entire flow regime at sufficiently large Reynolds numbers. Accordingly, the friction factor for a fully developed pipe flow is a function of the relative roughness, ε/D, in addition to the Reynolds number. The quantitative effect is clearly delineated in the Moody diagram; see Figure 3.17. Note that the fully-rough, or wholly-turbulent, zone designates the region where the friction factor stays essentially unchanged with increasing Reynolds number. This wholly-turbulent region commences at lower Reynolds numbers with increasing (relative) roughness. At the other extreme, a relative roughness, ε/D, of less than 10^{-6} has an inconsequential effect and, hence, the pipe with such small roughness is considered smooth.

Example 3.1 ***Smooth or rough pipe?***

Given

A 6.35 mm in diameter pipe is used for delivering water at around 5°C. The inner surface of the pipe has a roughness of 0.0635 mm.

Find

If the roughness is a concern when water is flowing at an average speed of (i) 0.3 m/s, and (ii) 1.5 m/s. The corresponding friction factor values.

Solution

The dynamic viscosity of water at 5°C, $\mu = 1.52\times10^{-3}$ N·s/m², and density, $\rho = 1000$ kg/m³.

Therefore, at 0.3 m/s,

$$\text{Re} = \rho DU/\mu = 1000\ (6.35/1000)\ (0.3) / \left(1.52 \times 10^{-3}\right) = 1253.$$

This is in the laminar flow regime and, thus, the effect of roughness is not a concern.

At 1.5 m/s,

$$\text{Re} = \rho DU/\mu = 1000\ (6.35/1000)\ (1.5) / \left(1.52 \times 10^{-3}\right) = 6266.$$

This is in the turbulent flow regime and, hence, the effect of roughness should be taken into account.

For a laminar pipe flow at Re = 1253, the Moody diagram gives

$$f = 0.051.$$

For a turbulent pipe flow at Re = 6266 and $\varepsilon/D = 0.01$, the Moody diagram gives

$$f = 0.047.$$

There are many software packages available to aid the determination of friction coefficients. However, good old back-of-the-envelope calculations are often enough, especially as a first estimate. The Moody diagram may be a little hard to read. Therefore, the Haaland equation,

$$1/\sqrt{f} = -1.8\ \log\ \left[(\varepsilon/D/3.7)^{1.11} + 6.9/\text{Re}\right] \tag{3.13}$$

is often resorted to for turbulent flow.

3.3.4 Head Loss along a Pipe Section

If we apply conservation-of-energy principles to a section of a pipe, as shown in Figure 3.18, we can obtain

$$P_1 + \tfrac{1}{2}\rho U_1^{\,2} + z_1\ \rho\ g = P_2 + \tfrac{1}{2}\rho U_2^{\,2} + z_2\ \rho\ g \tag{3.14}$$

This says that the sum of pressure (flow) energy plus kinetic energy plus potential energy at pipe section 1 is equal to that at pipe section 2. From the preceding section on friction that we have just mastered, we know that the flowing fluid loses some energy through friction, as it travels from section 1 to 2. We can denote this loss as *head loss*, h_L, which is in units of height or length. Dividing

Eq (3.14) by the specific weight, $\gamma = \rho g$, of the fluid involved, and accounting for the frictional loss, we have

$$P_1/\rho g + ½U_1^2/g + z_1 = P_2/\rho g + ½U_2^2/g + z_2 + h_L \tag{3.15}$$

If the pipe is of uniform cross-section, then U_1 is equal to U_2. We can simplify the head loss expression into

$$h_L = P_1/\rho g + z_1 - (P_2/\rho g + z_2) \tag{3.15a}$$

For a horizontal pipe, this is further reduced to

$$h_L = (P_1 - P_2)/\rho g \tag{3.15b}$$

The head loss is simply the pressure drop. Invoking Eq (3.12), we have head loss for a fully developed flow in a horizontal pipe,

$$h_L = \Delta P/\rho g = f\ (L_{pipe}/D)\ (½U^2/g) \tag{3.15c}$$

This head loss is called the *major head loss*, when it is associated with the straight section of the pipe. It is major because, in a typical piping system, most of the losses are due to pipe friction; less losses are caused by the fittings. We can also express the energy equation, Eq (3.14) or Eq (3.15), in terms of the friction factor as

$$P_1 + ½\rho U_1^2 + z_1\ \rho\ g = P_2 + ½\rho U_2^2 + z_2\ \rho\ g + f\ (L_{pipe}/D)\ (½\rho U^2) \tag{3.14a}$$

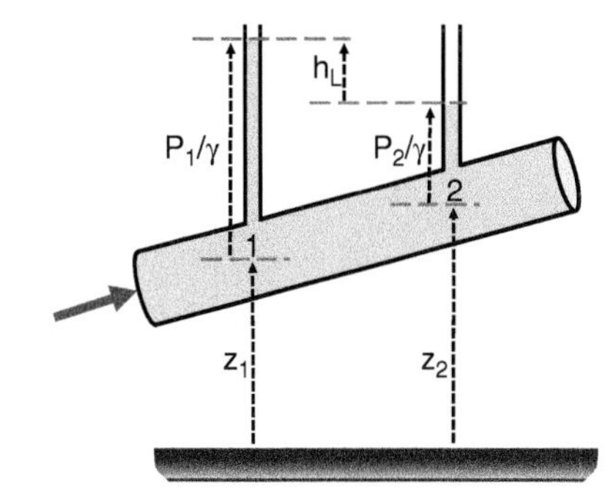

Figure 3.18 Energy conservation across a pipe section. Source: D. Ting.

The roughness values of a sample of commercial pipes are presented in Table 3.1. Let us re-examine Example 3.1 by employing a more realistic pipe. We will compare the associated head loss using pipes of different materials.

Table 3.1 Sample commercial pipe roughness values.

Pipe	[mm]	[in]
Plastic	0.0	0.0
Aluminum	0.001–0.002	$3.9 \times 10^{-8} - 7.9 \times 10^{-8}$
Drawn tubing	0.015	0.00006
Commercial steel	0.045	0.0018
Galvanized iron	0.15	0.006
Cast iron	0.26	0.010

Example 3.2 *Cast iron versus galvanized iron*

Given
A 6.35 mm in diameter pipe is called upon to deliver water at around 5°C over a distance of 120 m. You are asked to choose either a cast iron or galvanized iron pipe.

Find
The value of the friction factor for an average velocity of (i) 0.3 m/s, and (ii) 1.5 m/s.

Solution
The dynamic viscosity of water at 5°C, $\mu = 1.52 \times 10^{-3}\ \mathrm{N \cdot s/m^2}$, and density, $\rho = 1000\ \mathrm{kg/m^3}$.

At 0.3 m/s,

$$\mathrm{Re} = \rho DU/\mu = 1000\ (6.35/1000)\ (0.3) / \left(1.52 \times 10^{-3}\right) = 1253.$$

This corresponds to laminar flow and, hence, the effect of roughness is not a concern. The Moody diagram gives

$$f = 0.051.$$

At 1.5 m/s,

$$\mathrm{Re} = \rho DU/\mu = 1000\ (6.35/1000)\ (1.5) / \left(1.52 \times 10^{-3}\right) = 6266.$$

This signifies that the flow is fully turbulent flow; therefore, the effect of roughness needs to be taken into account. For cast iron, ε is 0.26 mm and, hence, $\varepsilon/D = 0.26/6.35$. The Moody diagram gives

$$f = 0.069.$$

If galvanized iron is used instead, ε is 0.15 mm and, hence, $\varepsilon/D = 0.15/6.35$. The Moody diagram gives

$$f = 0.057.$$

We can save approximately 17% of the pumping power when switching from cast iron to galvanized iron pipe.

Non-circular conduits are consistently employed to distribute the air associated with HVAC systems. To apply the Moody diagram to deduce the friction factor, the equivalent, hydraulic diameter can be called upon. For a rectangular duct, the hydraulic diameter is simply,

$$D_h = 2HW / (H + W) \tag{3.16}$$

where H is the height and W is the width.

Example 3.3 *Pressure loss in an air distribution duct*

Given
A 50.8 cm by 25.40 cm duct is used to distribute conditioned air at a volume flow rate of 5 m^3/s in a building. The total air-distributing duct work is 75 m in length.

Find
The major head loss.

Solution

For atmospheric air at about 27°C,

$$\mu = 1.8 \times 10^{-5}\ \mathrm{N \cdot s/m^2},$$

and

$$\rho = 1.2\ \mathrm{kg/m^3}.$$

The hydraulic diameter of the rectangular duct,

$$D_h = 2hw/(h + w) = 2\ (0.508)(0.2540)/(0.508 + 0.2540) = 0.3387\ \mathrm{m}.$$

Average air velocity,

$$U = 5\ \mathrm{m^3/s}/[0.508\,(0.254)] = 38.75\ \mathrm{m/s}$$

At 38.75 m/s,

$$Re = \rho DU/\mu = 1.2\ (0.3387)\ (38.75)/\left(1.8 \times 10^{-5}\right) = 8.7 \times 10^5.$$

This is in the fully turbulent flow regime and, hence, the effect of roughness has to be explicitly accounted for. For commercial steel, ε is 0.045 mm and, hence, $\varepsilon/D = 0.045/1000/0.3387 = 0.0001$. The Moody diagram gives

$$f = 0.019.$$

From Eq (3.15c),

$$h_L = \Delta P/\rho g = f\ (L_{pipe}/D)\ (\tfrac{1}{2}U^2/g)$$

$$= 0.019\ (75/0.3387)\ \tfrac{1}{2}(38.75)^2/9.81 = 3790\ \mathrm{Pa}/\rho g = 322\ \mathrm{m\ of\ air}.$$

3.3.5 Minor Head Loss

As discussed earlier, frictional loss in a pipe constitutes the largest pressure drop in typical, practical piping networks. Other than straight sections, various fittings are needed to realize a workable system. The most common fittings include elbows, bends, tees, and connectors. The additional pressure loss imposed by these fittings is call the *minor head loss*. It is minor because it is usually smaller than the major head loss due to the long sections of pipe. As the flows through the fittings are varied and complex, the value of the minor head loss associated with each fitting is deduced experimentally through standardized tests. As with the major head loss, Eq (3.15c), the pressure drop across a fitting is also proportional to the square of the velocity. Specifically, minor head loss,

$$h_{L,minor} = K_L\ \tfrac{1}{2}U^2/g \tag{3.17}$$

where K_L is the experimentally-deduced loss coefficient. This loss coefficient is a function of Reynolds number, fitting geometry, material, and make and model.

Physically, it is clear that the more abrupt a change in the flow is, the larger the pressure drop. For example, the prevailing air distribution duct work consists of many bends and smoothing the redirecting of the flow can significantly reduce the minor head loss. This is depicted in Figure 3.19, where the implementation of guide vanes largely eliminates flow reversals and/or energy-dissipating secondary flows.

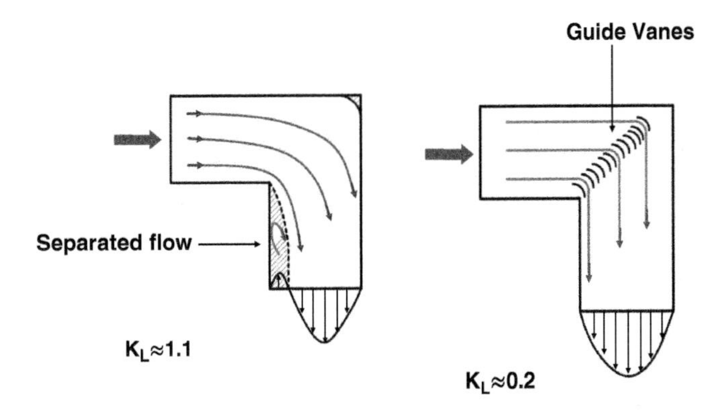

Figure 3.19 Reduce minor head loss in a bend via guide vanes. Source: X. Wang.

Example 3.4 *Reduce minor head loss across a 90-degree bend*

Given

A 50.8 cm square-cross-section air duct is fitted with a 90-degree bend. The efficacy of guide vanes is quantified in a laboratory by measuring the pressure drop across the 90-degree bend in the presence and absence of guide vanes. At 5 m^3/s atmospheric air at around 27°C, the pressure drops with and without the guide vanes are 180 Pa and 250 Pa, respectively.

Find

The corresponding values of loss coefficient for the minor head loss.

Solution

Let us derive the expression from the conservation of energy. We can reduce Eq (3.15),

$$P_1/\rho g + \tfrac{1}{2}U_1^2/g + z_1 = P_2/\rho g + \tfrac{1}{2}U_2^2/g + z_2 + h_L$$

into

$$P_1/\rho g = P_2/\rho g + h_{L,minor}.$$

This can be rewritten, along with Eq (3.17), as

$$h_{L,minor} = (P_1 - P_2)/\rho g = K_L \, \tfrac{1}{2}U^2/g$$

Therefore, the loss coefficient,

$$K_L = 2(P_1 - P_2)/(\rho U^2)$$

For atmospheric air at about 27°C,

$$\mu = 1.8 \times 10^{-5}\ \text{N·s/m}^2$$

and

$$\rho = 1.2\ \text{kg/m}^3$$

The hydraulic diameter,

$$D_h = 2hw/(h + w) = 2(0.508)(0.508)/(0.508 + 0.508) = 0.508\ \text{m}$$

The average velocity,

$$U = 5/\left(\pi D^2/4\right) = 24.67\ \text{m/s}$$

Substituting the above values into the loss coefficient expression gives

$$K_L = 2\Delta P/\left(\rho U^2\right)$$

In the presence of the guide vanes,

$$K_L = 2\,(180)\,/\,\left[(1.2)\,(24.67)^2\right] = 0.5$$

or

$$h_{L,minor} = \left(P_1 - P_2\right)/\rho g = K_L\ ½U^2/g = 15.3\ \text{m of air}$$

In the absence of the guide vanes,

$$K_L = 2\,(250)\,/\,\left[(1.2)\,(24.67)^2\right] = 0.7$$

or

$$h_{L,minor} = \left(P_1 - P_2\right)/\rho g = K_L\ ½U^2/g = 21.2\ \text{m of air}$$

3.4 Piping Network

The simplest piping network is a single pipe system for delivering a fluid from one spot to another. Single-pipe systems can be divided into three types. The most straightforward type is the deduction of the pressure drop, as we did in the examples we just covered. For this type of single-pipe problems, the flow rate, pipe length, and pipe diameter, among other details such as pipe material or roughness, are known. The solution process starts with the computation of the Reynolds number. This, along with the pipe roughness, enables the estimation of the friction factor with the help of the Moody diagram. The pressure drop falls out via the conservation of energy equation. Sizing of an air duct is frequently encountered in HVAC practice, so is the piping network for water distribution (McDonald and Magande, 2012). For this type of problem, the required flow rate, the pressure drop, which is dictated by the fan capacity, and the transport distance are known. The solution is obtained via an iterative approach. Let us go through an example on this.

Example 3.5 ***Air duct size for a given flow rate and pressure drop***

Given

Atmospheric air is to be transported through a 175 m-long commercial steel circular duct at 0.35 m^3/s. Based on the fan size, the head loss associated with the straight air duct section is to be no more than 20 m of air, and that due to fittings to be less than 5 m of air.

Find

The minimum duct size.

Solution

Assume the flow to be turbulent, the (major) head loss,

$$h_L = \Delta P/\rho g = f\ \left(L_{pipe}/D\right)\left(½U^2/g\right) \tag{E3.5.1}$$

Replace the velocity, U, with the volume flow rate,

$$\forall' = \pi(\tfrac{1}{2}D)^2\, U \tag{E3.5.2}$$

This gives

$$U = 4\forall' / \left(\pi D^2\right) \tag{E3.5.2a}$$

Substituting this into Eq (E3.5.1), we get

$$h_L = f\left(L_{pipe}/D\right) \tfrac{1}{2} \left(U^2/g\right) = 8\, f\, L_{pipe} \forall'^2 / \left(\pi^2\, D^5\, g\right) \tag{E3.5.1a}$$

First guess: D = 0.2 m
Then, from Eq (E3.5.2a), we have

$$U = 4\forall' / \left(\pi D^2\right) = 4\,(0.35) / \left[\pi (0.2)^2\right] = 11.14\ \text{m/s}$$

The Reynolds number, for atmospheric air at 20°C

$$Re = \rho U D/\mu = 1.204\,(11.14)\,(0.2)\,/1.825 \times 10^{-5} = 146{,}998$$

The relative roughness, $\varepsilon/D = 0.000\,045/0.2 = 0.000225$. From the Moody diagram,

$$f = 0.018$$

The corresponding head loss, Eq (E3.5.1a),

$$\begin{aligned} h_L &= 8\, f\, L_{pipe} \forall'^2 / \left(\pi^2\, D^5\, g\right) \\ &= 8\, f\ (175)\ (0.35)^2 / \left(9.81\, \pi^2\, D^5\right) \\ &= 1.771\ f/D^5 \end{aligned} \tag{E3.5.1b}$$

For f = 0.0180 and D = 0.2, we have

$$h_L = 1.771\ f/D^5 = 99.6\ \text{m}$$

This is way larger than 20 m. Let us try a larger pipe size to lower the head loss.

Second guess: D = 0.3 m
Then, from Eq. E3.5.2a, we have

$$U = 4\forall' / \left(\pi D^2\right) = 4\,(0.35) / \left[\pi (0.3)^2\right] = 4.951\ \text{m/s}$$

The Reynolds number, for atmospheric air at 20°C

$$Re = \rho U D/\mu = 1.204\,(4.951)\,(0.3)\,/1.825 \times 10^{-5} = 97{,}999$$

The relative roughness, $\varepsilon/D = 0.000\,045/0.2 = 0.000225$. From the Moody diagram,

$$f = 0.0188$$

For f = 0.0188 and D = 0.3, we have

$$h_L = 1.771\ f/D^5 = 13.7\ \text{m}$$

This is moderately less than 20 m. As commercial ducts usually come in standard sizes, this may be the closest to the available size which satisfies the requirement.

Problems

3.1 Pressure drop in an air distribution duct

Air at 1 atm and 37°C is distributed via a 30 cm by 30 cm square commercial steel duct at $0.6 \, m^3/s$. What are the pressure drop, head loss, and required pumping power, per unit length (meter) of duct?

3.2 Design an air distribution duct network

Conditioned air at 1 atm and approximately 12°C is to be transported via a circular duct at $0.3 \, m^3/s$. The distance between the air conditioner and the space to be conditioned is 50 m. The fan capacity limits the head loss to a maximum of 300 m of air. Find the minimum diameter of the duct.

Hint: Guess a diameter, from which you can deduce the average velocity. Then, based on the Reynolds number, and the roughness (based on the typical material, e.g. aluminum, commercial steel), obtain the friction factor. If the head loss is much less than the maximum, you have probably oversized the duct. In this case, reiterate with a smaller diameter. If the head loss is larger than the maximum, go up to the next duct size, and repeat the calculations.

3.3 Design a water delivery system

Water from a pond is to be pumped into a reservoir at an elevation of 17 m. Approximately $30 \, m^3$ of water is needed in the morning to irrigate a greenhouse over a period of one hour. The same amount and irrigation rate are needed again in the afternoon. The piping passage requires twelve 90-degree elbows and seven tees. Design a workable system.

3.4 Pumping water via two pumps in series or in parallel

Two identical pumps are available to pump water from a river into a tank at a higher elevation as a renewable energy storage system. Each pump can produce a pressure increase of ΔP and a volume flow rate of $\dot{\forall}$. The piping circuit is arranged such that these two pumps can either operate in parallel or in series. On a particular day the available renewable is abundant but brief and hence, it is desirable to pump the maximum amount of water over the brief period. Which of the following is the most appropriate arrangement?

a) Use only one pump.

b) Use both pumps in parallel.

c) Use both pumps in series.

3.5 Choose the better option for boosting flow rate

City water at 200 kPa gauge (above ambient pressure) is delivered through a 23 m pipe to the second floor of a building at 15 m above the water meter. The existing 1.27 cm diameter is old, resulting in a relative roughness ε/D of 0.05. There are two options to boost the flow rate. You can replace the old pipe with a newer one of near-zero roughness. Alternatively, you can install a pump to bring the water pressure up to 300 kPa gauge. Which option can lead to a higher water flow rate?

References

ASHRAE (2008). *ASHRAE Handbook, HVAC Systems and Equipment*. Atlanta, GA: ASHRAE.

AWWA (2006). *Distribution Valves: Selection, Installation, Field Testing, and Maintenance, M44*, 2nd ed. Denver, CO: American Water Works Association.

AWWA (2016). *Air-Valves: Air-Release, Air/Vacuum and Combination: Manual of Water Supply Practices, M51*, 2nd ed. Denver, CO: American Water Works Association.

Çengel, Y.A. and Cimbala, J.M. (2018). *Fluid Mechanics: Fundamentals and Applications*, 4th ed. New York: McGraw-Hill.

Franklin Institute (2019). Heart and blood vessels. https://www.fi.edu/heart/blood-vessels, accessed November 7, 2019.

Gerhart, P.M., Gerhart A.L., and Hochstein J.I. (2016). *Munon, Young and Okiishi's Fundamentals of Fluid Mechanics*, 8th ed. Hoboken, NJ: Wiley.

Hellemans, M. (2009). *The Safety Relief Valve Handbook: Design and Use of Process Safety Valves to ASME and International Codes and Standards*. Oxford: Elsevier.

McDonald A.G. and Magande, H.L. (2012). *Introduction to Thermo-Fluids Systems Design*. Chichester: Wiley.

Nesbitt, B. (2007). *Handbook of Valves and Actuators*. Oxford: Elsevier.

Smith, P. and Zappe, R.W. (2004). *Valve Selection Handbook: Engineering Fundamentals for Selecting the Right Valve Design for Every Industrial Flow Application*, 5th ed. London: Elsevier.

Ting, D.S-K. (2016). *Basics of Engineering Turbulence*. London: Academic Press.

4

Heat Exchangers

When I look at the solar system, I see the earth at the right distance from the sun to receive the proper amounts of heat and light. This did not happen by chance.

– Isaac Newton

Chapter Objectives

- Appreciate everyday heat exchangers.
- Differentiate direct contact heat exchangers from indirect ones.
- Classify (indirect) heat exchangers based on the flow configuration, i.e. parallel flow, counter-flow, and cross-flow.
- Categorize (indirect) heat exchangers according to the type of construction: shell-and-tube, finned-coil, and compact.
- Perform first law thermodynamic analysis on common heat exchangers.
- Understand LMTD (log mean temperature difference) and NTU (number of transfer units).

Nomenclature

A	area; A_c is the cross-sectional area, A_s is the surface area
C	the heat capacity rate; C_c is the heat capacity rate of the cold stream, C_h is the heat capacity rate of the hot stream
c_P	heat capacity at constant pressure; $c_{P,c}$ is the heat capacity of the cold stream, $c_{P,h}$ is the heat capacity of the hot stream
D	diameter; D_h is the hydraulic diameter
d	differential
F_C	correction factor
h	enthalpy; $h_{c,i}$ is the incoming cold stream enthalpy, $h_{c,o}$ is the enthalpy of the outgoing cold stream, h_{fg} is the enthalpy of vaporization, $h_{h,i}$ is the incoming hot stream enthalpy, $h_{h,o}$ is the outgoing hot stream enthalpy
h_{conv}	convective heat transfer coefficient
HEX	heat exchanger
k	thermal conductivity

L	length
LMTD	log mean temperature difference
m	mass; m' is mass flow rate, m_c' is the mass flow rate of the cold stream, m_h' is the mass flow rate of the hot stream, m_{phase}' is the rate of condensation or evaporation
n	number of tubes
NTU	number of transfer units
Nu	Nusselt number
p_L	perimeter
Pr	Prandtl number
Q	heat transfer; Q' is heat transfer rate, Q_{in}' is the rate of heat transfer going into the system, Q_{max}' is the maximum heat transfer rate, Q_{out}' is the rate of heat transfer going out of the system
R	thermal resistance
Re	Reynolds number
T	temperature; T_i is incoming fluid temperature, T_c is the cold stream temperature, $T_{c,i}$ is the incoming cold stream temperature, $T_{c,o}$ is the outgoing cold stream temperature, $T_{h,i}$ is the incoming hot stream temperature, $T_{h,o}$ is the outgoing hot stream temperature, T_o is the outgoing fluid temperature, T_{sat} is the saturation temperature, T_w is the wall temperature, ΔT_1 the temperature change in stream 1, ΔT_2 the temperature change in stream 2, ΔT_{lm} is the log mean temperature difference, $\Delta T_{lm,CF}$ is the log mean temperature difference of a counter-flow heat exchanger, ΔT_m is the mean temperature difference
t	temperature or time
U	overall heat transfer coefficient
V	velocity
x	x-direction; dx is a differential or infinitely small distance in the x-direction

Greek and Other Symbols

α	temperature change ratio
β	temperature change ratio
μ	dynamic viscosity
ν	momentum diffusivity or kinetic viscosity
κ	thermal diffusivity
ρ	density
ε	efficiency, heat exchanger effectiveness

4.1 Effective Exchange of Thermal Energy

Effective exchange of thermal energy is present in literally all creatures. For example, male dolphins, having their testis inside their warm bodies, require an effective counter-flow (counter-current) heat exchanger to keep their testis cool, preventing the extinction of the intelligent species; see Figure 4.1. As such, a *heat exchanger* is simply a system for exchanging heat (thermal energy) between media, which is usually two fluid streams. Note that the body of the dolphin is warmer than the optimal temperature for viable sperm production and growth. Therefore, the thermal energy is efficaciously transported away and dissipated elsewhere into the relatively cooler ambient.

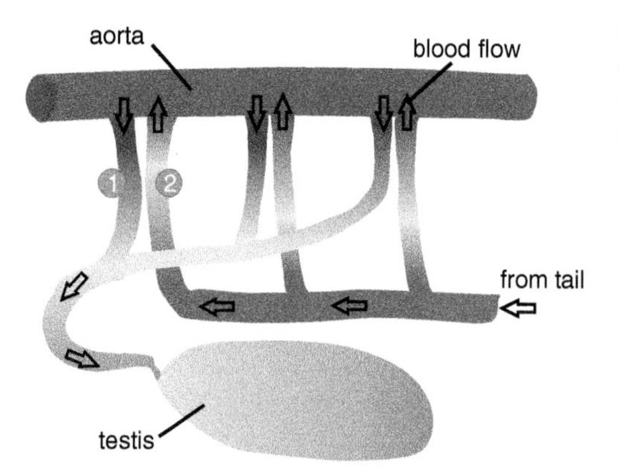

Figure 4.1 A natural vascular counter-flow heat exchanger effectively keeps the dolphin's testis cool, allowing the sperm to develop and flourish. Source: Y. Yang.

Another familiar example of heat exchangers in nature are the long and exposed legs of wading birds. As illustrated in Figure 4.2, to keep the naked skeletal legs from freezing in the cold, warm blood from the core of their bodies is continually delivered down to the very tips of their feet. Also, to keep chilled blood from stopping the heartbeat, the returning cold blood picks up thermal energy from the warm blood traveling in the opposite direction. It is interesting to note that by the time the warmer blood from the body reaches the cold feet, which may be in direct contact with icy cold surface, its temperature is significantly lower than that of the body's. In this way, the feet are kept from numbing, and yet, at the same time, the temperature gradient between the feet and the cold surface is moderate, to minimize heat loss.

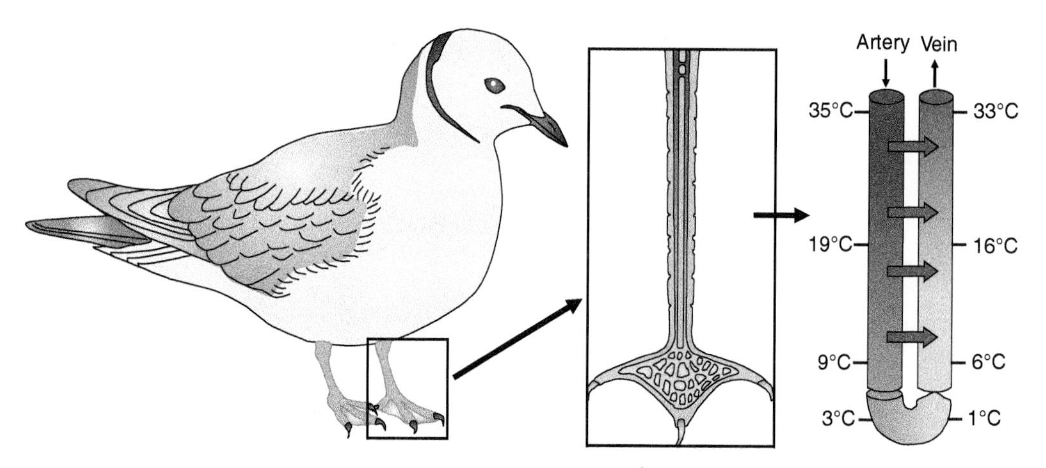

Figure 4.2 Counter-flowing blood vessels in a long-legged bird. Source: Y. Yang.

We see that heat management is meticulously designed in nature. This is also the case in engineering. One of the most recognizable man-made heat exchangers is the radiator of a car; see Figure 4.3. To efficiently remove some 60% of the thermal energy from the combustion of the fuel, the surface area for heat transfer has to be decidedly extended. This is accomplished using many large-surface-area thin fins that are closely spaced, and a fan is used to forcefully convect the heat into the atmosphere.

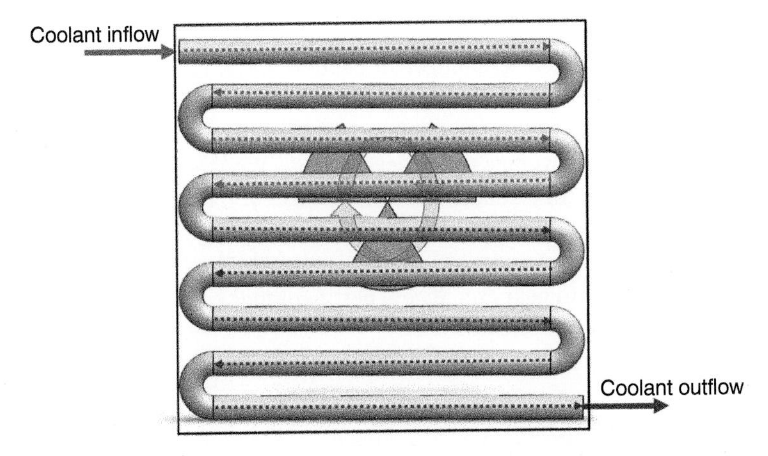

Figure 4.3 A car radiator as an effective heat exchanger. Source: S.K. Mohanakrishnan.

4.2 Types of Heat Exchangers

Heat exchange can take place without direct contact. A heat exchanger that operates without the warm and cold streams mixing or contacting each other is called an *indirect contact heat exchanger*, or simply *indirect heat exchanger*. All three examples given in Section 4.1 are indirect heat exchangers. It is, however, much more effective as far as the heat exchange is concerned, to have direct mixing of the working fluids. A cooling tower is one such *direct contact heat exchanger*. Cooling towers associated with power plants are easily noticeable by the visible water vapor plumes that rise from them. It is interesting to note that these harmless water vapor plumes have been mistaken as toxic pollution. Figure 4.4 depicts that the cooler atmospheric air entering from the lower part of the cooling tower picks up thermal energy from the hot water droplets in the spray. The thermal energy transport into the outgoing air drawn by the fan at the top of the cooling tower is made effective due to direct contact, where part of the water in the hot droplets is gathered by the upward and outward-bound atmospheric air. The latent heat of evaporation significantly augments the heat transfer rate.

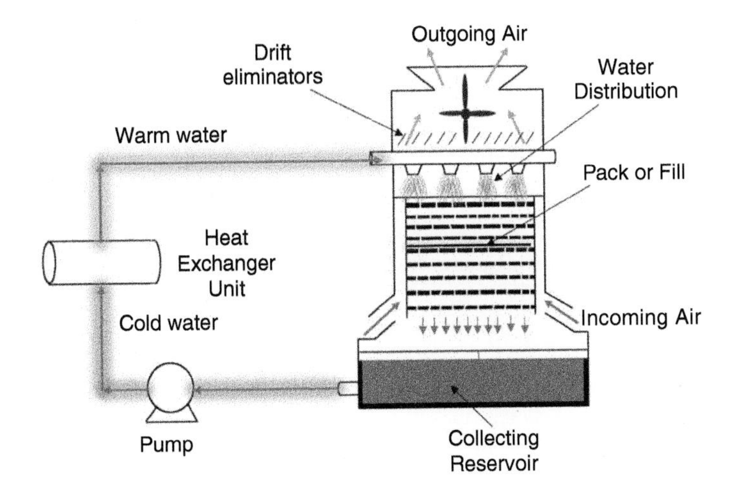

Figure 4.4 A cooling tower as a direct contact heat exchanger. Source: S.K. Mohanakrishnan.

More common is the occurrence of heat exchange without mixing or direct contact of two streams. Let us further classify this class of heat exchanger in Section 4.3. The intention here is to furnish a sufficient disclosure of heat exchangers, providing adequate background and understanding to equip engineers-in-training to execute the design and optimization of systems involving heat exchange. For more in-depth coverage of heat exchangers, the reader can refer to recognized books, such as Bergman et al (2018), Çengel and Ghajar (2020), Kakaç et al. (2012), and Thulukkanam (2013). The latter two are focused on heat exchanger design. The classical text on compact heat exchangers by Kays and London (1964) is also worth reading. McDonald and Magande (2012) dedicated two chapters to heat exchangers pertinent to thermofluid systems design.

4.3 Indirect-Contact Heat Exchangers

Indirect (contact) heat exchangers, where the two streams do not come into contact, can be categorized based on the flow arrangement or the construction type. With respect to the flow arrangement, there are parallel-flow, counter-flow, and cross-flow heat exchangers. We can infer from the intelligent design in nature that the counter-flow or counter-current arrangement is possibly the most forceful one. Physical settings, however, may call for a cross-flow heat exchanger arrangement. Parallel-flow heat exchangers are rarely seen in practice. Nevertheless, there is fundamental merit in understanding parallel-flow heat exchangers. In addition, comparing and contrasting the three types of heat exchangers can enhance our understanding and appreciation of all three types of indirect-contact heat exchanger based on flow arrangement.

4.3.1 A Single Fluid in a Conduit of Constant Temperature

Before we look in detail at the three common types of heat exchangers based on flow arrangement, let us first examine the simpler case of a single fluid flowing down a duct, as shown in Figure 4.5. To further simplify this single-fluid heat exchanger, we consider the case where the flow is steady and fully-developed and the wall is at a constant wall temperature, T_w. If the change in the fluid temperature is not too large, we may neglect the small changes in the fluid properties. Following this, we can assume that the overall heat transfer coefficient, U, is constant. Specifically, the assumptions are:

1) a fully-developed, steady flow in a uniform conduit;
2) the wall temperature, T_w, is constant throughout the entire conduit;
3) a moderate change in temperature and, hence, the fluid properties remain unchanged;
4) the overall heat transfer coefficient, U, is constant.

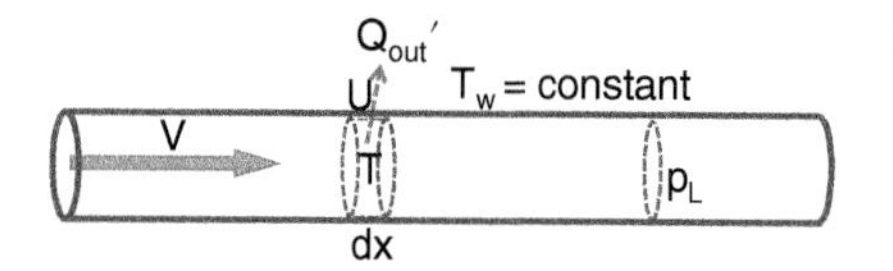

Figure 4.5 A single-fluid heat exchanger. Source: D. Ting.

Let us apply the conservation of energy principle to the system shown in Figure 4.5. The first law of thermodynamics says that

$$\text{energy storage rate} = \text{energy input rate} - \text{energy output rate} \tag{4.1}$$

Imagine we are in Ms. Frizzle's Magic School Bus following a fluid slug with mass m. The energy balance for a differential control volume of the fluid slug, with mass m, specific heat c_P, and heat transfer rate Q′,

$$m\ c_P\ dT/dt = Q_{in}' - Q_{out}' \tag{4.1a}$$

or

$$(\rho\ A\ dx)\ c_P\ dT/dt = 0 - U\ p_L\ dx\ (T - T_w) \tag{4.1b}$$

where T is temperature, t is time, ρ is the density of the fluid, and p_L is the perimeter of the conduit. The elapsed time from the entrance, t = x/V, where x is the distance downstream, and V is the average velocity of the flowing stream. The mass flow rate, m' = ρ V A_c, where A_c is the cross-sectional area. The heat transfer surface area, A = p_L L, where L is the length of the duct. Differentiating (T – T_w) gives

$$d\,(T - T_w) = dT \tag{4.2}$$

With V = dx/dt, we have dt = dx/V. Substituting this and Eq (4.2) into Eq (4.1b), we have

$$(\rho\ A\ dx\ c_P)\ d\,(T - T_w)\ /\ (dx/V) = -U\ p_L\ dx\ (T - T_w) \tag{4.3}$$

Rearranging, we get

$$d\,(T - T_w)\ /\ (dx/V) = -U\ p_L\ dx\ (T - T_w)\ /\ (\rho\ A\ dx\ c_P) \tag{4.3a}$$

or

$$d\,(T - T_w)\ /dx = -U\ p_L\ (T - T_w)\ /\ (\rho\ A\ c_P\ V) \tag{4.3b}$$

Multiplying both sides by the duct length, L, and noting that m' = ρAV, gives

$$L\ d\,(T - T_w)\ /dx = -U\ p_L\ L\ (T - T_w)\ /\ (m' c_P) \tag{4.3c}$$

which can be rewritten as

$$d\,(T - T_w)\ /\ (dx/L) = -UA\,(T - T_w)\ /\ (m' c_P) \tag{4.3d}$$

The above can be integrated from inlet to a distant x from the inlet, i.e.

$$\int_{T_i - T_w}^{T - T_w} \frac{1}{T - T_w} d\,(T - T_w) = \int_0^x -\frac{UA}{\dot{m} c_P} d\,(x/L) \tag{4.4}$$

This results in

$$T - T_w = (T_i - T_w)\ \exp\{-[UA/\ (m' c_P)]\ [x/L]\} \tag{4.5}$$

The fluid temperature at the outlet, x = L,

$$T_o - T_w = (T_i - T_w)\ \exp\{-[UA/\ (m' c_P)]\} \tag{4.6}$$

The term inside the square brackets is called the *number of transfer units*, which will be expounded shortly.

The heat transfer rate from the wall of the conduit into the flowing fluid,

$$Q' = m' c_P\ (T_o - T_i) \tag{4.7}$$

Recall the underlying assumptions listed at the beginning of this subsection, constant fluid properties, in particular. In the event that the fluid is warmer than the surface of the tube, the heat transfer rate into the flowing fluid, Q', will be negative. The maximum amount of heat transfer rate:

$$Q_{max}' = m'c_P\ (T_w - T_i) \tag{4.8}$$

The outgoing fluid temperature, T_o, reaches the wall temperature, T_w, as x approaches infinity.

Heat Exchanger Effectiveness and Number of Transfer Units

It is, however, unrealistic to have an infinitely-long heat exchange tube. Therefore, it is meaningful to express the actual heat transfer rate with respect to the maximum heat transfer rate, where the tube is infinitely long. The heat transfer rate with respect to its maximum value is called the *heat exchanger effectiveness*,

$$\varepsilon \equiv Q'/Q_{max}' = 1 - \exp(-NTU) \tag{4.9}$$

where the *number of transfer units*,

$$NTU = UA/m'c_P \tag{4.10}$$

We see that NTU is a non-dimensional parameter signifying the rate of heat transfer. It is commonly related to the Nusselt number, the Reynolds number, and the Prandtl number in the form,

$$NTU = UA/m'c_P = 4\ Nu/(Re\ Pr)\ (L/D_h) \tag{4.11}$$

where D_h is the hydraulic diameter of the tube.

The Prandtl Number, the Nusselt Number, and the Reynolds Number

Let us briefly recap Nu, Re, and Pr here. The Prandtl number, Pr, is named after the German engineer, Ludwig Prandtl (1875–1953). It stands for the molecular diffusivity of momentum with respect to the molecular diffusivity of heat. Pointedly, the *Prandtl number*,

$$Pr = \nu/\kappa = \mu/\rho/k/(\rho c_P) = \mu c_P/k \tag{4.12}$$

where ν is momentum diffusivity or kinetic viscosity, thermal diffusivity, $\kappa = k/(\rho c_P)$, where k is the thermal conductivity, and μ is dynamic viscosity. At atmospheric conditions, air has a Pr value of approximately 0.7, whereas water has a Pr of around 7.

Earlier than Prandtl, Osborne Reynolds (1842–1912), who has been referred to as the father of fluid mechanics, related the inertial force of a flowing fluid with respect to the viscous force of the fluid. See Section 3.3.1 for a general discussion of these forces. Specifically, the Reynolds number is,

$$Re = \rho VD/\mu = VD/\nu \tag{4.13}$$

where V is the velocity of the flowing fluid, and D is the characteristic length, i.e. diameter of the conduit. At a low velocity, in a small conduit, and/or for a viscous fluid, the corresponding Re is small and, thus, the viscous force dominates. At larger Re, the inertial force becomes larger than the underlying viscous force, and the flow becomes progressively more turbulent.

The Nusselt number developer, Wilhelm Nusselt (1882–1957), was a junior to Reynolds. The story has it that young Nusselt approached the then well-established Professor Reynolds, wanting to study under him. It is said that Reynolds looked at Nusselt and rejected Nusselt because he did not look smart enough in Reynolds' eyes. Instead of accepting this singular opinion and giving up, Nusselt went on and strived under the guidance of another advisor. Nusselt has proved that Eleanor Roosevelt was right when she said, "Nobody can make you feel inferior without your consent."

What is the Nusselt number? The Nusselt number is simply the ratio of convective heat transfer with respect to the underlying conductive heat transfer, i.e.

$$Nu = h_{conv}D/k \tag{4.14}$$

where h_{conv} is the convective heat transfer coefficient. When the involved fluid is completely stagnant, Nu is equal to one, signifying that there is no convection, and thermal energy is transferred by conduction.

Example 4.1 ***Heating water via condensing geothermal steam***

Given

Geothermal steam at 170°C condenses on the shell side of a heat exchanger over the tubes inside of each water flow at a rate of 0.7 kg/s. Water enters the 5 cm in diameter, 12 m-long tubes at 20°C.

Find

The exit temperature of the water.

Solution

Heat is transferred from the hot condensing steam on the outside of the tube into the water flowing inside the tube as shown in Figure 4.6.

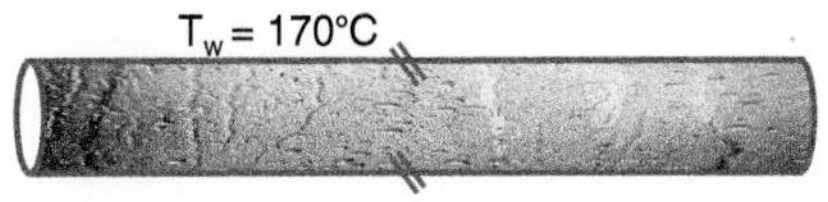

Figure 4.6 Transferring heat from condensing steam to moving fluid in a tube. Source: D. Ting.

The mass flow rate of the water is,

$$m' = \rho\, A\, V$$

where ρ is the density of the water, A is the heat transfer area, and V is the average velocity of the moving water. Substituting the values, we have

$$0.7\ kg/s = 998\ kg/m^3\ \left[\pi(0.05\ m)^2/4\right]\ V$$

Solving, we get V = 0.3572 m/s.

The corresponding Reynolds number of the flowing water in the tube is

$$Re = VD/\nu = 17825$$

This implies that the internal flow is turbulent, and the corresponding entrance length is about 10D. The tube is 12 m long, much longer than the entrance length and, therefore, we may assume fully developed turbulent flow throughout the entire tube for all practical purposes.

For a fully developed pipe flow, the Nusselt number can be related to the Reynolds number and the Prandtl numbers via the Dittus-Boelter equation. For the heating case, we have

$$Nu = h_{conv}D/k = 0.023\ Re^{0.8}\ Pr^{\ 0.4}$$

The thermal conductivity and the Prandtl number of water at 20°C are 0.5918 W/(m·K) and 6.99, respectively. Accordingly, we have

$$Nu = h_{conv}\,(0.05)\,/0.5918 = 0.023\ (17825)^{0.8}\ (6.99)^{0.4}.$$

This gives h_{conv} = 1491 W/(m²·K). Rearranging Eq (4.6) to solve for the exit water temperature, we have

$$T_o = (T_i - T_w)\exp\{-[UA/(m'c_P)]\} + T_w$$
$$= (20 - 170)\exp\{-[1491(\pi(0.05)(12))/(0.7(4182))]\} + 170.$$

This leads to T_o = 113°C.

4.3.2 Heat Transfer from a Hot Stream to a Cold Stream

Consider the adiabatic (no heat exchange with the surroundings) system under steady state conditions as shown in Figure 4.7. Conservation of energy says that energy gain by the cold fluid = energy loss by the hot fluid.

The total heat transfer rate,

$$Q' = m_c'(h_{c,o} - h_{c,i}) = m_h'(h_{h,i} - h_{h,o}) \tag{4.15}$$

where m_c' is the mass flow rate of the cold stream, $h_{c,o}$ is the enthalpy of the outgoing cold stream, $h_{c,i}$ is the incoming cold stream enthalpy, m_h' is the hot stream mass flow rate, $h_{h,i}$ is the incoming hot stream enthalpy, and $h_{h,o}$ is the outgoing hot stream enthalpy. When the temperature varies but only moderately, and without phase change, the specific heat capacity, c_P, can be assumed to remain constant. Under these conditions, the total heat transfer rate can be approximated fairly accurately from

$$Q' = m_c'c_{P,c}(T_{c,o} - T_{c,i}) = m_h'c_{P,h}(T_{h,i} - T_{h,o}) \tag{4.16}$$

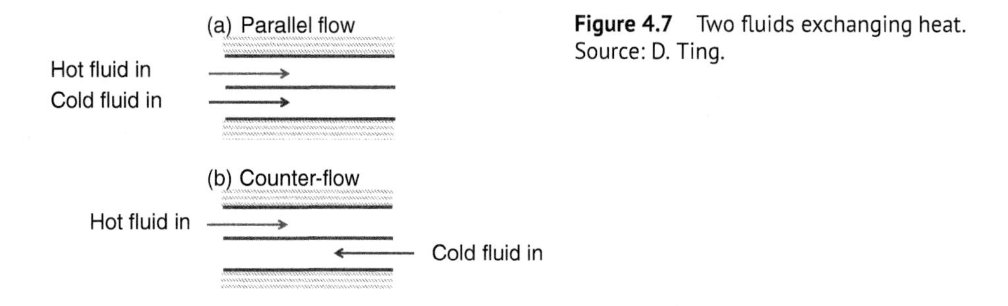

Figure 4.7 Two fluids exchanging heat. Source: D. Ting.

Newton's Law of Cooling

Newton's law of cooling is coined from Isaac Newton's 1701 work on convective cooling. For a moderate temperature difference between an object and its surrounding fluid, the heat transfer coefficient is relatively constant and, hence, the heat transfer rate may be expressed as

$$dQ/dt = -h\,A\,\Delta T(t) \tag{4.17}$$

where the negative sign signifies heat transfer in the negative temperature gradient direction, i.e. from high to low temperature, A is the heat transfer area, and $\Delta T(t)$ is the temperature difference as a function of time, t. Within the context of heat exchangers, it is more useful to express Newton's law of cooling as

$$Q' = U\,A\,\Delta T_m \tag{4.18}$$

where U is the overall heat transfer coefficient, and ΔT_m is the mean temperature difference.

Consider the two-fluid heat exchanger, say, for example, the parallel-flow heat exchanger shown in Figure 4.8. The rate of heat gained by a differentiable temperature difference is,

$$dQ' = m_c' c_{P,c} \left(T_{c,o} - T_{c,i}\right) = C_c \, dT_c \tag{4.19}$$

where $C_c = m_c' \, c_{P,c}$ is the *heat capacity rate*. According to the first law of thermodynamics, this amount gained by the cold stream is equal to that lost by the hot stream, i.e.

$$dQ' = m_c' c_{P,c} \left(T_{c,o} - T_{c,i}\right) = C_c \, dT_c = -m_h' c_{P,h} \left(T_{h,o} - T_{h,i}\right) = -C_h \, dT_h \tag{4.20}$$

Applying Newton's law of cooling, we have

$$dQ' = C_c dT_c = -C_h dT_h = U \Delta T \, dA \tag{4.21}$$

where the local temperature difference,

$$\Delta T \equiv T_h - T_c \tag{4.22}$$

Differentiating this, we have

$$d\Delta T = dT_h - dT_c = -dQ'/C_h - dQ'/C_c \tag{4.23}$$

This can be rewritten as

$$d\Delta T = -dQ' \left(1/C_h + 1/C_c\right) = -U \, \Delta T \, dA \, \left(1/C_h + 1/C_c\right) \tag{4.23a}$$

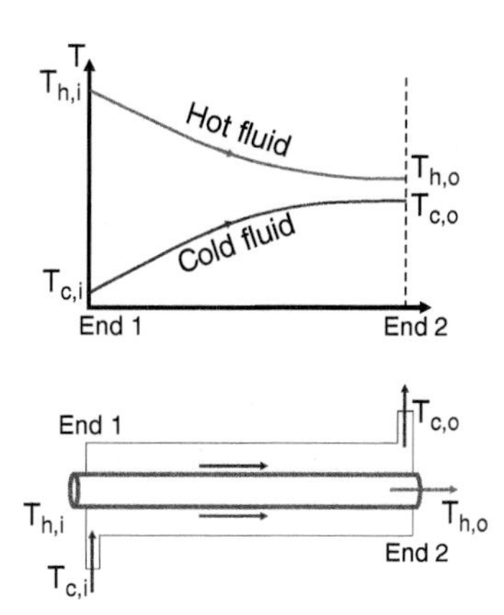

Figure 4.8 A parallel-flow heat exchanger. Source: D. Ting.

Heat Capacity Rate

The heat capacity rates of the hot and cold streams, respectively,

$$C_h = m_h' c_{P,h} \tag{4.24}$$

and

$$C_c = m_c' c_{P,c} \tag{4.25}$$

denote the required heat transfer rate for changing the temperature of the fluid through the heat exchanger by one degree. We note that:

1) The heat capacity rate, C, of liquid is typically much larger than that of gas, e.g. water versus air.
2) The temperature change of the hot stream is equal to that of the cold stream, $\Delta T_h = \Delta T_c$, only when the heat capacity rates of the two streams are equal, i.e. $C_h = C_c$.

Phase Change

Thermofluid systems such as condensers and boilers effectively operate by exploiting phase changes. In a condenser, water vapor condenses into liquid water and, in a boiler, liquid water vaporizes into steam. The corresponding heat transfer rate,

$$Q' = m_{phase}' h_{fg} \tag{4.26}$$

where m_{phase}' is the rate of condensation or evaporation, and h_{fg} is the enthalpy of vaporization. The condensing or boiling fluid is at a fixed temperature, i.e. the saturation temperature, T_{sat}. This is illustrated in Figure 4.9. Note that the heat capacity rate approaches infinity, as the temperature difference approaches zero.

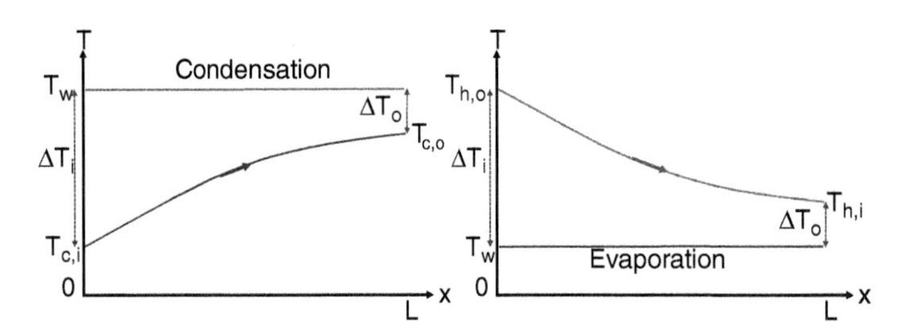

Figure 4.9 A two-fluid heat exchanger with one fluid (a) condensing, and (b) vaporizing or boiling. Source: D. Ting.

4.3.3 Log Mean Temperature Difference

We can rearrange Eq (4.23a) into

$$1/\Delta T \, d\Delta T = -U \left(1/C_h + 1/C_c\right) dA \tag{4.27}$$

Integrating from End 1 to End 2, with constant heat capacity rates and heat transfer coefficient,

$$\int 1/\Delta T \, d\Delta T = -U \left(1/C_h + 1/C_c\right) \int dA \tag{4.28}$$

we have

$$\ln \Delta T_2/\Delta T_1 = -UA \left(1/C_h + 1/C_c\right) \tag{4.29}$$

where ΔT_1 denotes the temperature difference between the two fluids at End 1, and ΔT_2 represents the temperature difference at End 2. This equation can be rewritten as

$$\ln \Delta T_2/\Delta T_1 = -UA[\left(T_{h,i} - T_{h,o}\right)/Q' + UA[\left(T_{c,o} - T_{c,i}\right)/Q' \tag{4.29a}$$

Rearranging, we get

$$\ln \Delta T_2/\Delta T_1 = -\left(UA/Q'\right) \left[\left(T_{h,i} - T_{c,o}\right) + \left(T_{h,o} - T_{c,o}\right)\right] \tag{4.29b}$$

which is

$$\ln \Delta T_2/\Delta T_1 = -\left(UA/Q'\right) \left(\Delta T_1 - \Delta T_2\right) \tag{4.29c}$$

This can be reorganized into

$$Q' = UA\left(\Delta T_1 - \Delta T_2\right)/\ln\left(\Delta T_1/\Delta T_2\right) \tag{4.29d}$$

We can introduce the *log mean temperature difference* (LMTD),

$$\Delta T_{lm} = (\Delta T_1 - \Delta T_2) / \ln(\Delta T_1/\Delta T_2) \tag{4.30}$$

Substituting this into Eq (4.29d), we have

$$Q' = UA\ \Delta T_{lm} \tag{4.31}$$

Note that the temperature variation along the heat exchanger is not linear, it is exponential in nature. Therefore, the arithmetic mean temperature difference would over-estimate the actual average temperature difference. The LMTD accounts for this exponential variation in temperature, giving the actual average temperature difference.

Parallel-Flow Heat Exchanger versus Counter-Flow Heat Exchanger

Combining Eqs. (4.30) and (4.31), we get

$$Q' = UA\ \Delta T_{lm} = UA\ (\Delta T_1 - \Delta T_2) / \ln(\Delta T_1/\Delta T_2) \tag{4.31a}$$

We note the following differences for parallel-flow and counter-flow heat exchangers:

1) For parallel-flow heat exchangers,

$$\Delta T_1 \equiv T_{h,1} - T_{c,1} = T_{h,i} - T_{c,i} \tag{4.32}$$

and

$$\Delta T_2 \equiv T_{h,2} - T_{c,2} = T_{h,o} - T_{c,o} \tag{4.33}$$

2) For counter-flow heat exchangers,

$$\Delta T_1 \equiv T_{h,1} - T_{c,1} = T_{h,i} - T_{c,o}, \tag{4.34}$$

and

$$\Delta T_2 \equiv T_{h,2} - T_{c,2} = T_{h,o} - T_{c,i} \tag{4.35}$$

It is clear, see Figure 4.10a, that for parallel-flow heat exchangers, there is a large temperature difference, ΔT, at the inlets. This large ΔT brings about large convection and conduction and, hence, a very high heat transfer rate at the inlets. The temperature difference, and thus also the heat transfer rate, decrease rapidly in a parallel-flow heat exchanger.

For a counter-flow heat exchanger, Figure 4.10b, a relatively large temperature difference, ΔT, is maintained throughout the passage. This large temperature difference induces large convection and conduction and, as a result, a large heat transfer rate throughout the entire passage. Consequently, a counter-flow heat exchanger gives rise to a larger overall heat transfer rate, compared to its parallel-flow counterpart. For any particular inlet and outlet temperatures, the LMTD for a counter-flow heat exchanger is always larger than that of a parallel-flow one. To put it another way, a smaller counter-flow heat exchanger can do the same desirable work as a relatively larger parallel-flow heat exchanger. It is thus unsurprising to see that parallel-flow heat exchangers are only employed when the limitations involved in the particular applications do not allow the utilization of the more-efficient, counter-flow counterpart.

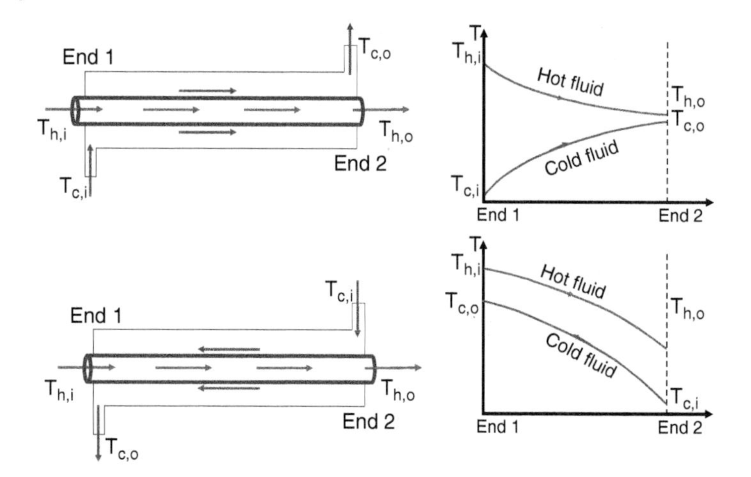

Figure 4.10 (a) Parallel-flow versus (b) counter-flow heat exchanger. Source: D. Ting.

It is worth reiterating the assumptions invoked; these include:

1) The heat exchanger is perfectly insulated from the surroundings, i.e. the system is adiabatic.
2) Axial conduction along the conduit is negligible.
3) The fluid properties, such as the specific heats, remain unchanged throughout the heat exchanger.
4) The overall heat transfer coefficient, U, is constant, which indirectly assumes that the flows are fully developed.
5) Potential and kinetic energy changes are negligible.

A couple of other points are also worth mentioning. For a counter-flow heat exchanger, when the heat capacity rates of the two streams are equal, the LMTD,

$$\Delta T_{lm} = \Delta T_2 = \Delta T_1 \tag{4.36}$$

When there is a phase change, such as that inside a condenser or a boiler, the parallel-flow and counter-flow heat transfer analyses give the same answer.

Example 4.2 *Heating water via condensing geothermal steam*

Given

Geothermal steam at 170°C condenses on the shell side of a heat exchanger over the tubes inside of each water flow at a rate of 0.7 kg/s. Water enters the 5 cm in diameter, 12 m-long tubes at 20°C.

Find

The rate of condensation of the geothermal steam.

Solution

From Example 4.1, $T_o = 113°C$ and $h_{conv} = 1491\ W/(m^2{\cdot}K)$. The heat transfer rate according to Eq (4.31) is

$$Q' = UA\ \Delta T_{lm}$$

where the log mean temperature difference,

$$\Delta T_{lm} = \left(\Delta T_1 - \Delta T_2\right) / \ln\left(\Delta T_1/\Delta T_2\right)$$

If we look at the example as a parallel-flow heat exchanger, then, according to Eqs. (4.32) and (4.44),

$$\Delta T_1 \equiv T_{h,1} - T_{c,1} = T_{h,i} - T_{c,i} = 170 - 20 = 150°C$$

and

$$\Delta T_2 \equiv T_{h,2} - T_{c,2} = T_{h,o} - T_{c,o} = 170 - 113 = 57°C$$

Substituting these for the heat transfer rate, we get

$$Q' = UA\ \Delta T_{lm} = 1491\ [\pi\,(0.05)\,(12)]\ (150 - 57) / \ln(150/57) = 270\ kW$$

If, instead, we consider it as a counter-flow heat exchanger, then, invoking Eqs. (4.34) and (4.35),

$$\Delta T_1 \equiv T_{h,1} - T_{c,1} = T_{h,i} - T_{c,o} = 170 - 113 = 57°C$$

and

$$\Delta T_2 \equiv T_{h,2} - T_{c,2} = T_{h,o} - T_{c,i} = 170 - 20 = 150°C$$

The corresponding heat transfer rate,

$$Q' = UA\ \Delta T_{lm} = 1491\ [\pi\,(0.05)\,(12)]\ (57 - 150) / \ln(57/150) = 270\ kW$$

We get the same answer whether we look at the problem as a parallel-flow or counter-flow heat exchanger. This is a uniqueness of heat exchange involving phase change.

4.3.4 Correction Factor

The LMTD derived thus far is applicable for counter-flow and parallel-flow heat exchangers. Similar LMTD expressions can be developed for cross-flow and multi-pass heat exchangers. These expressions can be quite complex. Therefore, it is more user-friendly to use the LMTD equations derived for counter-flow heat exchangers for cross-flow and multi-pass heat exchangers by introducing a correction factor. This correction factor is a function of the geometry of the heat exchangers and the inlet and outlet temperatures. Namely, the correction factor,

$$F_C = f\left(\text{geometry}, T_{h,i}, T_{h,o}, T_{c,i}, T_{c,o}\right) \tag{4.37}$$

With this correction factor, we have, for cross-flow and multi-pass heat exchangers, the equivalent temperature difference,

$$\Delta T_{lm} = F_C\ \Delta T_{lm,CF} \tag{4.38}$$

We see that the correction factor signifies the deviation of the LMTD from the corresponding values for the counter-flow case. As nature has consistently portrayed, counter-flow, or counter-current, heat exchange is the most forceful, therefore, the value of the correction factor is typically less than one. Figures 4.11a–d show the change in correction factor with respect to temperature and geometry of some of the most common heat exchangers.

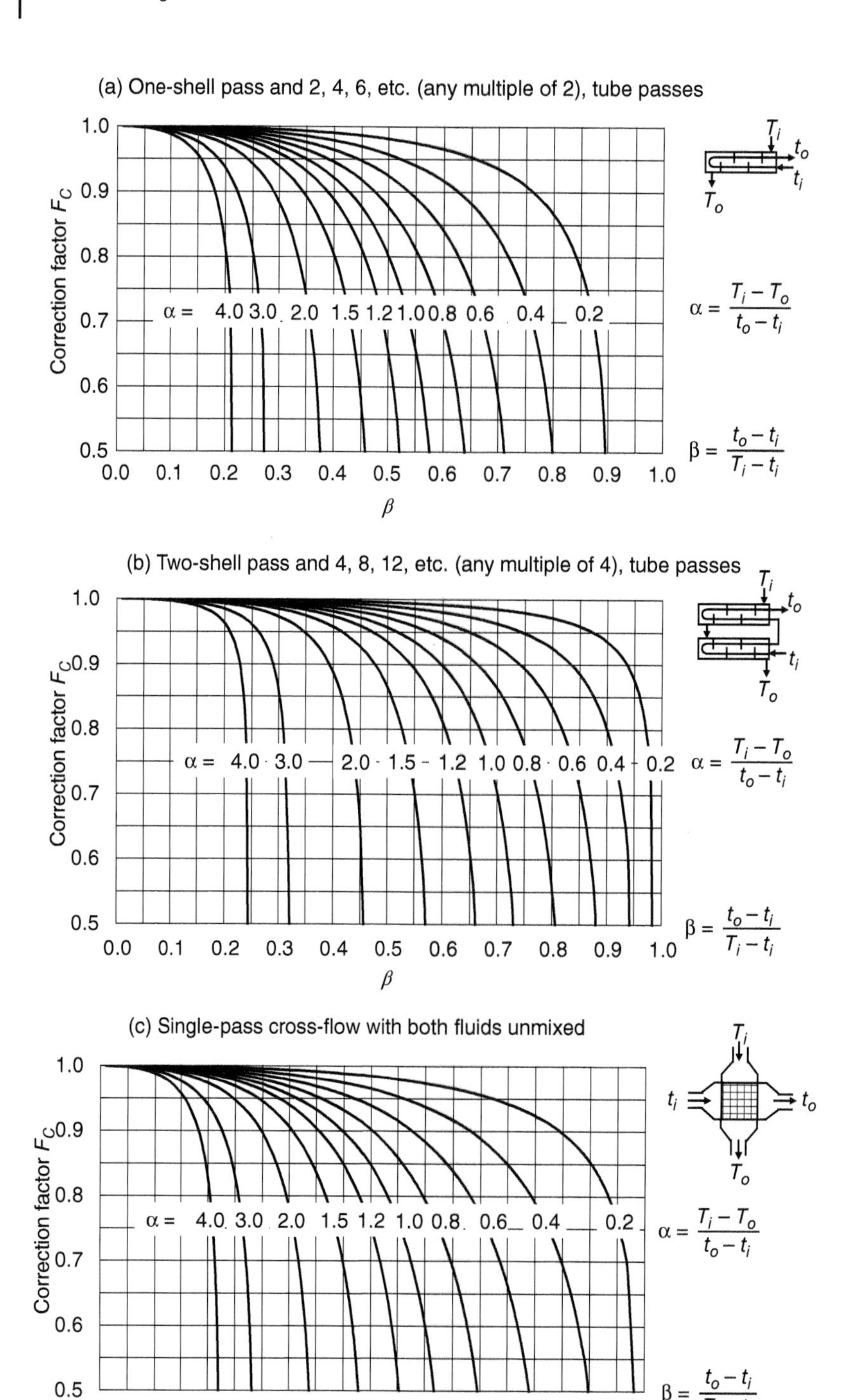

Figure 4.11a–d Correction factor versus temperature ratio for some common heat exchangers. Source: Y. Yang.

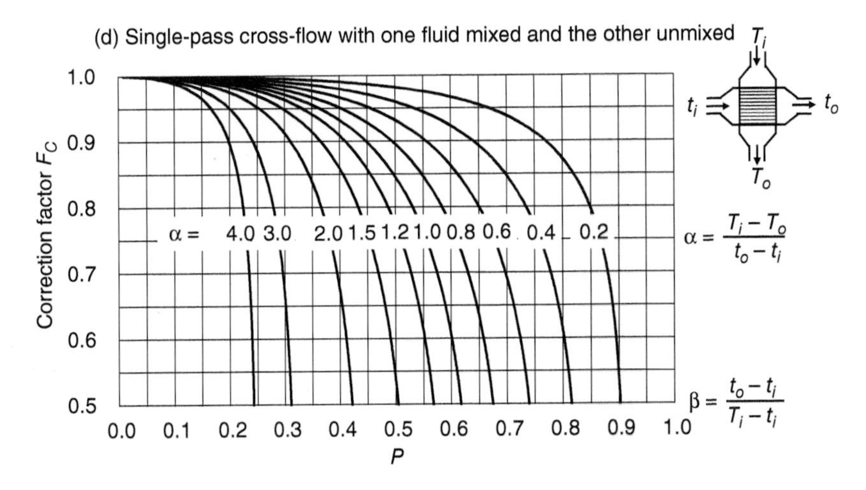

Figure 4.11 (*Continued*)

4.4 Comments on Heat Exchanger Selection

The thermofluid system design requirement can simply be changing the temperature of a stream from T_i to T_o. It can also removing or supplying a specific amount of thermal energy. As such, the design parameters include the heat transfer area, A, tube diameter, D, tube length, L, and the number of tubes, n. When the temperatures are known, it is conducive to use the LMTD method. It is more convenient to use the effective NTU method when one or more of the temperatures are not known *a priori*. Plots of heat exchanger effectiveness versus NTU for some common heat exchangers are shown in Figures 4.12a–c. Here are a few points worth highlighting:

1) NTU is a measure of heat transfer ability, that is, the higher the NTU, the higher the heat transfer rate.
2) As defined earlier, $NTU = UA/m'c_P$. With the minimum heat capacity rate dictating the amount of heat transfer, we can express this as $NTU = UA/C_{min}$. It is clear that NTU is also a yardstick for the area available for transferring heat. An increase in the area, A, increases NTU. Note, however, that the cost also increases with increasing A. This poses a practical upper limit for NTU.
3) From Figures 4.12a–c, we see that an initial increase in NTU rapidly improves the heat exchanger effectiveness, ε. The increase in ε slows down quickly with further increase in NTU. Beyond approximately 1.5, it is practically not cost-effective to further NTU, for the diminishing gain in ε cannot be justified by the additional cost.

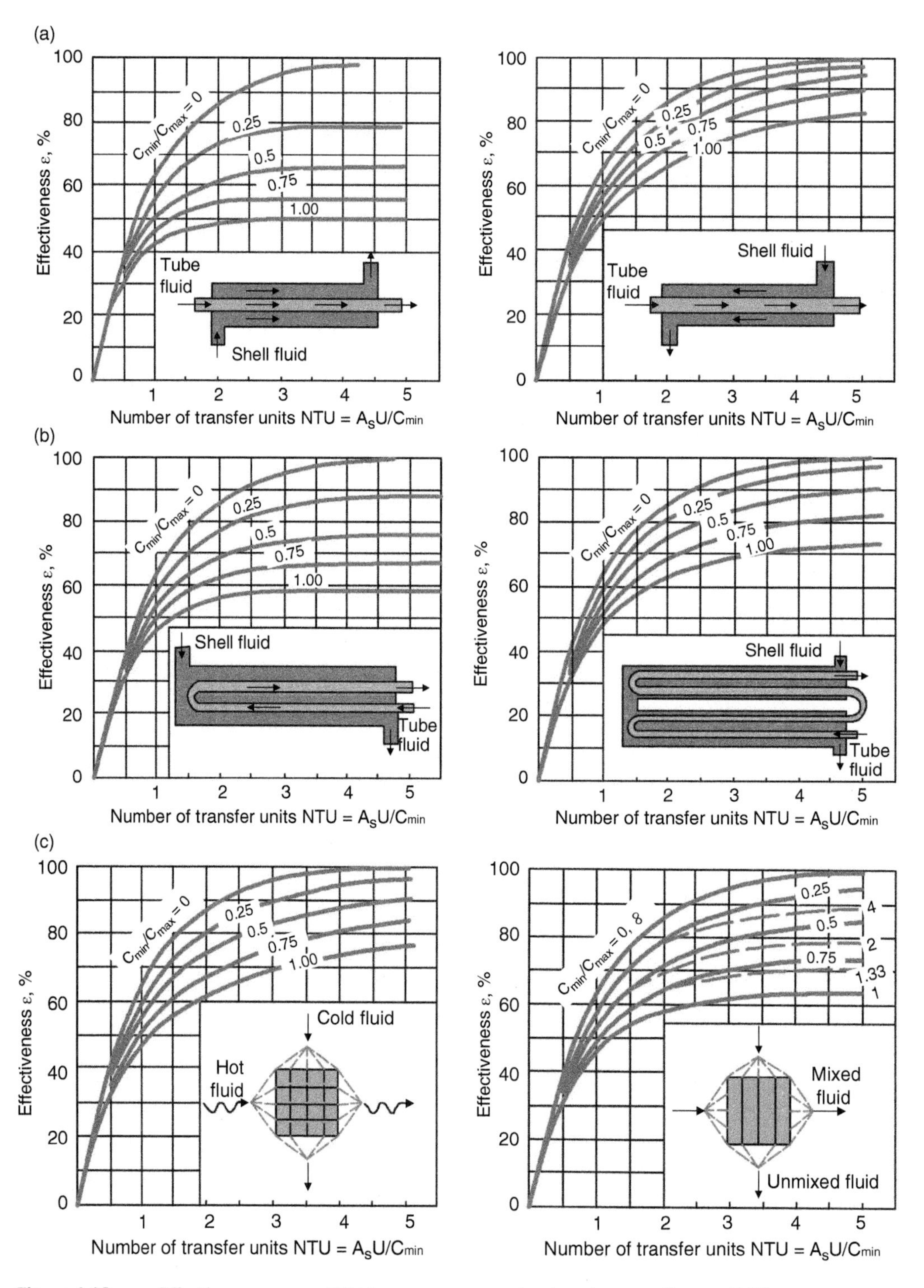

Figure 4.12a–c Effectiveness versus NTU for some common heat exchangers. Source: X. Wang.

Problems

4.1 Key parameters influencing the Nusselt number

In forced convection, the Nusselt number is typically a strong function of which *two* of the following non-dimensional parameters?

a) Biot number
b) da Vinci number
c) Eckert number
d) Fourier number
e) Grashof number
f) Jakob number
g) Lewis number
h) Prandtl number
i) Rayleigh number
j) Reynolds number
k) Schmidt number
l) Weber number

4.2 Maximum heat transfer rate

Cold water enters a counter-flow heat exchanger at 10°C and a rate of 8 kg/s, where it is heated by a hot-water stream that enters the heat exchanger at 70°C and a rate of 2 kg/s. Assume the specific heat of water remains constant at c_p=4.18 kJ/kg·K. Find the maximum heat transfer rate and the outlet temperatures of the cold- and the hot-water streams for this limiting case.

4.3 Choose the better counter-flow HEX

There are two counter-flow heat exchangers that can heat 230 kg/h of water from 35 to 93°C with 230 kg/h oil (c = 2.1 kJ/kg·C) entering at 175°C. These are:

HEX #1: U = 570 W/m^2 · °C, A = 0.47 m^2

HEX #2: U = 370 W/m^2 · °C, A = 0.94 m^2

Which of these is the more efficient heat exchanger? You can use the effectiveness versus NTU plot given in Figures 4.12 a–c.

4.4 Condenser

Cold water is used to condense steam in a condenser. The heat transfer rate, Q', and the temperature change in the cooling water, ΔT, are known. Write the appropriate expressions for which the steam condensation rate, the mass flow rate of the cooling water, and total thermal resistance, R, of the condenser can be determined.

4.5 Single-pass shell and tube heat exchangers

Water flowing at 100,000 lbm/h inside a 1.2-in diameter tube with a thickness of 0.12 in is heated from 60°F to 100°F. This is realized by condensing steam at 230°F on the shell side, where the convection heat transfer coefficient is 2000 Btu/h·ft^2·°F. The inlet velocity of the water is 4 ft/s and the inner surface convection heat transfer coefficient is 480 Btu/h·ft^2·°F. Find the required number and length of tubes.

4.6 Automobile radiator

Atmospheric air (c_p=1.01 kJ/kg·K) at 30°C is drawn across an automobile radiator at 10 kg/s. The radiator is equivalent to a cross-flow heat exchanger with UA = 10 kW/K. The coolant (c_p=4.00 kJ/kg·K) at 80°C flows through the radiator at 5 kg/s. The effectiveness of the radiator, $\varepsilon = 0.4$. Find the outgoing air temperature, $T_{a,o}$, and the heat transfer rate, Q'.

4.7 One large or two small heat exchangers?

Glycerin flowing at 1.5 kg/s is heated from 10°C to 50°C. This can be realized by a heat exchanger with an overall heat transfer coefficient of around 950 W/m^2·°C, where hot water enters at 120°C and leaves at 50°C. Alternatively, two small equal-surface-area heat exchangers in series, where 60% of the hot water goes through the first heat exchanger and the remaining 40% goes through the second heat exchanger, can be employed. The overall heat transfer coefficient of the two small heat exchangers system is the same as that of the large, single heat exchanger. Compare the options by deducing their heat transfer effectiveness, NTU, and surface area. What is the preferable option?

4.8 Geothermal double-pipe, parallel-flow HEX

A double-pipe, parallel-flow heat exchanger is used for heating water (c_p = 4.18 kJ/kg · K) from 25°C to 60°C at a rate of 0.2 kg/s. The heat is provided by 140°C geothermal water (c_p = 4.31 kJ/kg·K) that flows parallel at 0.3 kg/s in a thin-walled inner tube with a diameter of 0.8 cm. The overall heat transfer coefficient is 550 W/m^2·°C. Find the required length.

4.9 Car radiator

The jacket water (c_p = 1.0 Btu/lbm·F) of a car radiator at 190°F is cooled down to 140°F via a single-passage, cross-flow HEX with air (c_p = 0.245 Btu/lbm·F) at 90°F. The water is flowing at 92,000 lbm/h, while the air is flowing at 400,000 lbm/h. Find the log mean temperature difference, ΔT_{lm}.

References

Bergman, T.L., Lavine, A.S., Incropera, F.P., and DeWitt, D.P. (2018). *Fundamentals of Heat and Mass Transfer*, 8th ed, Hoboken, NJ: Wiley.

Çengel, Y.A. and Ghajar, A.J. (2020). *Heat and Mass Transfer: Fundamentals and Applications*, 6th ed. New York: McGraw-Hill.

Kakaç, S., Liu, H., and Pramuanjaroenkij, A. (2012). *Heat Exchangers: Selection, Rating, and Thermal Design*, 3rd ed. Boca Raton, FL: CRC Press.

Kays, W.M. and London, A.L. (1964). *Compact Heat Exchangers*, 2nd ed. New York: McGraw-Hill.

McDonald, A.G. and Magande, H.L. (2012). *Introduction to Thermo-Fluids Systems Design*. Chichester: Wiley.

Thulukkanam, K. (2013). *Heat Exchanger Design Handbook*, 2nd ed., Boca Raton, FL: CRC Press,

5

Equations

Life is a math equation. In order to gain the most, you have to know how to convert the negatives into positives.

– Anonymous

Chapter Objectives

- Understand what modeling is.
- Differentiate simulation from modeling.
- Categorize models into analog, mathematical, numerical, and physical types.
- Recognize the different forms of mathematical models.
- Become familiar with the common types of curve fitting.

Nomenclature

A	a variable or area; A_{flame} is the flame surface area, A_o is the original planar flame area
a	a coefficient, y-intercept
b	a coefficient, slope
C	a constant or (thermal) capacity; C_{air} is the thermal capacity of a volume of (indoor) air, $C_{u'}$ is an empirical coefficient
c	speed of light
CI	confidence interval
D	sum of the squared deviation
d	deviation
Da	Damköhler number = flow timescale with respect to chemical timescale
E	energy
h	enthalpy or heat transfer coefficient; h_{conv} is the convection heat transfer coefficient, h_{rad} is the linearized radiation heat transfer coefficient
HV	heating value
k	thermal conductivity
k_{tot}	total heat transmission coefficient
L	length or distance

m	number, order, or, mass; m_b' is the mass burning rate
N	total number of points
n	number of points, order of polynomial fit
P	percentage probability
Q	heat; Q' is heat transfer rate, Q_{conv}' is the convection heat transfer rate, Q_{rad}' is the radiation heat transfer rate, Q_{wall}' is the (conduction) heat transfer rate through a wall
R	range or resistance; R_{conv} is the convection thermal resistance, R_{elec} is the electrical resistance, R_{rad} is the radiation thermal resistance, R_{wall} is the thermal resistance of a wall
r^2	correlation coefficient or the coefficient of determination
S	speed; S_{flame} is the flame propagation speed, S_L is the laminar flame speed
$S_{x,y}$	standard error of estimate
SI	International System of Units
T	temperature; T_{ground} is the ground temperature, T_H is the higher temperature, T_i is indoor temperature, T_i' is the rate of change of indoor temperature, T_L is the lower temperature, T_o is outdoor temperature, T_{sur} is the temperature of the surroundings, T_{wall} is the wall temperature, T_∞ is the ambient (air) temperature
t	time; t_{vc} is the vortex consumption time, t_{wf} is the time for the flame to wrap around the vortex,
$t_{v,P}$	student-t value, for v degrees of freedom and P probability
u	velocity; u' is tangential velocity
v	degrees of freedom
X, x	a variable
Y, y	a variable
Z, z	a variable

Greek and Other Symbols

λ	size of the vortex
ρ	density; ρ_u is the density of the unburned gas
σ	Stefan-Boltzmann constant
τ	time constant or timescale; τ_{chem} is chemical timescale, τ_{flow} is flow timescale
ε	emissivity

5.1 Introduction

Equations are essential in engineering, as they are in life. Among others, Galileo Galilei speculated that "Mathematics is the language with which God has written the universe." We concern ourselves with equations in this chapter. An equation is a mathematical description equating two expressions. For example, the celebrated Einstein equation states that energy, E, is equal to mass, m, multiplied by the square of the speed of light, c^2, i.e.

$$E = m\,c^2 \tag{5.1}$$

If you have been told you have a "pea-sized brain," take heart. A 0.3 gram pea may give the impression that there is not much in it. Nothing is farther from the truth. If we multiply this mass with the speed of light ($c = 3\times10^8$ m/s) squared, we have 2.7×10^{13} J or 27 TJ of energy! This is approximately half the energy released (60 TJ) by the "Little Boy" that wiped out Hiroshima during World War II. The point is, even a pea-sized brain has great potential. The key is to harness it for good use.

On a lighter note, Albert Einstein inferred that a successful life is equal to working hard plus playing freely and keeping one's mouth shut. Pointedly, he said, "If A is a success in life, then A equals X plus Y plus Z. Work is X; Y is play; and Z is keeping your mouth shut." Mathematically, the equation is

$$A = X + Y + Z \tag{5.2}$$

An alternative, and more descriptive, title for this chapter is Mathematical Modeling. In plain English, equations are used to describe the thermofluid system, or sub-system, of interest. As such, the system is modeled mathematically to enable detailed analysis, frequently via simulation. Let us differentiate model and simulation.

5.1.1 Model Versus Simulation

Engineers, at times, have to build a model and perform simulations to study the real system of interest. The following descriptions concisely explain the related terms and, in so doing, allow easy differentiation among them.

Model. A model is a numerical or physical entity mimicking most of the salient features of the real system of concern.

Modeling. Modeling is the act of creating a model. It is the simplification of a practical problem into a system of equations.

Simulation. Simulation is the process of employing a model to examine the behavior of the real system of concern.

Simulating. Simulating is employing a model to study the characteristics of the real system of concern.

Example 5.1 ***Model a premixed planar flame interacting with a vortex***

Given

A premixed, where combustible fuel-oxidizer gas mixture is homogeneously mixed, planar flame approaches a two-dimensional vortex, as shown in Figure 5.1.

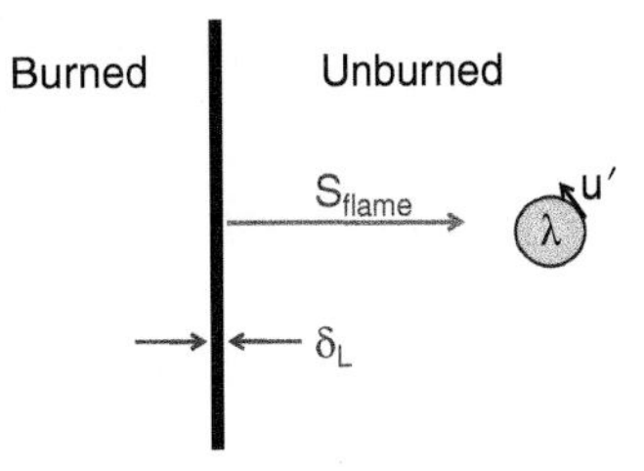

Figure 5.1 A premixed planar flame approaching a two-dimensional vortex. Source: D. Ting.

Find

A simple model describing the heat release rate of the vortex-induced flame.

Solution

Let us assume that the fluid dynamics is slow compared to the chemistry of combustion, which is typically the case as chemistry is generally much faster than flow motions. In other words, the characteristic flow time associated with the vortex, τ_{flow}, is large with respect to the characteristic chemical reaction time, τ_{chem}. This implies that the Damköhler number (Damköhler, 1940), which can be defined as

$$Da = \tau_{flow}/\tau_{chem} = (\lambda/u')\,/\,(\delta_L/S_L) \quad \text{(E5.1.1)}$$

is large. Here, λ is the size (diameter) of the vortex (see Figure 5.1), u' signifies the intensity of the vortex, i.e. its tangential velocity, δ_L is the thickness of the laminar flame, and S_L is the laminar flame speed.[1] The flame propagation speed,[2] S_{flame}, is proportional to the laminar flame speed. Assume the planar flame approaches the two-dimensional vortex, depicted in Figure 5.1 which takes place in a confined channel where the left-hand side is open and the right-hand side is closed. Then, the flame will advance to the right at the laminar flame speed, i.e. $S_{flame} = S_L$, the rate at which the flame consumes the unburned mixture. Note that the flame propagation speed is equal to the laminar flame speed because the burned gas is "freely expelled" through the open left side of the channel.

With the help of Figure 5.2, we can see that the flame surface corrugation increases with the intensity of the vortex, u', that is, the reacting surface area, A_{flame}, increases with u'. Namely, the faster the vortex rotates, the more affected is the flame as it passes through the vortex. Concisely, the relative intensity of the vortex with respect to the laminar flame speed, u'/S_L, signifies the degree of flame front corrugation and, thus, the reacting surface available. Consequently, the mass burning rate,

$$m_b' = \rho_u\, S_L\, A_{flame} \approx \rho_u\, S_L\, A_o\, (1 + C_{u'}\, u'/S_L) \quad \text{(E5.1.2)}$$

where ρ_u is the density of the unburned gas, A_o is the original planar flame area, and $C_{u'}$ is an empirically-deduced coefficient denoting the magnitude of the influence of the vortex intensity. The heat (energy) release rate is simply the mass burning rate multiplied by the heating value, HV, of the combustible mixture, i.e.

$$Q' = m_b'HV = \rho_u\, S_L\, A_{flame}\, HV \approx \rho_u\, S_L\, A_o\, (1 + C_{u'}\, u'/S_L)\, HV \quad \text{(E5.1.3)}$$

It is worth noting that the time it takes for the flame in Example 5.1 to consume the vortex, propagate across, and react with the vortex of unburned mixture is the vortex consumption time,

$$t_{vc} \approx \lambda/S_L \quad (5.3)$$

On the other hand, the time to wrap the flame around the vortex,

$$t_{wf} \approx \pi\lambda/u' \quad (5.4)$$

Beyond one revolution of wrapping, the immediate flame area around the vortex does not increase any further. Accordingly, $\pi\lambda/u'$ is roughly the longest time for the maximum vortex influence, as far

1 The laminar flame speed is defined as the speed at which the unburned gas enters the reacting flame front, behind which it exists as burned gas. By definition, it signifies the (volumetric) combustion rate (of the unburned mixture) per unit reacting flame surface.

2 The flame propagation speed is the speed at which the flame moves with respect to a fixed frame of reference, e.g. the channel enclosing the flame.

as increasing the reacting surface area, the flame front, is concerned. In plain English, the asymptote of vortex intensity influence, u'/S_L, is on the order of π, for the ideal case. In reality, however, the upper limit can be larger than π; as the vortex wraps the flame around more than one revolution, it can influence the nearby flame surface, i.e. creating more reacting area. This is particularly the case when there is more than one vortex present, making wrinkles of the surface in the neighborhood.

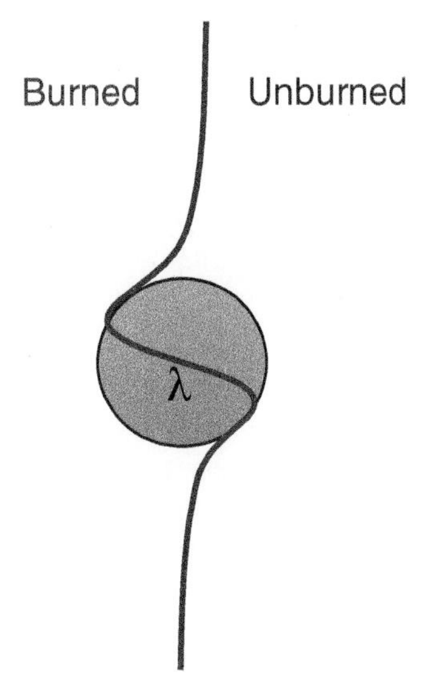

Figure 5.2 A premixed flame interacting with a vortex. Source: D. Ting.

5.1.2 Simulation

Once the system of interest is described by a set of equations, it can be analyzed or experimentally scrutinized. For the planar flame interacting with a two-dimensional vortex example, the mass burning rate is modeled by Eq (E5.1.2). The fundamental laminar flame speed, S_L, can be numerically deduced via chemical kinetics and the associated transport equations for heat and the various involved species. There are a few reliable software packages, and for well-defined combustible mixture conditions, such as near-stoichiometric methane-air mixtures combusting around atmospheric conditions, the involved chemical kinetics is quite well established.

In short, the set of equations created via modeling can be analyzed (i) numerically or analytically, and/or (ii) experimentally. Numerical analysis typically invokes a powerful computer. While less computer-intensive, a computer is also generally required for many analytical analyses. Experimental analysis is usually the last resort, as it tends to be very time-consuming and expensive. In other words, physical experimentation is generally only executed when there is no analytical or numerical solution for the problem at hand. By and large, most physical models are scaled models. Therefore, proper scaling and the limits of scaling must be explicitly followed and observed. Dimensional analysis, when correctly executed, can be viable and robust. Curve fitting is commonly employed to describe the experimentally acquired results.

Simulations can thus be considered to be the evaluation of a model in predicting the real system performance. They estimate the behavior of the system of interest prior to building an expensive system, which may or may not work well. As the actual system is yet to be built, there is a varying amount of missing information. Therefore, numerous assumptions and/or idealizations have to be invoked in simulations. Extreme, dangerous, very fast, very slow, etc. conditions of interest can also be simulated to understand the limits of the operation of the system. The simulation results obtained from numerical, analytical, or experimental simulations are used for design and optimization. To formulate a good design, the system behavior is simulated over wide ranges of operating conditions. For optimization of the design, the simulation is extended to evaluate different design options over extended conditions. This enables the engineer to achieve the optimal design and/or operating conditions.

5.2 Types of Models

One way to categorize models is to divide them into the following two kinds.

Descriptive model. A descriptive model describes and explains the system behavior. As such, the system already exists and real data of the running system are available. The model is created to describe the data. A detailed descriptive model can also reveal the inter-working of the various involved components. Descriptive models are useful for revealing the basic mechanics and for explaining the underlying principles.

Predictive model. As the name says, a predictive model predicts the system behavior. The system is either not in existence or has not been run. A model is created to predict how the system would perform under a given condition.

We note that good physical insight is needed for a model to appropriately capture the essential physics involved in a real system. For the planar flame-two-dimensional-vortex interaction example, realizing that the burning rate varies with the flame surface area is a necessary insight to create the analytical model. Note that a descriptive model, upon appropriate validation (typically by experimental results), can be used as a predictive model, within (the validation) limits. Figure 5.3 depicts how the value (reward) of the model increases with respect to the ability of the model. A descriptive model describes what an existing information or data set means. Invoking scientific knowledge to diagnose the data set can shed light on why it happened. The understanding gathered can then be used to predict what will happen under different scenarios, such as variation of particular variables. For instance, we could apply the planar flame-two-dimensional-vortex model to predict, within limits, what would happen if the vortex size or intensity changes. Ultimately, we wish to be able to come up with prescriptive analytics. For example, how can we make a spark-ignition engine run at a desirable speed? This may be achieved by generating the appropriate number of vortices of particular strength and size. Note that the value or reward furnished by the model increases as we progress from hindsight to insight to foresight, but so do the difficulty and risk or uncertainty.

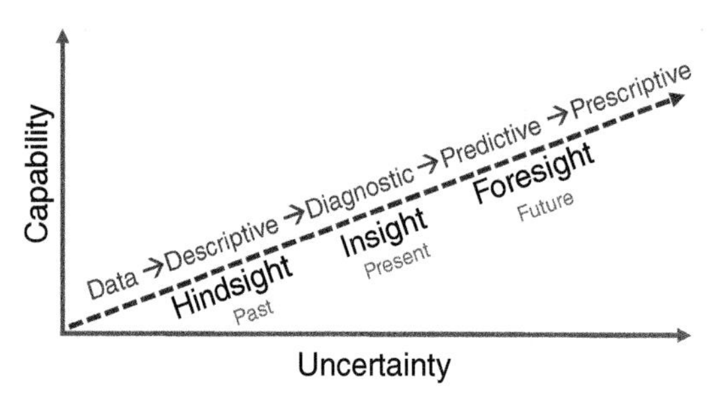

Figure 5.3 Capability of model versus its uncertainty. Source: D. Ting.

More specifically, we can classify models into four main types:

1) Analog models
2) Mathematical models
3) Numerical models
4) Physical models

Let us delineate these four types of models in the following subsections.

5.2.1 Analog Models

Analog models are based on analogy or similarity. The knowledge from a solved problem that is similar to the yet-to-be-solved one can sometimes be accurately adopted.

A classic example is the heat transfer problem which borrowed from the electric circuit analysis. Specifically, temperature is equivalent to voltage, thermal resistance to electrical resistance, and heat flow to electric current; see Figure 5.4. The conduction heat flow rate through a homogeneous wall, such as that shown in Figure 5.4, is

$$Q_{wall}' = k\,A\,\left(T_H - T_L\right)/L \tag{5.5}$$

As we are using conduction as an illustration, k is the thermal conductivity, A is the area, T_H is the "high" temperature, T_L is the "low" temperature, and L is the thickness or distance. Compared with an electric circuit, we see that the temperature potential (difference), T_H - T_L, is analogous to voltage potential, and the equivalent thermal resistance,

$$R_{wall} = L/\left(k\,A\right) \tag{5.6}$$

Intuitively, we can visualize that thickening the wall, or reducing the conductivity or area, restricts the heat flow rate, that is, increases the thermal resistance. Therefore, the heat transfer rate through the wall can also be expressed as

$$Q_{wall}' = \left(T_H - T_L\right)/R_{wall} \tag{5.7}$$

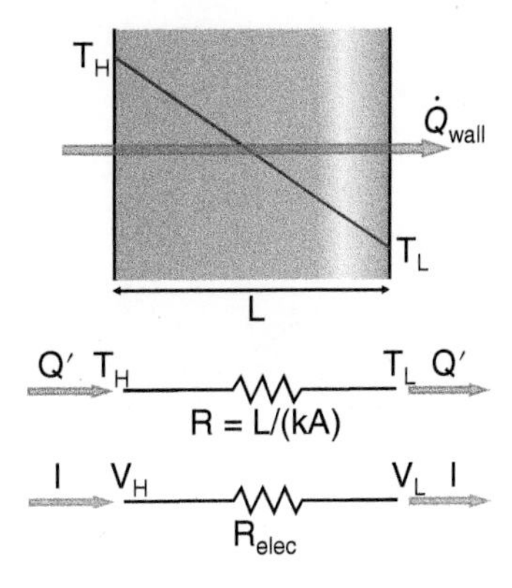

Figure 5.4 Heat conduction through a wall described equivalent to an electric circuit. Source: D. Ting.

The same analogy can be applied to convection heat transfer, as portrayed in Figure 5.5. The convection heat transfer rate,

$$Q_{conv}' = \left(T_{wall} - T_{\infty}\right)/R_{conv} \tag{5.8}$$

where T_{wall} is the wall temperature and T_{∞} is the ambient (air) temperature. The convection thermal resistance,

$$R_{conv} = 1/\left(h_{conv}\,A\right) \tag{5.9}$$

where h_{conv} is the convection heat transfer coefficient. In reality, there usually is radiation heat transfer in addition to convection heat transfer, as illustrated in Figure 5.5. The rate of radiation heat transfer can be expressed as

$$Q_{rad}' = \left(T_{wall} - T_{sur}\right)/R_{rad} \tag{5.10}$$

The radiation thermal resistance,

$$R_{rad} = 1/\left(h_{rad}\, A\right) \tag{5.11}$$

In typical engineering practice, the "linearized" radiation heat transfer coefficient,

$$h_{rad} = \varepsilon\, \sigma\, \left(T_{wall} - T_{sur}\right)\, \left(T_{wall}^{\ 2} - T_{sur}^{\ 2}\right) \tag{5.12}$$

over a narrow range of temperatures, may be invoked. Here, ε is emissivity, σ is the Stefan-Boltzmann constant, and T_{sur} is the temperature of the surroundings. Note that when dealing with radiation, the temperatures have to be in an absolute scale, which is Kelvin in SI units.

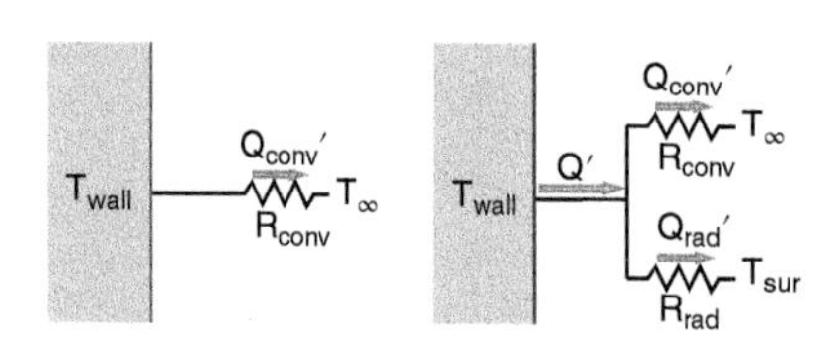

Figure 5.5 Heat convection from a wall, (a) without and (b) with radiation heat transfer. Source: D. Ting.

Time Constant of a Building

The time constant of a building can be deduced from the equivalent electric resistance with a capacitance. For a simple, single-zone building, the corresponding thermal network may be as illustrated in Figure 5.6. For example, the heat transfer from indoor to outdoor air through the roof, the top series of heat transfer resistors, passes through the convection resistor under the ceiling, the conduction resistor of the ceiling board, the convection resistor above the ceiling, the mass of attic air, the convection resistor under the roof, the conduction resistor of plywood, the conduction resistor of the shingle, and the above-shingle convection resistor. The thermal masses associated with the ceiling board, attic air, plywood, and shingle are represented by the capacitors.

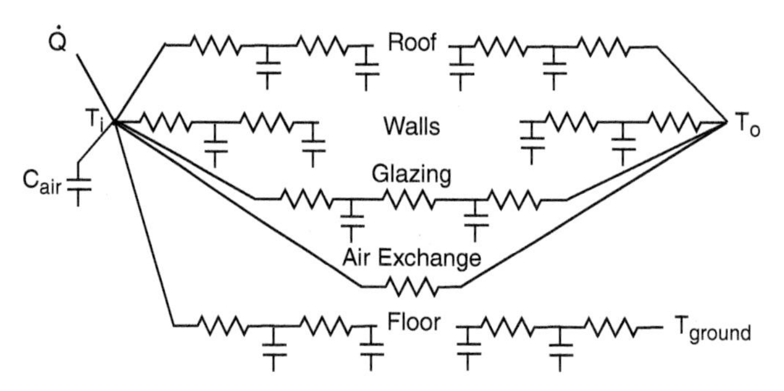

Figure 5.6 The thermal network of a simple building. A pair of short parallel lines designates a capacitor. Source: Y. Yang, edited by D. Ting.

To model the thermal inertia of the building, an equivalent capacitance, which denotes the storage effect, can be used. If we consider the entire indoor air as a single node, we can express the system under consideration as sketched in Figure 5.7. The associated thermal resistance,

$$R = 1/k_{tot} \tag{5.13}$$

where k_{tot} is the building total heat transmission coefficient.

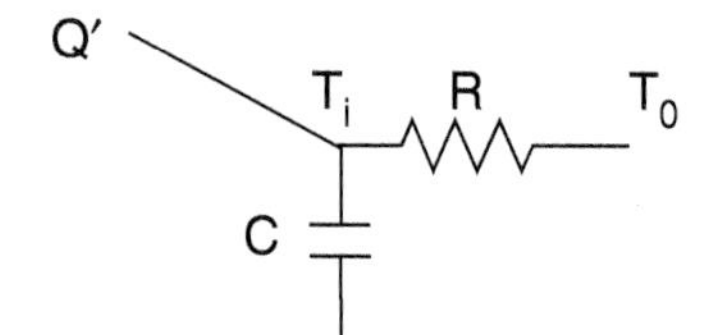

Figure 5.7 An equivalent electric model of a single-node building with thermal inertia. Source: D. Ting.

The corresponding heat balance of the single-node building can be expressed as

$$C\,T_i' = (T_o - T_i)/R + Q' \tag{5.14}$$

where C is the building thermal capacity, T_i' is the change in indoor temperature with respect to time, T_o is the outdoor temperature, which is assumed to be a constant, and Q' is the steady heat input rate, i.e. from the heating system. If we multiply Eq (E5.2.2) by R and define the time constant,

$$\tau = RC \tag{5.15}$$

we have

$$\tau\,T_i' + T_i = T_o + R\,Q' \tag{5.16}$$

Let

$$T(t) = T_i - T_o - R\,Q' \tag{5.17}$$

Differentiating this with respect to time, we get

$$T' = T_i' \tag{5.18}$$

Substituting these into Eq (5.16), we obtain,

$$\tau\,T' + T = 0 \tag{5.19}$$

The solution of this first-order differential equation is

$$T(t) = T(t = 0)\ \exp(-t/\tau) \tag{5.20}$$

We note that after one time constant, τ, the temperature, T(t), decays to 1/e, or 36.8% of its initial value, T(0).

Example 5.2 *Time constant of a single-zone building*

Given

A building can be approximated as a single zone via the RC network, with R = 5 K/kW and C = 10 MJ/K. $T_o = 0°C$, and T_i and the heat input rate, Q', have been constant until time, t = 0, when the thermostat is set back.

Find

The time it takes for T_i to drop from 20°C to 15°C.

Solution

The exponential decrease of the indoor temperature can be described, see Figure 5.8, as

$$T(t) = T(t = 0)\ \exp(-t/\tau) \tag{5.20}$$

That is,

$$15°C - 0°C = (20°C - 0°C)\ \exp(t/\tau) \tag{E5.2.1}$$

where $\tau = RC = 5K/kW\ (10\ MJ/K) = 50{,}000s$. Substituting, we get

$$t = 14{,}384\ s = 4\ hr$$

This value falls within a reasonable range for typical buildings. Practically speaking, it is often worthwhile setting the thermostat back when the space or building is not occupied over an extended time period, e.g. a couple of hours or more.

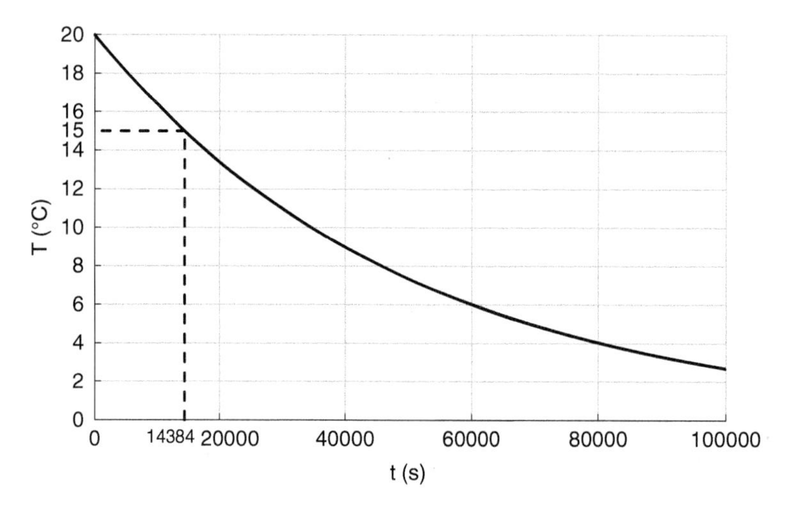

Figure 5.8 Exponential decay of indoor temperature following a thermostat setback. Source: Y. Yang.

We note that analog models frequently end up in one or more mathematical equations. When this occurs, they are equivalent to mathematical models.

5.2.2 Mathematical Models

A system behavior can frequently be described by one or more equations. Once appropriately described, a numerical simulation can be conducted. As such, mathematical models form the basis for numerical simulation. The quantitative results obtained are needed for the realization of a viable design.

Mathematical models are typically derived from either theories or physical experiments. Mathematical models based on theory are referred to as theoretical models. The heat release rate deduced based on vortex-induced flame area increase is a theoretical model. The effect of an engine valve opening on the engine power output can be investigated by conducting a physical experiment using a real engine and measuring the output power as the valve opening is varied. The obtained results may be expressed mathematically as $Q' = f(\text{valve opening})$. Mathematical equations such as that form empirical models.

5.2.3 Numerical Models

Other than very simple mathematical models, which can be solved analytically, most mathematical models require the equations to be solved numerically via a computer. In thermofluids, a classical example of analytical execution is the elegant manipulation of the equations associated with potential flow to obtain the exact solutions for inviscid (frictionless) flow.

When a mathematical model is solved numerically, via discretization, it becomes a numerical model. The problem is solved with the help of a computer by invoking an appropriate numerical method. The two most common numerical methods are (i) the finite-difference method, and

(ii) the finite-element method. The finite-difference method is primarily used to solve fluid mechanic problems. The finite-element method is typically employed to solve solid mechanic problems. The most common finite-difference method is called computational fluid dynamics (CFD). Finite element analysis (FEA) is the best-known finite-element method.

Two necessary checks are required in numerical analysis. The grid size, time step, boundary conditions, etc. need to be properly chosen or refined. This first check is called the *verification* of the model. Software is verified when appropriate results are produced without any bugs. Once the model is verified, the full extent of the required simulations is carried out. The accuracy of the simulated results is then validated with limited experimental or other known results. In a nutshell, *validation* is checking to see if the model is physically correct.

5.2.4 Physical Models

When there is no reliable theory, or no known mathematical equations for a problem at hand, a physical model is sought. In such cases, a scale model resembling the real system is designed and constructed. It is employed to conduct experiments to find out the behavior of the system of interest. Dimensional analysis is invoked to appropriately scale the physical models based on the scaling laws. One of the many fascinating scale model studies is to examine the inflation of a 30-ft parachute in a relatively small water tunnel; see Desabrais (2002), for example. Other than allowing a significant reduction in the size of the parachute, the deployment speed is also substantially reduced in water, easing the analysis of the inflation process. Sometimes, due to the nature of the system of interest, a full-scale prototype may need to be built and tested. Physical modeling is typically the most expensive modeling approach, but it is necessary when relatively unknown, costly and/or expensive systems are to be realized.

5.3 Forms of Mathematical Models

Mathematical models can be expressed in many forms. One way to categorize them is depicted in Figure 5.9. The differential model is presumably the most exploited model.

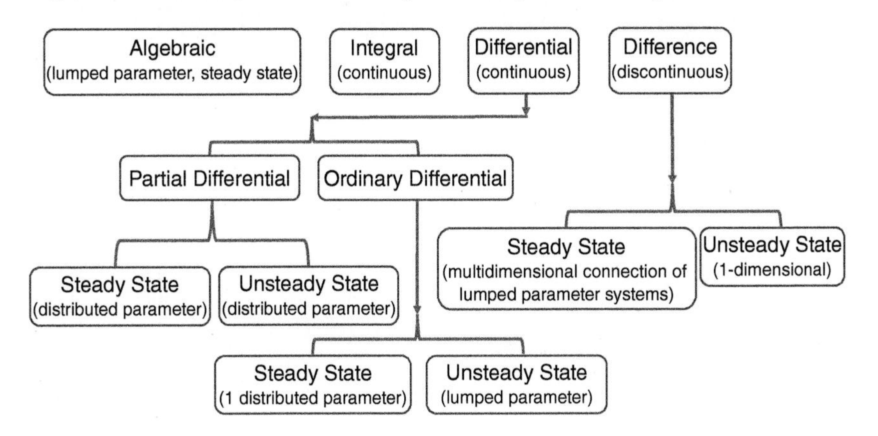

Figure 5.9 Categorization of mathematical models. Source: D. Ting.

Algebraic Mathematical Models

The earlier-discussed building thermal mass (inertia) model is an algebraic model. The thermal mass of the entire building is lumped into one capacitor. Accordingly, lumped-parameter models are algebraic models. The overall heat transmission resistance, consisting of heat loss through the wall, ceiling-roof, floor, basement, windows, doors, infiltration/exfiltration, etc., is lumped into one resistor. The resistance of this overall heat transmission resistor, $R = 1/k_{tot}$, where k_{tot} is the total heat transmission coefficient of the building. Furthermore, the net heat gain rate is Q'. Accordingly, the heat balance can be expressed as $C\,T_i' = (T_o - T_i) / R + Q'$.

5.4 Curve Fitting

Most engineering analyses involve data. Curve fitting is the most common way to describe the available data mathematically for the intended analysis. The set of data is often obtained from experiments or actual measurements of an existing, operating system. Thermofluid system behavior and characteristics are typically described by sets of data or performance lines.

5.4.1 Least Error Linear Fits

All experimental data have some amount of scatter. Therefore, uncertainty analysis is a critical element when reporting any experimental undertaking. When it comes to curve fitting, the standard definition of a best fit is the one where the sum of the absolute deviations is the smallest. This is commonly realized by minimizing the sum of the squares of the deviation.

In the absence of prior knowledge, unless the data clearly show otherwise, a linear best fit is a good starting point. This first-degree polynomial, or linear fit, can be written as

$$Y = a + bx \tag{5.22}$$

where Y is value of the dependent variable from the fit, and x is the independent variable. This linear fit requires a minimum of two data points. It is noted that no deviation, from the fit, can be deduced when there are only two data points. In other words, we need more than two data points to have a sense of the scatter of the empirical data set. For each value of the independent variable, x_i, the deviation, d_i, of the actual dependent variable, y_i, from the corresponding value on the linear fit, Y_i, is

$$d_i = y_i - Y_i \tag{5.23}$$

To emphasize, y_i is the value of the data point, while Y_i is the value of the line, at x_i; see Figure 5.10.

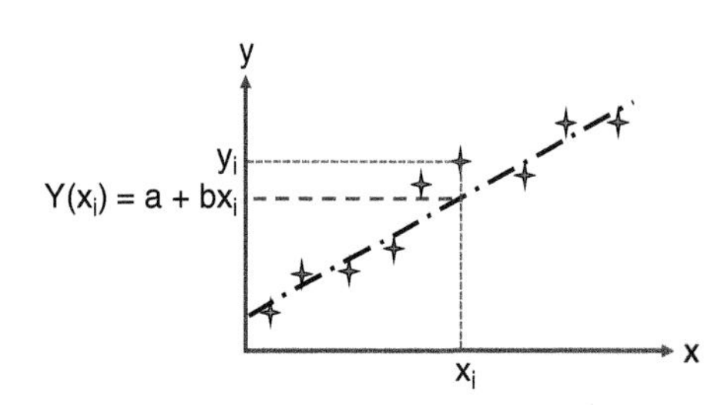

Figure 5.10 A least error linear fit. Source: D. Ting.

Following Figure 5.10, the square of the deviation,

$$d_i^2 = (y_i - Y_i)^2 = (y_i - a - b\,x_i)^2 \tag{5.24}$$

The sum of the squared deviations for all the data points,

$$D = \sum_{i=1}^{n} [y_i - Y(x_i)]^2 = \sum_{i=1}^{n} [y_i - a - bx_i]^2 \tag{5.25}$$

To minimize the sum of the squared deviations, we set

$$\frac{\partial D}{\partial a} = 0 = -\sum_{i=1}^{n} 2(y_i - a - bx_i) \tag{5.26}$$

and

$$\frac{\partial D}{\partial b} = 0 = -\sum_{i=1}^{n} 2x_i(y_i - a - bx_i) \tag{5.27}$$

Solving for a and b,

$$a = \frac{\Sigma y_i \Sigma x_i^2 - \Sigma x_i \Sigma x_i y_i}{n\Sigma x_i^2 - (\Sigma x_i)^2} \tag{5.28}$$

$$b = \frac{n\Sigma x_i y_i - \Sigma x_i \Sigma y_i}{n\Sigma x_i^2 - (\Sigma x_i)^2} \tag{5.29}$$

Standard Error of Estimate

To quantify how well the fit represents the data trend, we can estimate the error associated with fitting the line through the set of data points. The standard error of estimate,

$$S_{y,x} = \sqrt{\frac{\Sigma y_i^2 - a\Sigma x_i - b\Sigma x_i y_i}{n-2}} \tag{5.30}$$

is a measure of how well the least-squares line represents the data. It has the same units as y and is a measure of the data scatter about the best-fit line.

Alternatively, the correlation coefficient, or, the coefficient of determination

$$r^2 = 1 - \frac{\Sigma(y_i - a - bx_i)^2}{\Sigma(y_i - \bar{y})^2} \tag{5.31}$$

where $r^2 = 1$ signifies a perfect fit. A low r^2 value may indicate that there are other important variables, which affect the result y, that have not been considered.

Example 5.3 ***Truck air deflector***

Given

A group of engineers are working on improving the aerodynamics of the air deflector of an 18-wheel truck. The performance of two air deflector designs are to be evaluated with respect to the no-deflector reference case; see Figure 5.11. Before that can be executed, a spring used to measure the aerodynamic load on a model truck in a wind tunnel has to be calibrated using dead weights. A set of calibration results is presented in Table 5.1.

Figure 5.11 A model truck with different air deflectors. Source: Photos by D. Yang.

Table 5.1 Spring extension versus force calibration.

Weight [lb]	0.00	0.50	1.00	1.50	2.00	2.50
Displacement [in]	0.05	0.52	1.03	1.50	2.00	2.56

Table 5.2 Calculations of the least-squares linear fit coefficients.

							sum, $\sum$
y	0.00	0.50	1.00	1.50	2.00	2.50	7.5
x	0.05	0.52	1.03	1.50	2.00	2.56	7.66
xy	0	0.26	1.03	2.25	4.00	6.4	13.94
x^2	0.0025	0.2704	1.0609	2.25	4.00	6.5536	14.1374
y^2	0	0.25	1	2.25	4	6.25	13.75

Find

i) the best linear fit, (ii) the standard error of estimate, (iii) the correlation coefficient (coefficient of determination).

Solution

By plotting the weight (or force) versus displacement, it is quite clear that a linear fit

$$Y = a + bx$$

is in order. The coefficients, a and b can be solved using Eqs. (5.28) and (5.29),

$$a = \frac{\Sigma y_i \Sigma x_i^2 - \Sigma x_i \Sigma x_i y_i}{n\Sigma x_i^2 - (\Sigma x_i)^2},$$

$$b = \frac{n\Sigma x_i y_i - \Sigma x_i \Sigma y_i}{n\Sigma x_i^2 - (\Sigma x_i)^2}.$$

Calculations of the individual terms of coefficients a and b are systematized in Table 5.2.

Substituting,

$$a = [7.5\,(14.1374) - 7.66\,(13.94)] \,/\, \left[6\,(14.1374) - (7.66)^2\right] = -0.02868,$$

$$b = [6\,(13.94) - 7.66\,(7.5)] \,/\, \left[6\,(14.1374) - (7.66)^2\right] = 1.002.$$

Therefore,

$$Y = -0.02868 + 1.002x.$$

The standard error of estimate can be deduced from Eq (5.30),

$$S_{y,x} = \sqrt{\frac{\Sigma y_i^2 - a\Sigma x_i - b\Sigma x_i y_i}{n - 2}}.$$

Substituting,

$$S_{y,x} = \sqrt{\{[13.75 - (-0.02868)(7.5) - 1.002(13.94)]/(6-2)\}} = 0.00078.$$

The coefficient of determination can be determined using Eq (5.31),

$$r^2 = 1 - \frac{\Sigma(y_i - a - bx_i)^2}{\Sigma(y_i - \bar{y})^2}.$$

Substituting, we get $r^2 = 0.97$.

5.4.2 Least Error Polynomial Fits

The method of least squares can be extended to higher-order polynomials. When the trend of the data appears nonlinear, the choice of the actual higher-order fit to be invoked is a compromise between

1) reducing the random error of the fit;
2) the physics of the problem;
3) the convenience of using a lower-order polynomial.

For an mth order polynomial fit, we have,

$$Y = a_0 + a_1 x + a_2 x^2 + \ldots a_m x^m \tag{5.32}$$

The sum of the squared deviation,

$$D = \sum [Y_i - (a_0 + a_1 x + a_2 x^2 + \ldots a_m x^m)]^2 \tag{5.33}$$

Minimizing the sum of the squared deviation,

$$\partial D/\partial a_0 = 0,$$
$$\partial D/\partial a_1 = 0,$$
$$\ldots$$
$$\partial D/\partial a_m = 0.$$

Solving the m + 1 equations gives the values of $a_0, a_1, a_2, \ldots a_m$. The goodness of the fit, the standard error of fit, is

$$S_{y,x} = \sqrt{\frac{\sum_{i=1}^{n}(Y_i - y_i)^2}{n-(M+1)}} \tag{5.34}$$

In short, the best order is the lowest order with an acceptable $S_{y,x}$ value that retains the physics between the dependent and independent variables. The rule of thumb is to have at least two independent data points for each order of polynomial attempted. The *confidence interval* of curve fit at any value of x is

$$CI = \pm t_{v,P} S_{y,x}\left[\frac{1}{N} + \frac{(x-\bar{x})^2}{\sum_{i=1}^{N}(x_i - \bar{x})^2}\right]^{1/2} \tag{5.35}$$

If the independent variable is a known and controlled value, we may assume that the principal source of curve fit is due to the random error in the measured dependent variable. In this case, the confidence interval can be simplified to

$$CI = \pm t_{v,P}\, S_{y,x}/\sqrt{n} \tag{5.36}$$

where values of $t_{\nu,P}$ can be obtained from the student-t table. The curve fit can thus be described by

$$Y \pm t_{\nu,P}\, S_{y,x}/\sqrt{n} \tag{5.37}$$

See, for example, Lipson and Sheth (1973) and Miller and Freund (1985), for regression analysis of multiple variables of the form $y = f(x_1, x_2, \ldots)$.

Dealing with Small Coefficients of the High-Degree Terms

At times, the coefficients of the high-degree terms in a polynomial may be small. This is particularly challenging when the independent variable is large. This is commonly the case when the thermodynamic properties derived from experimental measurements are expressed over a wide range of temperatures using a high-order polynomial. For example, the saturated water vapor enthalpy,

$$h = a_0 + a_1\, T + \ldots + a_{n-1}\, T^{n-1} + a_n\, T^n \tag{5.38}$$

where T goes up to hundreds of degrees. As such, the values of the coefficients, a_{n-1}, a_n,..., are very small. One way to mitigate this is to lower the independent variable by dividing the independent parameter by a factor. For example,

$$h = a_0 + a_1\, (T/100) + \ldots + a_{n-1}\, (T/100)^{n-1} + a_n\, (T/100)^n \tag{5.39}$$

Simplification When the Independent Variable Is Uniformly Spaced

If the spacing between consecutive values of the independent variable is constant, we can simplify the polynomial expression. Consider a fourth-degree polynomial with a constant spacing Δx. Then, we can simplify

$$y = a_0 + a_1\, x + a_2\, x^2 + a_3\, x^3 + a_4\, x^4 \tag{5.40}$$

into

$$\begin{aligned} y - y_0 = {} & a_1\, [(n/R)\,(x - x_0)] + a_2\, [(n/R)\,(x - x_0)]^2 \\ & + a_3\, [(n/R)\,(x - x_0)]^3 + a_4\, [(n/R)\,(x - x_0)]^4 \end{aligned} \tag{5.41}$$

where n is the order of the polynomial, and R is the range of the independent variable x under study. To find the values of the coefficients, substitute for (x_1, y_1),

$$\Delta y_1 = a_1\, [4\,(x_1 - x_0)\,/R] + a_2\, [4\,(x_1 - x_0)\,/R]^2 + a_3\, [4\,(x_1 - x_0)\,/R]^3 + a_4\, [4\,(x_1 - x_0)\,/R]^4 \tag{5.42}$$

With $n = 4$ and uniform spacing, Δx,

$$n\,(x_1 - x_0)\,/R = 4\Delta x/\,(x_4 - x_0) = 4\Delta x/\,(4\Delta x) = 1 \tag{5.43}$$

Therefore,

$$\Delta y_1 = a_1 + a_2 + a_3 + a_4 \tag{5.44}$$

Similarly, substituting (x_2, y_2), and noting that $n(x_2 - x_0)/R = 2$, we have

$$\Delta y_2 = 2a_1 + 4a_2 + 8a_3 + 16a_4 \tag{5.45}$$

In the same way, for (x_3, y_3), and (x_4, y_4), we have

$$\Delta y_3 = 3a_1 + 9a_2 + 27a_3 + 81a_4 \tag{5.46}$$

$$\Delta y_4 = 4a_1 + 16a_2 + 64a_3 + 256a_4 \tag{5.47}$$

Solving the four Δy_i equations, we obtain the values of the coefficients, a_1, a_2, a_3, and a_4. See Table 5.3 for a summary.

Table 5.3 Values of the coefficients for uniformly-spaced independent variable.

Order of polynomial				
1st	Δy_1			
2nd	Δy_1-a_2	$\frac{1}{2}\Delta y_2$-Δy_1		
3rd	Δy_1-a_2 -a_3	$\frac{1}{2}\Delta y_2$-Δy_1-3a_3	$\frac{1}{2}\Delta y_1$-$\frac{1}{2}\Delta y_2$+Δy_3/6	
4th	Δy_1-a_2 -a_3 -a_4	$\frac{1}{2}\Delta y_2$-Δy_1-3a_3-7a_4	$\frac{1}{2}\Delta y_1$-$\frac{1}{2}\Delta y_2$+Δy_3/6-6a_4	-Δy_1/6 - Δy_2/4 - Δy_3/6 - Δy_4/24

Which Fit Is Appropriate?

When it comes to equations, notably curve-fitting equations, one would have to agree with Paul Dirac that, "It is more important to have beauty in one's equations than to have them fit experiments." Also, Albert Einstein proclaimed, "Everything should be made as simple as possible, but not simpler." Combining these two quotes, we end up with the eloquent saying by Johann Wolfgang von Goethe, "The ideal of beauty is simplicity and tranquility." Therefore, the answer concerning curve fitting is always to choose the lowest order of polynomial which can adequately describe the data.

When the data points form a straight line, it is obvious that the first-order polynomial or linear fit is the most beautiful and appropriate choice. For y = f(x), i.e.

$$y = a_0 + a_1\, x \tag{5.48}$$

For the specific case shown in Figure 5.12a, $a_0 = 1$, $a_1 = 2/3$, i.e. $y = 1 + (2/3)\,x$. For Figure 5.12b

$$z)_{x=c} = a_0 + a_1\, y \tag{5.49}$$

In this particular case, $a_1 < 0$, i.e. negative slope.

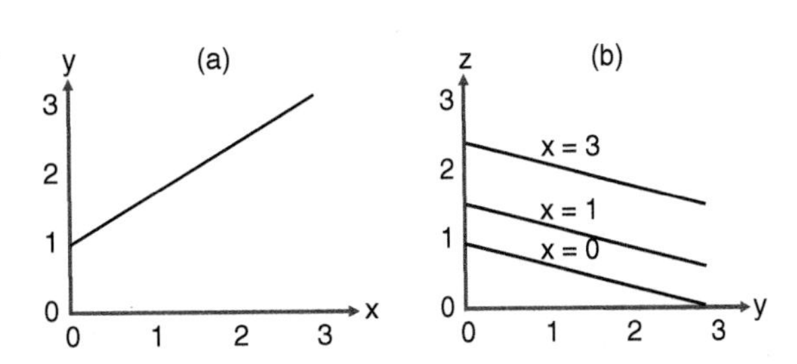

Figure 5.12 Sample first-order polynomial or linear fits. Source: D. Ting.

When the data points follow a single curvature trend, such as that depicted in Figure 5.13, a second-order polynomial is likely an appropriate fit. Explicitly, for Figure 5.13a,

$$y = a_0 + a_1\, x + a_2\, x^2 \tag{5.50}$$

For the decreasing y with increasing x case plotted in Figure 5.13b, we have

$$y = a_0 + a_1\, x^{-1} + a_2\, x^{-2} \tag{5.51}$$

Note that the exponents, rather than the coefficients, are negative.

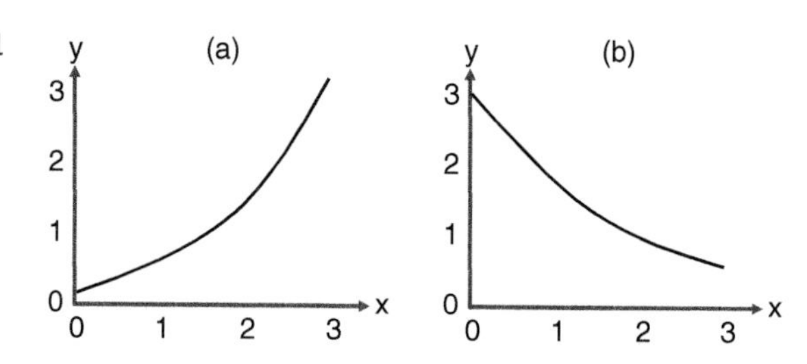

Figure 5.13 Sample second-order polynomial fits. Source: D. Ting.

Consider the case when there are two independent variables. For the z = f(x, y) case shown in Figure 5.14, we have

$$z = a_0 + a_1 x + a_2 y + a_3 x^2 \tag{5.52}$$

As the highest power is two, this is also a second-order polynomial.

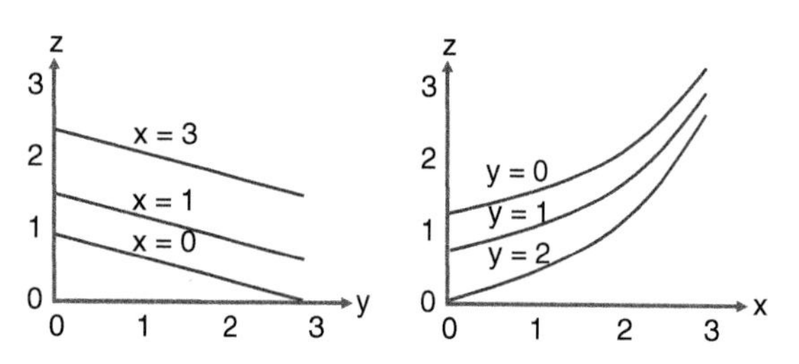

Figure 5.14 Sample multi-independent-variable second-order polynomial fits. Source: D. Ting.

When the data points follow a curve with a reverse curvature, i.e. an inflection point, a third-order polynomial is invoked. For the example shown in Figure 5.15, we have

$$y = a_0 + a_1 x + a_2 x^2 + a_3 x^3 \tag{5.53}$$

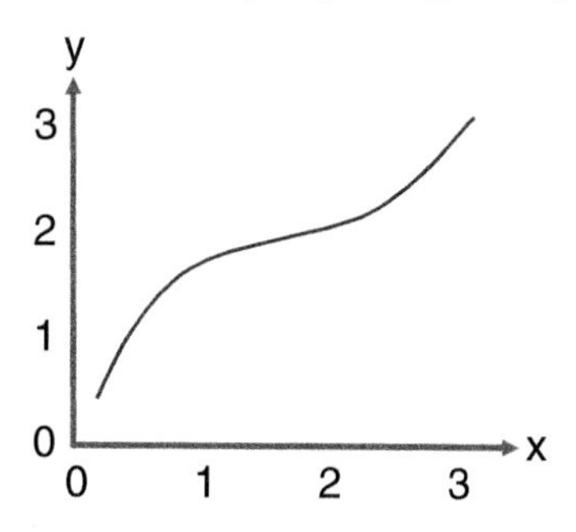

Figure 5.15 A sample third-order polynomial fit. Source: D. Ting.

5.4.3 Non-Polynomial into Polynomial Functions

Many other non-polynomial functions can also apply the least squares method to fit the data. When applicable, linearizing the equation can significantly ease its handling.

Exponential Functions

Exponential functions in the form,

$$y = a\,e^{bx} = a\ \exp(bx) \tag{5.54}$$

Taking the logarithm, we have

$$\ln\ y = \ln\ a + bx \tag{5.55}$$

This can be linearized as

$$Y = A + bx \tag{5.56}$$

where

$$Y = \ln\ y, \text{and } A = \ln\ a.$$

Power Law

The power law fit,

$$y = a\,x^b \tag{5.57}$$

can also be transformed into a polynomial fit. Taking the logarithm, we have

$$\ln\ y = \ln\ a + b\ \ln\ x \tag{5.58}$$

This can be linearized into

$$Y = A + b\,X \tag{5.59}$$

where $Y = \ln y$, $A = \ln a$, and $X = \ln x$.

Polynomial in Terms of 1/x

Polynomial functions in terms of $1/x$ such as

$$y = a_0 + a_1/x + a_2/x^2 + \ldots + a_m/x^m \tag{5.60}$$

can also be converted into the typical polynomial form. Let $X = 1/x$, then Eq (5.60) becomes

$$y = a_0 + a_1\,X + a_2\,X^2 + \ldots + a_m\,X^m \tag{5.61}$$

Function of the Form ax/(b+x)

Functions in the form,

$$y = ax/\,(b + x) \tag{5.62}$$

can be rewritten as

$$1/y = 1/a + (b/a)\ (1/x) \tag{5.63}$$

The corresponding linear equation is

$$Y = A + BX \tag{5.64}$$

where

$$Y = 1/y, A = 1/a, B = b/a, X = 1/x.$$

5.4.4 Multiple Independent Variables

The same error minimization process can be applied when there is more than one independent variable. Consider

$$y = f\left(x_1, x_2\right) = a_0 + a_1\,x_1 + a_2\,x_2 \tag{5.65}$$

where a_0, a_1 and a_2 are constants. The square of the deviation,

$$d_i^{\,2} = \left(y_i - Y_i\right)^2 = \left(y_i - a_0 - a_1\,x_{1,i} - a_2\,x_{2,i}\right)^2 \tag{5.66}$$

The sum of the squared deviations for all the data points,

$$D = \sum_{i=1}^{n} [y_i - Y(x_i)]^2 = \sum_{i=1}^{n} [y_i - a_0 - a_1 x_{1,i} - a_2 x_{2,1}]^2. \tag{5.67}$$

To minimize the sum of the squared deviations, we set the partial derivative with respect to each coefficient to zero, that is,

$$\frac{\partial D}{\partial a_0} = 0 = -\sum_{i=1}^{n} 2\left(y_i - a_0 - a_1 x_{1,i} - a_2 x_{2,i}\right) \tag{5.68}$$

$$\frac{\partial D}{\partial a_1} = 0 = -\sum_{i=1}^{n} 2x_{1,i}\left(y_i - a_0 - a_1 x_{1,i} - a_2 x_{2,i}\right) \tag{5.69}$$

$$\frac{\partial D}{\partial a_2} = 0 = -\sum_{i=1}^{n} 2x_{2,i}\left(y_i - a_0 - a_1 x_{1,i} - a_2 x_{2,i}\right) \tag{5.70}$$

These three equations can be used to solve for a_0, a_1, and a_2. Specifically, Eqs. (5.68–5.70) can be rewritten as

$$na_0 + a_1 \sum_{i=1}^{n} x_{1,i} + a_2 \sum_{i=1}^{n} x_{2,i} = \sum_{i=1}^{n} y_i \tag{5.71}$$

$$a_0 \sum_{i=1}^{n} x_{1,i} + a_1 \sum_{i=1}^{n} \left(x_{1,i}\right)^2 + a_2 \sum_{i=1}^{n} x_{1,i} x_{2,i} = \sum_{i=1}^{n} x_{1,i} y_i \tag{5.72}$$

$$a_0 \sum_{i=1}^{n} x_{2,i} + a_1 \sum_{i=1}^{n} x_{1,i} x_{2,i} + a_2 \sum_{i=1}^{n} \left(x_{2,i}\right)^2 = \sum_{i=1}^{n} x_{2,i} y_i \tag{5.73}$$

Nonlinear functions such as exponential and power-law variations can easily be linearized. For example,

$$y = a_0\, x_1^{a1}\, x_2^{a2}\, x_3^{a3} \ldots x_m^{am} \tag{5.74}$$

where a_0, a_1, a_3, etc. are constants. This can be linearized into

$$\ln(y) = \ln\left(a_0\right) + a_1 \ln\left(x_1\right) + a_2 \ln\left(x_2\right) \ldots + a_m \ln\left(x_m\right) \tag{5.75}$$

Problems

5.1 Spark locations for fast combustion

You are asked to pick the best locations to place three spark plugs for the fastest combustion of a four-cylinder, spark-ignition engine. The engine cylinders can be approximated as circular disks such as that shown in Figure 5.16. The three spark plugs can be placed at the spots marked A, B, or C. Specifically you can place the three plugs at A1, A2, and A3; B1, B2, and B3; or C1, C2, and C2. Where would you place the three spark plugs?

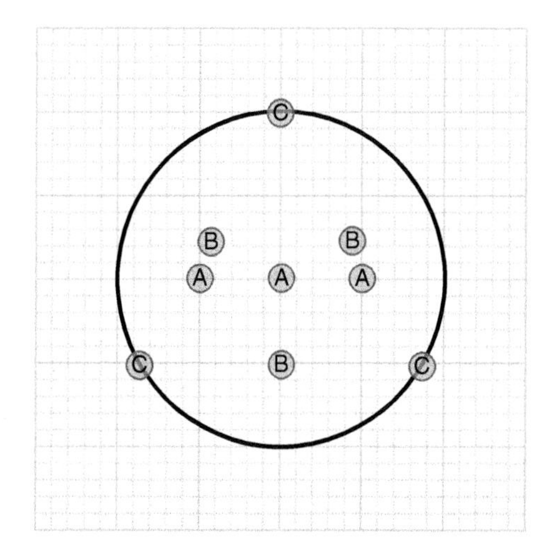

Figure 5.16 A circular cylinder head of a spark-ignition engine. Source: D. Ting.

5.2 Fitting three data points with a second-order polynomial equation

The dependent variable, y, is known to depend on x. From an experiment, three (x, y) points have been obtained. They are: (1, 3), (2, 4), and (2, 6). Find if $y = a + bx + cx^2$ is an appropriate curve fit. If not, propose a suitable curve fit.

5.3 Power law curve fitting

The line, $y = c\,x^m$, passes through (100, 50) and (1000, 10). Note that it is a straight line on a log-log plot. Find the values of c and m.

5.4 Least sum error best fit

The line, $y = a\,x + b/x$ passes through (1, 10.5), (3, 8), and (8, 18). Find the values of a and b.

5.5 Express a system behavior with an appropriate equation

The Engineering Design and Optimization of Thermofluid Systems class is interested in the effect of temperature on the behavior of a spring. Their experiment resulted in:
At temperature, $T = 0°C$, the load in Newtons, $y = 10x$, where x is the displacement in cm.
At $T = 10°C$, $y = 5 + 15x$.
At $T = 20°C$, $y = 10 + 20x$.
Express the load, y [N], in terms of T [°C] and x [cm].

References

Damköhler, G. (1940).Der Einfluss der Turbulenz auf die Flammengeschwindigkeit in Gasgemischen. *Physikalische Chemie* 46(11): 601–652, 1940. English translation: NACA Technical Memorandum, *No.* 1112, 1947.

Desabrais, K.J. (2002). *Velocity field measurements in the near wake of a parachute canopy*. PhD dissertation, Mechanical Engineering, Worcester Polytechnic Institute.

Lipson, C. and Sheth, N.J. (1973). *Statistical Design and Analysis of Engineering Experiments*. New York: McGraw-Hill.

Miller, J.E. and Freund, I. (1985). *Probability and Statistics for Engineers*. Englewood Cliffs, NJ: Prentice-Hall.

6

Thermofluid System Simulation

... in real life, mistakes are likely to be irrevocable. Computer simulation, however, makes it economically practical to make mistakes on purpose. If you are astute, therefore, you can learn much more than they cost.

– John H. Mcleod

Chapter Objectives

- Understand what system simulation is.
- Appreciate information flow diagrams.
- Differentiate sequential from simultaneous solution approaches.
- Recap common matrix approaches for solving a set of equations.
- Learn to apply successive substitution.
- Employ the Newton-Raphson method to solve an equation and a set of equations.

Nomenclature

A area or variable

a variable or coefficient

b variable or coefficient

C coefficient

c_p heat capacity

g gravity

h height or enthalpy; h_f is the enthalpy of saturated liquid, h_g is the enthalpy of saturated vapor

m mass; m' is mass flow rate

P pressure; P_{atm} is the atmospheric pressure, P_{static} is the static pressure

T temperature

U overall heat transmission coefficient

x variable or parameter

y variable or parameter

z variable or parameter

Greek and Other Symbols

ρ density
∀ volume; ∀' is volume flow rate

6.1 What Is System Simulation?

Within the context of this book, *system simulation* is solving a set of equations describing a system. Recall that a system is a collection of interrelated components. For example, a conventional air conditioning system consists of a condenser, a compressor, an evaporator, a throttling valve, along with the piping network. System simulation is the deduction of operating variables such as pressure, temperature, volume flow rate, and heat transfer rate. The simulation is typically conducted under steady-state operation of the concerned system, where the equations describing

1) the component performance characteristics
2) the thermodynamic properties at the operating conditions
3) the mass and energy conservation

are solved simultaneously. We will focus on continuous, deterministic, steady-state, thermofluid systems. Closely related references for this chapter include Burmeister (1998), Dhar (2016), Jaluria (2019), Penoncello (2018), Suryanarayana and Arici (2003), and of course, Stoecker (1989). To effectively execute the design and optimization of a thermofluid system, we call for "systems thinking." What Peter Senge said, "Systems thinking is a discipline for seeing wholes. It is a framework for seeing interrelationships rather than things, for seeing 'patterns of change' rather that static 'snapshots,'" very much applies.

Returning to system simulation, we note that the simulated system is a synthetic system which imitates a real one. In design and optimization, the simulation aims at designing a viable system for the particular application. The simulation is also extended to explore prospective modifications to improve the system, and to search for an optimum system.

Systems can be categorized into

1) steady-state versus dynamic
2) continuous versus discrete
3) deterministic versus stochastic.

As mentioned above, a thermofluid system is most often simulated under steady-state condition where its operation is unchanging with respect to time. When the transient start-up, shut-down, and varying operations are of interest, the system needs to be simulated as a dynamic system. The flow of discrete objects such as vehicles can be modeled as a discrete system in transportation. In thermofluids, we are concerned mostly with moving fluid and/or energy, which are simulated as a continuous system. By and large, the involved variables are precisely specified or calculated and, thus, the system is deterministic. A stochastic system is utilized when the input conditions are uncertain. One such system is a wind-energy harnessing system in which the amount of wind tends to follow some probability distribution, with completely random variations at times.

Let us remind ourselves of the big picture of thermofluid system design and optimization. First, there is a need to solve an engineering problem. Namely, we need a system with particular requirements to solve the problem at hand; see Figure 6.1. Spell out the involved design variables and the unavoidable constraints. Once the concept is properly developed, we have a prospective system. This system is modeled via a set of descriptive equations. Simulation is subsequently performed, solving the governing equations. The simulated results are evaluated to see if the modeled system is acceptable. If not, the model, and sometimes the prospective system or even the concept, will be revised until the solution is sound.

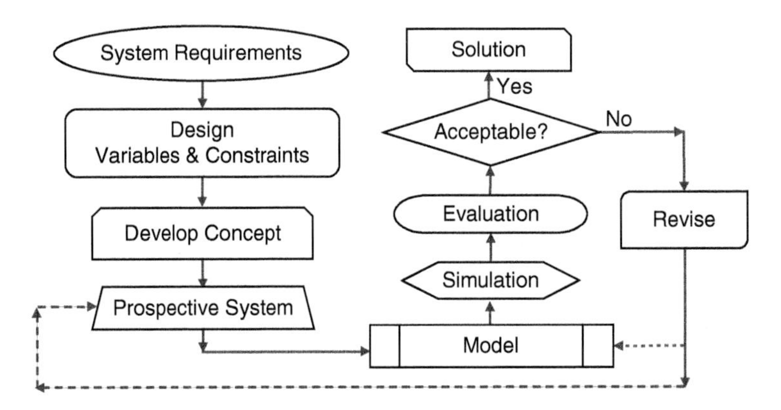

Figure 6.1 An overview of the design and optimization process involving simulation. Source: D. Ting.

There are two approaches to solving the set of equations: (i) direct and (ii) iterative:

Direct methods: Direct methods solve the equations exactly, except for the round-off error, in a finite number of operations. Matrix decomposition, matrix inversion, Gaussian elimination, and Gauss-Jordan elimination are some of the most common direct methods.

Iterative methods: Iterative methods are particularly useful for solving large sets of linear equations that are generally sparse, large sets of equations that arise in the finite difference and finite element solutions of partial differential equations and nonlinear equations. Gauss-Seidel iteration, successive substitution, and Newton-Raphson methods are some of the prominent iterative methods.

We will cover both approaches, and compare and contrast them, in this chapter.

6.2 Information-Flow Diagram

An *information-flow diagram* is a diagram illustrating the flow of information. In the thermofluid system context, it is versatile in depicting the connection between inputs and outputs. As such, it is a visual way to convey the equations that describe the system or component. Figure 6.2 is a schematic of a water pump under steady-state operation. Water at a mass flow rate of m' enters the pump at pressure P_1 and it leaves the pump at pressure P_2. The relationship between these three parameters is the pump characteristic equation, which can be expressed generically as

$$f\left(P_1, P_2, m'\right) = 0 \tag{6.1}$$

To solve this, we can make use of the information-flow diagram shown in Figure 6.3, where the blocks are used to signify the equation. We see in Figure 6.3 that with a known pump characteristic equation, we can deduce the outlet pressure, P_2, from known mass flow rate, m', and inlet pressure, P_1. Alternatively, we can consider inlet and outlet pressures as inputs and the mass flow rate as the output of the equation.

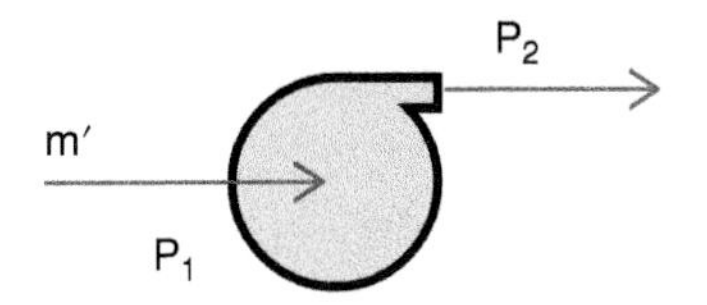

Figure 6.2 A water pump under steady-state operation. Source: D. Ting.

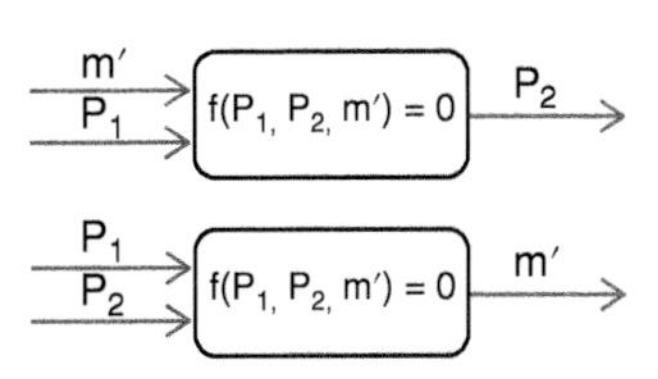

Figure 6.3 A pump information-flow diagram. Source: D. Ting.

Let us look at the fire-water system depicted in Figure 6.4. The mass flow rate via Hydrant A,

$$m_A' = C_A \surd\left(P_3 - P_{atm}\right) \tag{6.2}$$

where C_A is the flow coefficient of Hydrant A, P_3 is the inlet pressure, and the water is released into the open atmosphere, which is at the atmospheric pressure, P_{atm}. We note that the square root is from the relationship between flow rate and pressure, i.e. the Bernoulli equation.

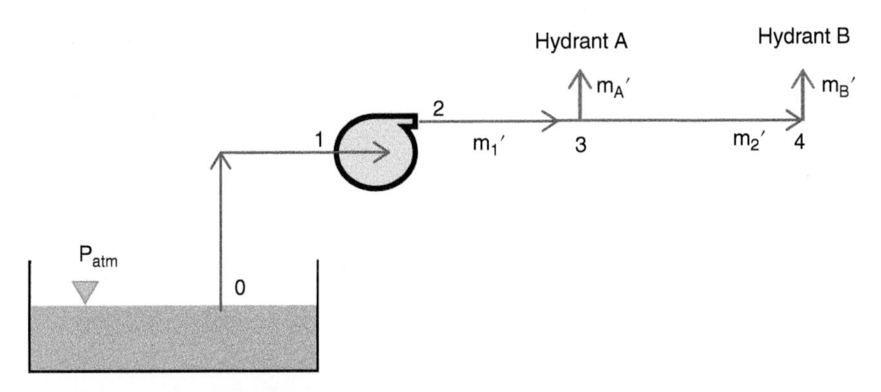

Figure 6.4 A fire-water system. Source: D. Ting.

Similarly, for Hydrant B,

$$m_B' = C_B \surd\left(P_4 - P_{atm}\right) \tag{6.3}$$

It is understandable that the flow coefficient varies from hydrant to hydrant and, thus, C_B is unlikely to be equal to C_A. Also, Hydrant B is farther downstream and, hence, P_4 is less than P_3.

For pipe Section 0–1, we have

$$P_{atm} - P_1 = C_1\, m_1' + \rho\, g\, h \tag{6.4}$$

The water is drawn from atmospheric pressure against the elevation increase from Point 0 to Point 1. The first term on the right denotes the pressure loss (head drop) caused by the friction imposed by the section of pipe.

For the horizontal pipe section, Section 2–3,

$$P_2 - P_3 = C_2\, m_1' \tag{6.5}$$

It is clear that there is only friction-caused pressure (head) loss. The same is true for Section 3–4, that is,

$$P_3 - P_4 = C_3\, m_2' \tag{6.6}$$

From the above, the five equations are:

$$F_1\left(m_A{}', P_3\right) = m_A{}' - C_A\sqrt{\left(P_3 - P_{atm}\right)} = 0 \tag{6.7}$$

$$F_2\left(m_B{}', P_4\right) = m_B{}' - C_B\sqrt{\left(P_4 - P_{atm}\right)} = 0 \tag{6.8}$$

$$F_3\left(m_1{}', P_1\right) = P_{atm} - P_1 - C_1\, m_1{}' - \rho\, g\, h = 0 \tag{6.9}$$

$$F_4\left(m_1{}', P_2, P_3\right) = P_2 - P_3 - C_2\, m_1{}' = 0 \tag{6.10}$$

$$F_5\left(m_2{}', P_3, P_4\right) = P_3 - P_4 - C_3\, m_2{}' = 0 \tag{6.11}$$

We note that m_A', m_B', m_1', m_2', P_1, P_2, P_3 and P_4 are the eight unknowns. Therefore, in addition to the above five equations, we need three more independent equations.

The sixth equation is the pump characteristic equation,

$$F_6\left(m_1{}', P_1, P_2\right) = 0 \tag{6.12}$$

A typical pump characteristic curve is shown in Figure 6.5. This equation is determined experimentally following the appropriate standard for deducing the pump characteristic curve under a controlled condition.

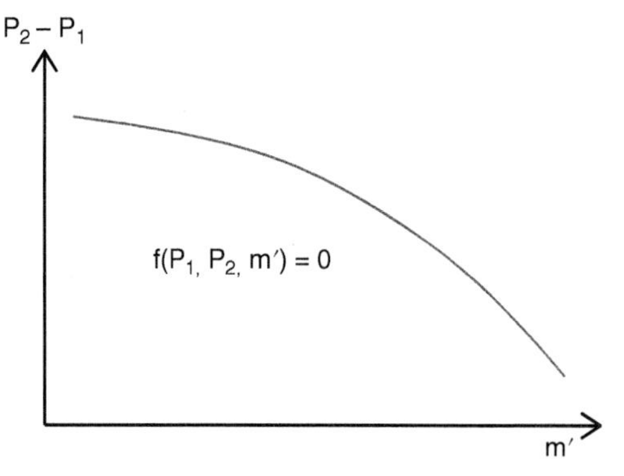

Figure 6.5 A pump characteristic curve. Source: D. Ting.

The remaining equations required are from mass conservation. At the Hydrant A junction, we have

$$m_1{}' = m_A{}' + m_2{}' \tag{6.13}$$

We can express this in a consistent format as that for the other six equations, i.e.

$$F_7\left(m_1{}', m_2{}', m_A{}'\right) = m_1{}' - m_A{}' - m_2{}' = 0 \tag{6.14}$$

Similarly, at the Hydrant B junction,

$$m_2{}' = m_B{}' \tag{6.15}$$

and thus,

$$F_8\left(m_1{}', m_B{}'\right) = m_2{}' - m_B{}' = 0 \tag{6.16}$$

6.3 Solving a Set of Equations via the Matrix Approach

It is clear that a thermofluid system model invariably consists of a set of equations. The set of equations can be expressed in terms of matrices, so that they can be systematically solved. Matrices form a convenient approach to solve the set of equations. An m×n matrix can be written as

$$\begin{bmatrix} a_{1,1} & \cdots & a_{1,n} \\ \vdots & \ddots & \vdots \\ a_{m,1} & \cdots & a_{m,n} \end{bmatrix} \tag{6.17}$$

The *transpose* of the matrix is

$$\begin{bmatrix} a_{1,1} & \cdots & a_{m,1} \\ \vdots & \ddots & \vdots \\ a_{1,n} & \cdots & a_{n,m} \end{bmatrix} \tag{6.18}$$

For example, the transpose of Matrix A,

$$[A] = \begin{bmatrix} 1 & 2 \\ 3 & 4 \\ 5 & 6 \end{bmatrix} \tag{6.19}$$

is

$$[A]^T = \begin{bmatrix} 1 & 3 & 5 \\ 2 & 4 & 5 \end{bmatrix} \tag{6.20}$$

When appropriate, we can multiply one matrix with another one. For example, we can multiply the transpose of Matrix A above with Matrix B,

$$[B] = \begin{bmatrix} 11 & 12 \\ 13 & 14 \\ 15 & 16 \end{bmatrix} \tag{6.21}$$

to give

$$\begin{bmatrix} a_{1,1}b_{1,1} + a_{1,2}b_{2,1} + a_{1,3}b_{3,1} & a_{1,1}b_{1,2} + a_{1,2}b_{2,2} + a_{1,3}b_{3,2} \\ a_{2,1}b_{1,1} + a_{2,2}b_{2,1} + a_{2,3}b_{3,1} & a_{2,1}b_{1,2} + a_{2,2}b_{2,2} + a_{2,3}b_{3,2} \end{bmatrix} \tag{6.22}$$

Substituting

$$\begin{bmatrix} 1(11) + 3(13) + 5(15) & 1(12) + 3(14) + 5(16) \\ 2(11) + 4(13) + 6(15) & 2(12) + 4(14) + 6(16) \end{bmatrix} = \begin{bmatrix} 125 & 134 \\ 164 & 176 \end{bmatrix} \tag{6.23}$$

Matrix Determinant

Many matrix solution approaches invoke the determination of the determinant. For a 1×1 matrix, $[a_{1,1}]$, its determinant,

$$|a_{1,1}| = a_{1,1} \tag{6.24}$$

For a 2×2 matrix, its determinant,

$$\begin{vmatrix} a_{1,1} & a_{1,2} \\ a_{2,1} & a_{2,2} \end{vmatrix} = +\backslash - / = a_{1,1}a_{2,2} - a_{2,1}a_{1,2} \tag{6.25}$$

where \ signifies multiplication of terms from top left to bottom right, which is positive, and / denotes going from bottom right to top left and, thus, it is negative.

We can extend the above determinant computation to a 3×3 matrix. The corresponding determinant,

$$\begin{vmatrix} a_{1,1} & a_{1,2} & a_{1,3} \\ a_{2,1} & a_{2,2} & a_{2,3} \\ a_{3,1} & a_{3,2} & a_{3,3} \end{vmatrix} = +\backslash + \backslash + \backslash - / - / - / \tag{6.26}$$

This is illustrated in Figure 6.6, which is equal to

$$a_{1,1}a_{2,2}a_{3,3} + a_{1,2}a_{2,3}a_{3,1} + a_{1,3}a_{2,1}a_{3,2} - a_{3,1}a_{2,2}a_{1,3} - a_{3,2}a_{2,3}a_{1,1} - a_{3,3}a_{2,1}a_{1,2} \tag{6.27}$$

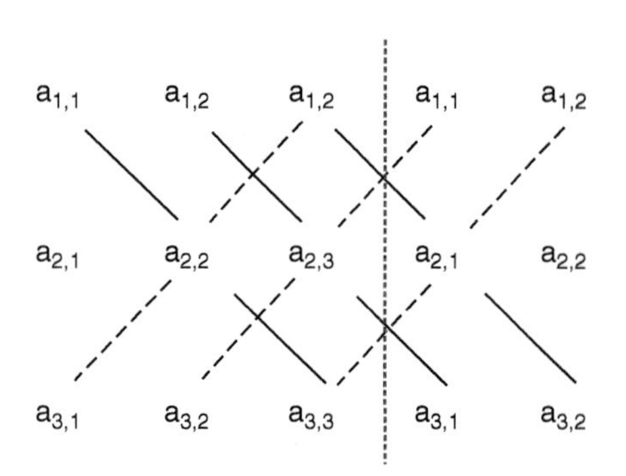

Figure 6.6 Deduction of a 3×3 matrix by adding the negatively-sloped terms and subtracting the positively-sloped terms. Source: D. Ting.

There is a commoner way of determining the determinant. For a 3×3 matrix, the determinant is

$$a_{1,1}\begin{vmatrix} \blacksquare & \blacksquare & \blacksquare \\ \blacksquare & a_{2,2} & a_{2,3} \\ \blacksquare & a_{3,2} & a_{3,3} \end{vmatrix} - a_{1,2}\begin{vmatrix} \blacksquare & \blacksquare & \blacksquare \\ a_{2,1} & \blacksquare & a_{2,3} \\ a_{3,1} & \blacksquare & a_{3,3} \end{vmatrix} + a_{1,3}\begin{vmatrix} \blacksquare & \blacksquare & \blacksquare \\ a_{2,1} & a_{2,2} & \blacksquare \\ a_{3,1} & a_{3,2} & \blacksquare \end{vmatrix} \tag{6.28}$$

Expanding this, we have

$$a_{1,1}\begin{vmatrix} a_{2,2} & a_{2,3} \\ a_{3,2} & a_{3,3} \end{vmatrix} - a_{1,2}\begin{vmatrix} a_{2,1} & a_{2,3} \\ a_{3,1} & a_{3,3} \end{vmatrix} + a_{1,3}\begin{vmatrix} a_{2,1} & a_{2,2} \\ a_{3,1} & a_{3,2} \end{vmatrix} \tag{6.29}$$

When the determinant is zero, there are infinite solutions to the set of equations. That system of equations is called a dependent system of equations. Let us use Example 6.1 to illustrate this.

Example 6.1 *A set of dependent equations*

Given

Four variables are found to give the following set of four linear equations, i.e.

$$x_1 + 2x_2 - 2x_3 + 3x_4 = 5 \tag{E6.1.1}$$

$$2x_1 - x_2 + 3x_3 - 2x_4 = 18 \tag{E6.1.2}$$

$$-x_1 + 3x_2 + x_3 - 4x_4 = -6 \tag{E6.1.3}$$

$$x_1 - 3x_2 + 5x_3 - 5x_4 = 13 \tag{E6.1.4}$$

We can express these equations in terms of the more familiar dependent variable, y, as a function of the independent variables, x_1, x_2, x_3, and x_4. For example, Eq (E6.1.1) can be expressed as

$$y = x_1 + 2x_2 - 2x_3 + 3x_4 - 5. \tag{E6.1.1a}$$

Find

The values of x_1, x_2, x_3, and x_4. Note that the set of equations are suspected to be dependent on each other, therefore, make sure this is not the case before solving them.

Solution

Test the matrix by evaluating the determinant,

$$\begin{vmatrix} 1 & 2 & -2 & 3 \\ 2 & -1 & 3 & -2 \\ -1 & 3 & 1 & -4 \\ 1 & -3 & 5 & -5 \end{vmatrix} = (1)(-1)^{1+1}\begin{vmatrix} -1 & 3 & -2 \\ 3 & 1 & -4 \\ -3 & 5 & -5 \end{vmatrix} + (2)(-1)^{2+1}\begin{vmatrix} 2 & -2 & 3 \\ 3 & 1 & -4 \\ -3 & 5 & -5 \end{vmatrix}$$

$$= (-1)(-1)^{3+1}\begin{vmatrix} 2 & -2 & 3 \\ -1 & 3 & -2 \\ -3 & 5 & -5 \end{vmatrix} + (1)(-1)^{4+1}\begin{vmatrix} 2 & -2 & 3 \\ -1 & 3 & -2 \\ 3 & 1 & 4 \end{vmatrix}$$

$$= (1)(30) - (2)(30) + (1)(0) - (1)(-30)$$

$$= 0$$

Therefore, the equations are dependent, i.e. there are infinite number of solutions.

More often, the set of equations are derived based on physics and sound measurements. As such, it is an independent system of equations with one definite solution. Consider a set of linear equations,

$$a_{1,1}\, x + a_{1,2}\, y - a_{1,3}\, z = c_1 \tag{6.30}$$

$$a_{2,1}\, x + a_{2,2}\, y - a_{2,3}\, z = c_2 \tag{6.31}$$

and

$$a_{3,1}\, x + a_{3,2}\, y - a_{3,3}\, z = c_3 \tag{6.32}$$

In matrix form, we have

$$[A]\,[X] = \begin{bmatrix} a_{1,1} & a_{1,2} & a_{1,3} \\ a_{2,1} & a_{2,2} & a_{2,3} \\ a_{3,1} & a_{3,2} & a_{3,3} \end{bmatrix} \begin{bmatrix} x \\ y \\ z \end{bmatrix} = \begin{bmatrix} c_1 \\ c_2 \\ c_3 \end{bmatrix} = [C] \tag{6.33}$$

With known values of $a_{i,j}$ and c_i, we can solve for x, y and z. Let us look at Cramer's Rule followed by the Gaussian elimination methods.

Cramer's Rule

From the above, we see that a system of equations can be represented as

$$[A_{i,j}]\,[X_i] = [C_i] \tag{6.34}$$

The set of equations described by Eqs (6.30–6.32) is expressed in such a manner by Eq (6.34). According to Cramer's Rule, the solution, values of X_i, can be deduced from the determinant of matrix [A] with [C] substituted in the ith column, divided by the determinant of matrix [A]. Specifically,

$$X_i = |\,[A_{i,j}] \text{ with } [C_i] \text{ substituted in the } i^{th} \text{ column}\,| \,/\, |\,A\,| \tag{6.35}$$

Let us illustrate this in Example 6.2.

Example 6.2 ***Solving a 3x3 matrix using Cramer's Rule***

Given

The optimum performance of a car is related to its (normalized) engine speed, x, the incoming wind velocity, y, and the fuel flow rate, z. It has been experimentally determined that

$$2x + y - z = 3 \quad \text{(E6.2.1)}$$

$$x - 2y + 2z = 9 \quad \text{(E6.2.2)}$$

$$-x + 3z = 0 \quad \text{(E6.2.3)}$$

Find

The engine speed, incoming wind velocity, and fuel flow rate at which the optimum performance of the engine occurs using Cramer's Rule.

Solution

In matrix form, we have

$$\begin{bmatrix} 2 & 1 & -1 \\ 1 & -2 & 2 \\ -1 & 0 & 3 \end{bmatrix} \begin{bmatrix} x \\ y \\ z \end{bmatrix} = \begin{bmatrix} 3 \\ 9 \\ 0 \end{bmatrix}$$

Applying Cramer's Rule, we have

$$x = \begin{vmatrix} 3 & 1 & -1 \\ 9 & -2 & 2 \\ 0 & 0 & 3 \end{vmatrix} \Bigg/ \begin{vmatrix} 2 & 1 & -1 \\ 1 & -2 & 2 \\ -1 & 0 & 3 \end{vmatrix} = \frac{3(-6-0) - 1(27-0) - 1(0-0)}{2(-6-0) - 1(3 - -2) - 1(0-2)} = \frac{-45}{-15} = 3,$$

and

$$y = \begin{vmatrix} 2 & 3 & -1 \\ 1 & 9 & 2 \\ -1 & 0 & 3 \end{vmatrix} \Bigg/ \begin{vmatrix} 2 & 1 & -1 \\ 1 & -2 & 2 \\ -1 & 0 & 3 \end{vmatrix} = \frac{2(27-0) - 3(3 - -2) - 1(0 - -9)}{2(-6-0) - 1(3 - -2) - 1(0-2)} = \frac{30}{-15} = -2.$$

Substituting these into Eq (E6.2.1), we get z = 1.

In short, x = 3, y = −2 (wind on the back of the car), and z = 1 at optimum performance.

More commonly used for solving a set of linear equations is the method called Gaussian elimination. The two major stages in Gaussian elimination are:

1) conversion of the coefficient matrix into a triangular matrix
2) back substitution to obtain the solution.

Let us solve the car performance example via Gaussian elimination.

Example 6.3 ***Solving a 3x3 matrix using Gaussian elimination***

Given

The optimum performance of a car is related to its (normalized) engine speed, x, the incoming wind velocity, y, and the fuel flow rate, z. It has been experimentally determined that

$$2x + y - z = 3 \quad \text{(E6.3.1)}$$

$$x - 2y + 2z = 9 \quad \text{(E6.3.2)}$$

$$-x + 3z = 0 \quad \text{(E6.3.3)}$$

Find

The engine speed, incoming wind velocity and fuel flow rate at which the optimum performance of the engine occurs via Gaussian elimination.

Solution

In matrix form, we have

$$\begin{bmatrix} 2 & 1 & -1 \\ 1 & -2 & 2 \\ -1 & 0 & 3 \end{bmatrix} \begin{bmatrix} x \\ y \\ z \end{bmatrix} = \begin{bmatrix} 3 \\ 9 \\ 0 \end{bmatrix}$$

The idea is to convert the coefficient matrix [A] into a triangular matrix. To do so, we proceed to remove the first term in the second row, i.e. make it zero. To achieve that, we add Eqs. (E6.3.2) and (E6.3.3). Doing so, we get

$$0 - 2y + 5z = 9. \qquad \text{(E6.3.2a)}$$

The matrix becomes

$$\begin{bmatrix} 2 & 1 & -1 \\ 0 & -2 & 5 \\ -1 & 0 & 3 \end{bmatrix} \begin{bmatrix} x \\ y \\ z \end{bmatrix} = \begin{bmatrix} 3 \\ 9 \\ 0 \end{bmatrix}$$

The next step is to remove the first term in the third row. This can be realized by adding Eq (E6.3.1) with two times Eq (E.6.3.3), which gives

$$0 + y + 5z = 3 \qquad \text{(E6.3.3a)}$$

In matrix form, this is

$$\begin{bmatrix} 2 & 1 & -1 \\ 0 & -2 & 5 \\ 0 & 1 & 5 \end{bmatrix} \begin{bmatrix} x \\ y \\ z \end{bmatrix} = \begin{bmatrix} 3 \\ 9 \\ 3 \end{bmatrix}.$$

To make it into a triangular matrix, we only need to set the second term of the third row to zero. This can be accomplished by adding Eq (E6.3.2a) with two times Eq (E6.3.3.a). Executing this gives

$$\begin{bmatrix} 2 & 1 & -1 \\ 0 & -2 & 5 \\ 0 & 0 & 15 \end{bmatrix} \begin{bmatrix} x \\ y \\ z \end{bmatrix} = \begin{bmatrix} 3 \\ 9 \\ 15 \end{bmatrix}$$

In equation format, we have

$$2x + y - z = 3 \qquad \text{(E6.3.1)}$$

$$-2y + 5z = 9 \qquad \text{(E6.3.2a)}$$

$$15z = 15 \qquad \text{(E6.3.3b)}$$

From Eq (E6.3.3b), we get $z = 1$. Substituting this into Eq (E6.3.2a) leads to

$$-2y + 5\,(1) = 9$$

or $y = -2$. With this along with $z = 1$, Eq (E6.3.1) can be written as

$$2x + (-2) - (1) = 3$$

which gives $x = 3$.

To sum things up, the car performs best when $x = 3$, $y = -2$, and $z = 1$.

6.4 Sequential versus Simultaneous Calculations

Let us examine and compare solving the set of equations one at a time versus solving all of them simultaneously. Sequential calculations are favored when the problem can be solved sequentially. This approach is more straightforward to execute and is not subject to convergence issues. It starts with an input, from which the output is sequentially fed as an input to the subsequent unit of the concerned system. For example, a conventional air conditioning system consists of an evaporator, a compressor, a condenser, and an expansion (throttling) valve. Suppose the inlet condition of the evaporator is known, then the outlet can be calculated. The evaporator outlet may be assumed to be equal to the inlet into the compressor, which enables the deduction of the compressor's outlet. Subsequently, we can use it to determine the condenser outlet, which is equal to the expansion valve inlet. Another example of sequential calculation is a power generation system. The input is the fuel flow rate, it is used to calculate the combustor output. The combustor output is employed to deduce the boiler output, which is fed as the input to determine the steam turbine output.

More frequently, however, the system under study is more complex than a one-path cycle, such as that of a conventional air conditioning referenced above. In such a case, simultaneous calculations are commonly invoked. Which approach is preferable largely depends on the problem at hand and the familiarity of the analyst.

We will look closely at some of the standard simultaneous calculation methods. In Section 6.5, successive substitution is delineated. Then, the Newton-Raphson method, based on Taylor series expansion, is detailed.

6.5 Successive Substitution

Successive substitution is closely associated with simultaneous calculations. The calculation is initiated by guessing the value of one or more variables of concern. With the guessed values, the series of calculations are carried through until the originally-guessed variables are recalculated. The process is iterated until satisfactory convergence is realized. The convergence criterion is typically a change of no more than a small set value, less than 1%, for example, between subsequent iterations for all interested variables. Let us use a simple water pumping system in Example 6.4 to demonstrate this.

Example 6.4 ***A water pumping system***

Given

A water pumping system is shown in Figure 6.7. The friction loss in the pipe is $6.3\,m_2'^2$ in kPa, where m' is in kg/s. The characteristic equation of Pump 1 is

$$\Delta P = 720 - 20\ m_1' - 2.5\ m_1'^2 \tag{E6.4.1}$$

in kPa. For Pump 2, it is

$$\Delta P = 800 - 60\ m_2' - 25\ m_2'^2 \tag{E6.4.2}$$

in kPa.

Find

ΔP, m_1', m_2', m' using Successive Substitution.

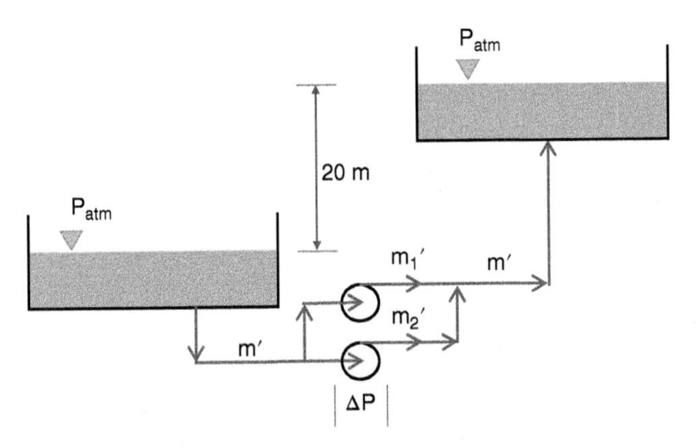

Figure 6.7 A water pumping system. Source: D. Ting.

Solution

For consistency, all pressures are in kPa and all mass flow rates are in kg/s. Other than Eqs. (E6.4.1) and (E6.4.2), we have the total head loss and mass conservation equations.

The total head loss,

$$\Delta P = 6.3\, m'^2 + \rho\, g\, h = 6.3\, m'^2 + 196.20 \qquad \text{(E6.4.3)}$$

Mass conservation,

$$m' = m_1' + m_2' \qquad \text{(E6.4.4)}$$

In short, we have four equations with four unknowns, ΔP, m', m_1', and m_2'.

Let us assume m_2' to be 1 kg/s to get the calculations started according to the information flow diagram depicted in Figure 6.8, and the obtained values are summarized in Table 6.1. Substituting this into Eq (E6.4.2), we get

$$\Delta P = 800 - 60\, m_2' - 25\, m_2'^2 = 800 - 60\,(1) - 25\,(1)^2 = 715$$

Substitute this pressure drop into Eq (E6.4.1),

$$2.5\, m1'^{\,2} + 20\, m1' + (\Delta P - 720) = 0$$

which gives

$$m_1' = \left\{-20 \pm \sqrt{\left[(20)^2 - 4\,(2.5)\,\,(715 - 720)\right]}\right\} / \,[2\,(2.5)]$$
$$= \{-20 \pm 21.2\} \,/\, [2\,(2.5)]$$
$$= 0.24$$

From Eq (E6.4.3),

$$m' = \sqrt{\left[(\Delta P - 196.2)\,/6.3\right]} = 9.07$$

From Eq (E6.4.4),

$$m_2' = m' - m_1' = 9.07 - 0.24 = 8.83$$

Iterate

$$\Delta P = 800 - 60\, m_2' - 25\, m_2'^2 = 800 - 60\left(8.89\right) - 25\left(8.89\right)^2 = -1680.$$

It diverged!

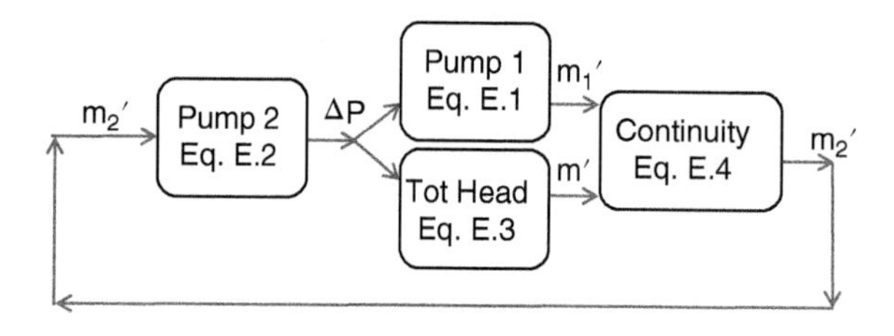

Figure 6.8 Information-flow diagram showing one calculation process for a water pumping system. Source: D. Ting.

Table 6.1 Iterations of Example 6.4.

m_2' [kg/s]	ΔP [kPa]	m_1' [kg/s]	m' [kg/s]
1	715	0.2426	9.0746
8.8320	−1680		

We encounter the prevailing challenge of successive substitution method, that is, it tends to diverge. For Example 6.4, the specific process always leads to divergence, irrespective of the initial guess. Fortunately, there are methods for minimizing divergence. Arranging the largest terms diagonally on the matrix consisting of the set of equations, relaxation, etc. can mitigate this divergence challenge.

Example 6.5 ***Solve the water pumping system via a different sequence***

Given

Example 6.4.

Find

ΔP, m_1', m_2' and m', by following the calculation order that is depicted in Figure 6.9.

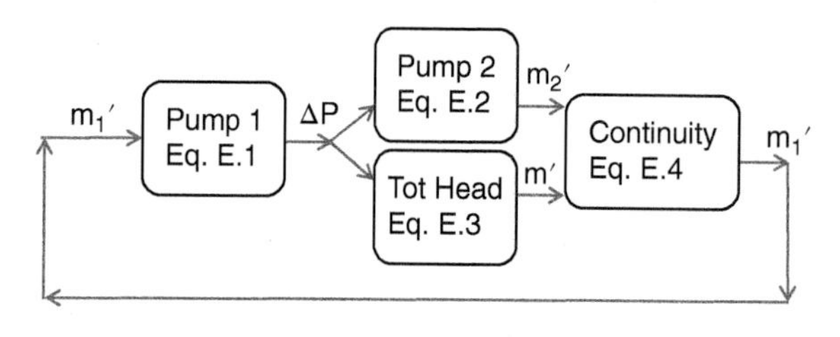

Figure 6.9 A converging iteration process for a water pumping system. Source: D. Ting.

Solution

For Pump 1,

$$\Delta P = 720 - 20\ m_1' - 2.5\ m_1'^2. \tag{E6.5.1}$$

For Pump 2,

$$\Delta P = 800 - 60\ m_2' - 25\ m_2'^2 \tag{E6.5.2}$$

The total head loss,

$$\Delta P = 6.3\ m'^2 + \rho\ g\ h = 6.3\ m'^2 + 196.2. \tag{E6.5.3}$$

Mass conservation,

$$m' = m_1' + m_2' \tag{E6.5.4}$$

The four unknowns of the above four equations are ΔP, m', m_1', and m_2'.

Table 6.2 Iterations of Example 6.5.

Iteration	ΔP [kPa]	m_2' [kg/s]	m' [kg/s]	m_1' [kg/s]
1	637.5	1.6178	8.3694	6.7516
2	471.0053	2.6210	6.6045	3.9836
3	600.6566	1.8682	8.0125	6.1443
⋮				
22	545.8335	2.2069	7.4497	5.2428
⋮				
52	546.1696	2.2049	7.4532	5.2484
53	546.1698	2.2049	7.4532	5.2484
54	546.1697	2.2049	7.4532	5.2484

Following Figure 6.9, we guess $m_1' = 3$.
Equation (E6.5.1),

$$\Delta P = 720 - 20\,m_1' - 2.5\,m_1'^2 = 720 - 20\,(3) - 2.5\,(3)^2 = 637.5.$$

Equation (E6.5.2),

$$25\,m_2'^2 + 60\,m_2' + (\Delta P - 800) = 0$$

which gives

$$m_2' = \left\{-60 \pm \sqrt{\left[(60)^2 - 4\,(25)\,\left(637.5 - 800\right)\right]}\right\} / \left[2\,(25)\right] = 1.618.$$

Equation (E6.5.3),

$$m' = \sqrt{\left[(\Delta P - 196.2)/6.3\right]} = 8.369.$$

Equation (E6.5.4),

$$m_1' = m' - m_2' = 6.751.$$

We see from Table 6.2 that within 22 iterations, the solutions largely converged. If a stricter convergence criterion is imposed, it can take slightly over 50 iterations.

Once again, Example 6.5 demonstrates the importance of a viable sequence when it comes to the successive substitution method. With experience, one can have a sense of which particular sequence is more likely to lead to convergence. Let us acquaint ourselves with successive substitution a little better with another example.

Example 6.6 ***A fan-and-duct system***

Givven

A fan-and-duct system with the following duct and fan characteristics.

For the duct,

$$P_{static} = 90 + 11.82\forall'^2 \quad \text{(E6.6.1)}$$

in Pa, where volume flow rate, ∀', is in m^3/s.
For the fan,

$$\forall' = 17 - 75.8 \times 10^{-6}\, P_{static}{}^2 \quad \text{(E6.6.2)}$$

Find

P_{static} and ∀' using Successive Substitution, with $P_{static} = 250$ and ∀' = 15 as initial guesses.

Solution

Follow Table 6.3 as the iterative process proceed.
First attempt: solve for ∀' first using $P_{static} = 250$.

$$\forall' = 17 - 75.8 \times 10^{-6}\, P_{static}{}^2 = 12.26$$

$$P_{static} = 90 + 11.82\forall'^2 = 1867$$

$$\forall' = 17 - 75.8 \times 10^{-6}\left(1867\right)^2 = -247$$

Diverged!

Table 6.3 Iterations of Example 6.6.

Iteration	P_{static} [Pa]	∀' [m^3/s]
1	250	3.6792
2	419.2092	5.2775
3	393.2563	5.0652
4	396.8013	5.0947
5	396.3103	5.0906
6	396.3781	5.0912
7	396.3687	5.0912
8	396.3700	5.0911
9	396.3699	5.0911
10	396.3699	5.0911

Note: Extra significant figures kept for computation purpose.

Second attempt: solve for P_{static} first using ∀' = 15.

$$P_{static} = 90 + 11.82\forall'^2 = 2749.5$$

$$\forall' = 17 - 75.8 \times 10^{-6}\left(2749.5\right)^2 = -556$$

Diverged!
Third attempt: solve for ∀' via Eq (E6.3.1) first.
From Eq (E6.6.1), we have

$$\forall' = \sqrt{\left[\left(P_{static} - 90\right)/11.82\right]} = 3.679.$$

From Eq (E6.6.2), we can write

$$P_{\text{static}} = \sqrt{[(17 - \forall') / 75.8 \times 10^{-6}]} = 419.2.$$

Iterate

$$\forall' = \sqrt{\left[\left(419.2 - 90\right) / 11.82\right]} = 5.277.$$

$$P_{\text{static}} = \sqrt{\left[\left(17 - 5.277\right) / 75.8 \times 10^{-6}\right]} = 393.3.$$

$$\forall' = \sqrt{\left[\left(393.3 - 90\right) / 11.82\right]} = 5.065.$$

It converged rapidly!

Since we are on a roll, let us look at a little more complex thermofluid system, a steam boiler. We will examine the part of the system called the steam boiler blowdown. It is employed to bleed out water from a boiler to avoid a buildup of impurities.

Example 6.7 *Steam boiler blowdown*

Given

A steam boiler blowdown is shown in Figure 6.10. Water at 400°C is blown down at 0.4 kg/s to heat the feed water at 90°C at 5 kg/s, where UA = 15 kW/K.

Saturated liquid water enthalpy,

$$h_f = 4.19\ T \tag{E6.7.1}$$

where temperature T is in °C.

Saturated water vapor enthalpy,

$$h_g = 2530 + 0.4\ T \tag{E6.7.2}$$

where temperature T is in °C.

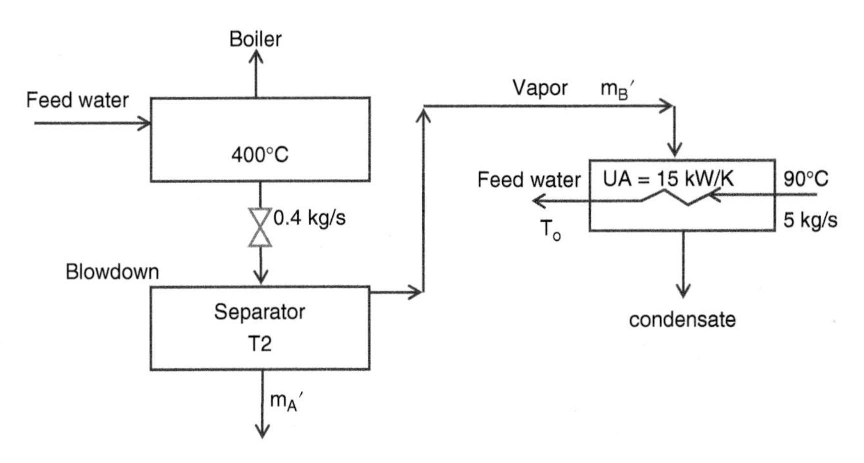

Figure 6.10 A steam boiler with blowdown. Source: D. Ting.

Find

(i) an information-flow diagram, (ii) values of T_2, T_o, m_A', and m_B' via successive substitution.

Solution

To solve the problem, the relevant equations need to be obtained. Figure 6.10 eases the determination of the following equations.

Mass balance at the separator,

$$m_A' + m_B' = 0.4 \quad \text{(E6.7.3)}$$

Here the mass flow rates are in kg/s.

Energy balance at the separator,

$$m' c_P \, \Delta T = m_A' \Delta h_A + m_B' \Delta h_B \quad \text{(E6.7.4)}$$

Substituting known values and Eqs. (E6.7.1) and (E6.7.2), we get

$$0.2\ (4.19)\ (400) = m_A'\,(4.19)\ T_2 + m_B'\,(2530 - 0.4\ T2) \quad \text{(E6.7.4a)}$$

Energy balance at the condenser,

$$m'\left(c_P\ \Delta T\right)_{\text{Feedwater}} = \left(m_B' \Delta h\right)_{\text{condenser}} \quad \text{(E6.7.5)}$$

Substituting known values and Eqs. (E6.7.1) and (E6.7.2), we get

$$5\ (4.19)\ \left(T_o - 90\right) = m_B'\left[\left(2530 - 0.4\ T_2\right) - 4.19\ T_2\right] \quad \text{(E6.7.5a)}$$

For the condensing-steam HEX (heat exchanger),

$$T_o = T_i + \left(T_{const} - T_i\right)\ \left\{1 - \exp\left[-UA/\left(m' c_P\right)\right]\right\} \quad \text{(E6.7.6)}$$

Substituting for known values, we get

$$T_o = 90 + \left(T_2 - 90\right)\ (0.5088) \quad \text{(E6.7.6a)}$$

In short, we have four equations, i.e.

$$m_A' + m_B' = 0.4 \quad \text{(E6.7.3)}$$

$$670.4 = 4.19\ m_A' T_2 + m_B'\left(2530 - 0.4\ T_2\right) \quad \text{(E6.7.4b)}$$

$$20.95\ \left(T_o - 90\right) = m_B'\left(2530 - 3.39\ T_2\right) \quad \text{(E6.7.5b)}$$

$$T_o = 90 + 0.5088\ \left(T_2 - 90\right) \quad \text{(E6.7.6b)}$$

The four unknowns are m_A', m_B', T_o, and T_2.

A possible information-flow diagram is given in Figure 6.11, and the iteration progress is summarized in Table 6.4. Let us guess $T_o = 110°C$ and $m_B' = 0.2$ kg/s. Then, from Eq (E6.7.6b), we have

$$T_2 = \left(T_2 - 90\right)/0.5088 + 90 = 129.31$$

Equation (E6.7.5b) can be rearranged into

$$T_o = m_B'\left(2530 - 3.79\ T_2\right)/20.95 + 90 = 109.47$$

Equation (E6.7.3) can be recast as

$$m_A' = 0.4 - m_B' = 0.2$$

Finally, Eq. (E6.7.4b) can be rewritten as

$$m_B' = \left(670.4 - 4.59\ T_2\right)/\left(2530 - 0.4\ T_2\right) = 0.2177$$

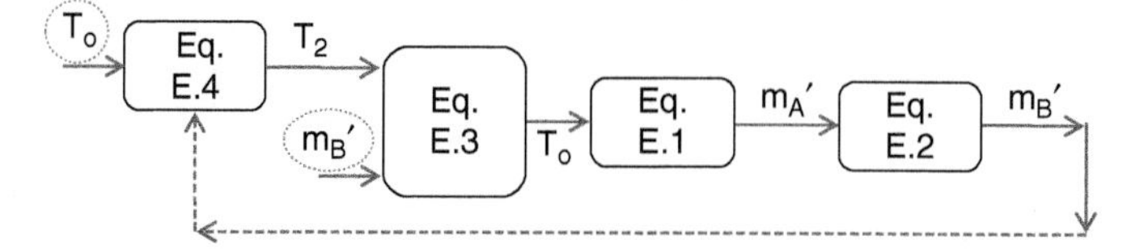

Figure 6.11 A possible information-flow diagram for the steam boiler with blowdown. Source: D. Ting.

Iterating, we have

$$T_o = 111.24$$
$$m_A' = 0.1823$$
$$m_B' = 0.2218$$

It more-or-less converged in only two iterations!

Table 6.4 Iterations of Example 6.7.

Iteration	T_2 [°C]	T_o [°C]	m_A' [kg/s]	m_B' [kg/s]
		110		0.2
1	129.3082	109.4742	0.2	0.2177
2	128.2748	111.2383	0.1823	0.2218
3	131.742	111.4949	0.1782	0.2215
4	132.2462	111.4478	0.1785	0.2213
5	132.1536	111.4299	0.1787	0.2212
6	132.1185	111.4296	0.1788	0.2212
7	132.1179	111.4303	0.1788	0.2212
8	132.1194	111.4305	0.1788	0.2212
9	132.1196	111.4304	0.1788	0.2212
10	132.1196	111.4304	0.1788	0.2212

6.6 Taylor Series Expansion and the Newton-Raphson Method

A time-tested, versatile method to solve a set of equations is the Newton-Raphson method. The Newton-Raphson method is based on the Taylor series expansion. Accordingly, we will review the Taylor series expansion in context before going through the Newton-Raphson method.

6.6.1 Taylor Series Expansion

The idea behind the Taylor series is that as the order of differentiation is increased, the importance of the term diminishes. If $y = y(x)$, as shown in Figure 6.12, one can expand about the point x=a. The corresponding Taylor series is

$$y = c_o + c_1 (x - a) + c_2 (x - a)^2 + c_3 (x - a)^3 + \dots \tag{6.36}$$

At $x = a$, we only have the first term, i.e. the 0th-order term, and all higher order terms are zero. Specifically,

$$y)_{x=a} = y(a) = c_o \tag{6.37}$$

The first derivative with respect to x is

$$dy/dx = c_1 + 2\,c_2\,(x - a) + 3\,c_3\,(x - a)^2 + \ldots \tag{6.38}$$

At x = a, the first derivative is

$$dy/dx)_{x=a} = dy\,(a)\,/dx = y'\,(a) = c_1 \tag{6.39}$$

Differentiating Eq. (6.38, we get the second derivative, i.e.

$$d^2y/dx^2 = 2\,c_2 + 6\,c_3\,(x - a) + \ldots \tag{6.40}$$

The corresponding value for x = a is

$$d^2y/\,dx^2)_{x=a} = d^2y\,(a)\,/dx^2 = y"\,(a) = 2\,c_2 \tag{6.41}$$

Differentiating again we get the third derivative,

$$d^3y/dx^3 = 6\,c_3 + 24\,c_4\,(x - a) + \ldots \tag{6.42}$$

The value at x = a is

$$d^3y/\,dx^3)_{x=a} = d^3y\,(a)\,/dx^3 = y'''\,(a) = 6\,c_3 \tag{6.43}$$

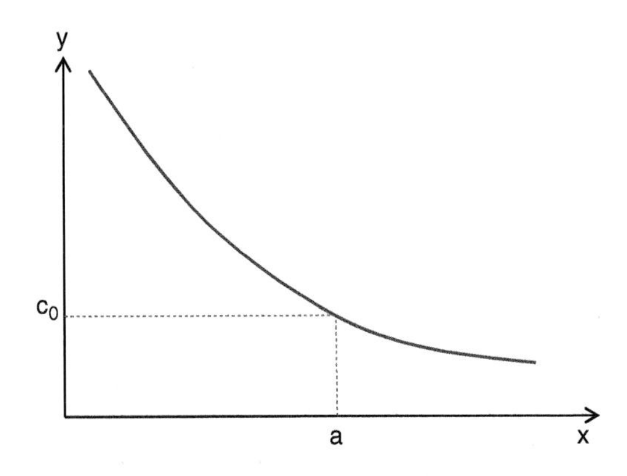

Figure 6.12 Taylor series expansion for y = y(x) at x = a. Source: D. Ting.

Summary

In summary, we have:

$$y = c_o + c_1\,(x - a) + c_2\,(x - a)^2 + c_3\,(x - a)^3 + c_4\,(x - a)^4 + \ldots$$

$$y\,(a) = c_o$$

$$y'\,(a) = c_1$$

$$y"\,(a) = 2c_2$$

$$y'''\,(a) = 6c_3, \ldots$$

Therefore, we can write

$$y = y\,(a) + y'\,(a)\,(x - a) + \tfrac{1}{2}\,y"\,(a)\,(x - a)^2 + (1/6)\,y'''\,(a)\,(x - a)^3 + \ldots \tag{6.44}$$

Let us move on to the two independent variables case, where $y = y(x_1, x_2)$, as shown in Figure 6.13. Expanding about the point $x_1 = a$ and $x_2 = b$ gives

$$y = c_o + \left[c_1\,(x_1 - a) + c_2\,(x_2 - b)\right] + \left[c_3\,(x_1 - a)^2 + c_4\,(x_1 - a)\,(x_2 - b) + c_5\,(x_2 - b)^2\right] + \ldots \tag{6.45}$$

The corresponding first derivatives are

$$\partial y/\partial x_1 = c_1 + 2c_3\ (x_1 - a) + c_4\ (x_2 - b) + \ldots \tag{6.46}$$

and

$$\partial y/\partial x_2 = c_2 + c_4\ (x_1 - a) + 2c_5\ (x_2 - b) + \ldots \tag{6.47}$$

Substituting for $x_1 = a$ and $x_2 = b$, we have

$$y\,(a, b) = c_o \tag{6.48}$$

$$\partial y\,(a, b)\,/\partial x_1 = c_1 \tag{6.49}$$

$$\partial y\,(a, b)\,/\partial x_2 = c_2 \tag{6.50}$$

Differentiating Eq (6.45) with respect to x_1 again gives

$$\partial^2 y\,(a, b)\,/\partial {x_1}^2 = 2\ c_3 \tag{6.51}$$

Similarly,

$$\partial^2 y\,(a, b)\,/\partial x_1\ \partial x_2 = c_4 \tag{6.52}$$

$$\partial^2 y\,(a, b)\,/\partial {x_2}^2 = 2\ c_5 \tag{6.53}$$

and so on.

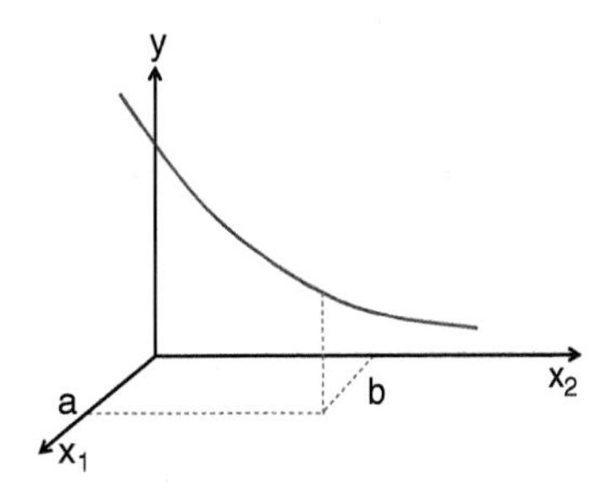

Figure 6.13 Taylor series expansion for $y = y(x_1, x_2)$ at $x_1 = a$ and $x_2 = b$. Source: D. Ting.

The General Expression

If $y = y(x_1, x_2, x_3, \ldots x_n)$, we can expand about the point $x_1 = a_1, x_2 = a_2, x_3 = a_3, \ldots x_n = a_n$. Doing so leads to

$$y\,(x_1, x_2, x_3, \ldots x_n) = y\,(a_1, a_2, a_3, \ldots a_n) + \sum \{[\partial y\ (a_1, a_2, a_3, \ldots a_n)\,/\partial x_j]\ (x_j - a_j)\}$$

$$+ \tfrac{1}{2} \sum \sum \{[\partial^2 y\ (a_1, a_2, a_3, \ldots a_n)\,/\partial x_i\ \partial x_j]\ (x_i - a_i)\ (x_j - a_j)\} + \ldots \tag{6.54}$$

For completeness, let us walk through the Taylor series expansion of a two-variable function in Example 6.8.

Example 6.8 ***Taylor series expansion of a two-variable function***

Given

$$y = \ln\left(x_1^{\,2}/x_2\right)$$

Find

The Taylor series at $y(x_1 = 2, x_2 = 1)$.

Solution

From Eq (6.45), we have

$$y = c_o + \left[c_1\ (x_1 - a) + c_2\ (x_2 - b)\right] + \left[c_3\ (x_1 - a)^2 + c_4\ (x_1 - a)\ (x_2 - b) + c_5\ (x_2 - b)^2\right] + \ldots$$

Also,

$$y\,(a, b) = c_o$$
$$\partial y\,(a, b)\,/\partial x_1 = c_1$$
$$\partial y\,(a, b)\,/\partial x_2 = c_2$$
$$\partial^2 y\,(a, b)\,/\partial x_1^{\,2} = 2\ c_3$$
$$\partial^2 y\,(a, b)\,/\partial x_1\ \partial x_2 = c_4$$
$$\partial^2 y\,(a, b)\,/\partial x_2^{\,2} = 2\ c_5, \ldots$$

Substituting for $a = 2$ and $b = 1$, we get

$$c_o = y\,(a, b) = \ln\left(2^2/1\right) = 1.386$$
$$c_1 = \partial y\,(a, b)\,/\partial x_1 = \left(2x_1/x_2\right)\,/\,\left(x_1^{\,2}/x_2\right) = 2/x_1 = 1$$
$$c_2 = \partial y\,(a, b)\,/\partial x_2 = -\left(x_1^{\,2}/x_2^{\,2}\right)\ \left(x_1^{\,2}/x_2\right) = -1/x_2 = -1$$
$$c_3 = ½\ \partial^2 y\,(a, b)\,/\partial x_1^{\,2} = ½\ \left(-2/x_1^{\,2}\right) = -1/4$$
$$c_4 = \partial^2 y\,(a, b)\,/\partial x_1\ \partial x_2 = 0$$
$$c_5 = \partial^2 y\,(a, b)\,/\partial x_2^{\,2} = ½\ \partial^2 y\,(a, b)\,/\partial x_2^{\,2} = ½\ \left(1/x_2^{\,2}\right) = ½$$

Therefore,

$$y = 1.386 + \left(x_1 - 2\right) - 1\left(x_2 - 1\right) - (1/4)\left(x_1 - 2\right)^2 + 0 + ½\ \left(x_2 - 1\right)^2\Big] + \ldots$$

6.6.2 The Newton-Raphson Method

The Newton-Raphson method is based on the first two terms of the Taylor series. Recall from Figure 6.12 and Eq. (6.36) that the Taylor series expansion about $x = a$ is

$$y = c_o + c_1\ (x - a) + c_2\ (x - a)^2 + c_3\ (x - a)^3 + \ldots \tag{6.36}$$

From which we get Eq (6.44),

$$y = y\,(a) + y'\,(a)\ (x - a) + ½\,y''\,(a)\ (x - a)^2 + (1/6)\ y'''\,(a)\ (x - a)^3 + \ldots \tag{6.44}$$

As x approaches a, the higher-order terms decrease; therefore,

$$y \cong y\,(a) + y'\,(a)\ (x - a) \tag{6.45}$$

This is the basis of the Newton-Raphson iterative technique for solving a nonlinear algebraic equation. The Newton-Raphson iterative technique is based on Eq (6.55).

The Newton-Raphson Method with One Equation and One Unknown

In the simplest case, the objective function is only a function of one independent variable, i.e. $y = f(x)$. In this case, only one equation is needed to solve the problem. An example within the context of optimization of a thermofluid system is the engine efficiency, y, as a function of the engine speed, x. More often than not, the engine efficiency increases with engine speed until it reaches a maximum value, beyond which its efficiency decreases with further increases in engine speed. For a typical internal combustion automobile, the engine efficiency peaks at around 80 km/h. Mathematically, the equation is roughly of the form,

$$y = -ax^2 + bx + c \tag{6.56}$$

where a, b and c are coefficients deduced using a curve fit. To solve for the maximum y using the Newton-Rapson method, we introduce a dependent variable, Y, where

$$Y = -ax^2 + bx + c - y_{opt} \tag{6.57}$$

where y_{opt} is the optimum y, which is the maximum y for engine efficiency. This optimum y occurs when $Y = 0$. Let us use a real numerical example to illustrate how the Newton-Raphson method solves such a problem.

Example 6.9 ***Solve a single-variable equation using the Newton-Raphson method***

Given

$$x + 2 = e^x$$

Find

x via the Newton-Raphson method

Solution

Let $y = x + 2 - e^x$ and x_c is the correct value that solves the equation, giving $y(x_c) = 0$.

If we start with an initial guess, $x = 2$, we have

$$y = x + 2 - e^x = (2) + 2 - e^{(2)} = -3.39$$

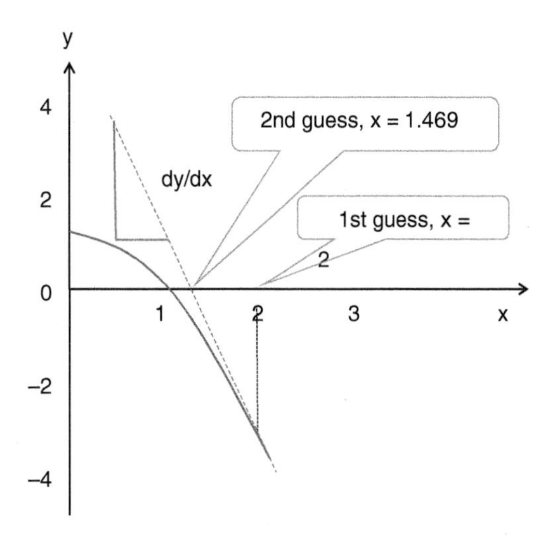

Figure 6.14 An illustration of the rapid convergence of the Newton-Raphson method for a single variable function. Source: D. Ting.

From Eq. (6.55),

$$y(x) \cong y(x_c) + y'(x_c)\ (x - x_c) \cong 0 + y'(x_c)\ (x - x_c)$$

This can be expressed as

$$x_c \cong x - y(x)/y'(x)$$

This is illustrated in Figure 6.14, where the new guess, x_c, is the x-axis intercept by the tangent from the current guess, x. The use of the slope, y'(x), significantly hastens the improvement of the guess. Accordingly, we can improve our guess,

$$x_c \cong (2) - (-3.39) / \left(1 - e^{(2)}\right) = 1.469$$

$$y = x + 2 - e^x = (1.469) + 2 - e^{(1.469)} = -0.876$$

This new y is much closer to zero, the solution.

Iterate,

$$x_c \cong (1.469) - (-0.876) / \left(1 - e^{(1.469)}\right) = 1.208$$

$$y = x + 2 - e^x = (1.208) + 2 - e^{(1.208)} = -0.132$$

The calculation is converging, i.e. y approaching zero.

Iterate again,

$$x_c \cong (1.208) - (-0.132) / \left(1 - e^{(1.208)}\right) = 1.152$$

$$y = x + 2 - e^x = (1.152) + 2 - e^{(1.152)} = -0.018$$

To the first decimal, the calculation has converged. Nevertheless, students are encouraged to iterate one more time, to convince themselves that the solution is indeed $x = 1.15$.

The Newton-Raphson Method with Multiple Equations and Unknowns

In practice, we are often faced with more than one independent variable. By its very nature, we need as many equations as the number of unknowns we have to solve. Let us consider a generic case where there are three variables, x_1, x_2 and x_3. The three equations are

$$f_1(x_1, x_2, x_3) = 0 \tag{6.58}$$

$$f_2(x_1, x_2, x_3) = 0 \tag{6.59}$$

and

$$f_3(x_1, x_2, x_3) = 0 \tag{6.60}$$

Applying the Newton-Raphson method, we start with guessed values of x_1, x_2 and x_3. Then, we calculate f_1, f_2, and f_3, based on these guesses. The corresponding first derivatives, f_1', f_2', and f_3', are computed to improve our estimates of x_1, x_2, and x_3. Only the first two terms of the Taylor series are employed, i.e.

$$\begin{aligned} f_1(x_1, x_2, x_3) &\cong f_1(a_1, a_2, a_3) \\ &+ \left[\partial f_1(x_1, x_2, x_3)/\partial x_1\right] / (x_1 - a_1) \\ &+ \left[\partial f_1(x_1, x_2, x_3)/\partial x_2\right] / (x_1 - a_2) \\ &+ \left[\partial f_1(x_1, x_2, x_3)/\partial x_3\right] / (x_1 - a_3) \end{aligned} \tag{6.61}$$

where a_i are the correct values (solutions) for x_i. Similarly, this can be invoked for f_2 and f_3. The three equations can be short-handed in matrix form,

$$\begin{bmatrix} \partial f_1/\partial x_1 & \partial f_1/\partial x_2 & \partial f_1/\partial x_3 \\ \partial f_2/\partial x_1 & \partial f_2/\partial x_2 & \partial f_2/\partial x_3 \\ \partial f_3/\partial x_1 & \partial f_3/\partial x_2 & \partial f_3/\partial x_3 \end{bmatrix} \begin{bmatrix} x_1 - a_1 \\ x_2 - a_2 \\ x_3 - a_3 \end{bmatrix} = \begin{bmatrix} f_1 \\ f_2 \\ f_3 \end{bmatrix} \tag{6.62}$$

This is used to improve the guessed values of x_1, x_2 and x_3. The calculation is repeated until satisfactory convergence is reached. Let us apply this method to solve the water pumping problem in Example 6.10.

Example 6.10 ***Resolve the water pumping system using the Newton-Raphson method***

Given

A water pumping system, as shown in Figure 6.7. The friction loss in the pipe is 6.3 $m_2'^2$ in kPa, where m' is in kg/s. The characteristic equation of Pump 1 is

$$\Delta P = 720 - 20\ m_1' - 2.5\ m_1'^2 \quad (E6.10.1)$$

in kPa. For Pump 2, it is

$$\Delta P = 800 - 60\ m_2' - 25\ m_2'^2 \quad (E6.10.2)$$

in kPa.

Find

ΔP, m_1', m_2', m' using the Newton-Raphson method.

Solution

The total head loss,

$$\Delta P = 6.3\ m'^2 + \rho\ g\ h = 6.3\ m'^2 + 196.20 \quad (E6.10.3)$$

Mass conservation,

$$m' = m_1' + m_2' \quad (E6.10.4)$$

There are four equations, Equations (E6.10.1) to (E6.10.4), with four unknowns, ΔP, m', m_1', and m_2'.

Step 1. Cast the equations in the form $f_i(x_j) = 0$

$$y_1\left(\Delta P, m_1', m_2', m'\right) = \Delta P - 6.3\ m'^2 - 196.20 = 0 \quad (E6.10.5)$$

$$y_2\left(\Delta P, m_1', m_2', m'\right) = \Delta P - 720 + 20\ m_1' + 2.5\ m_1'^2 = 0 \quad (E6.10.6)$$

$$y_3\left(\Delta P, m_1', m_2', m'\right) = \Delta P - 800 + 60\ m_2' + 25\ m_2'^2 = 0 \quad (E6.10.7)$$

$$y_4\left(\Delta P, m_1', m_2', m'\right) = m' - m_1' - m_2' = 0 \quad (E6.10.8)$$

Step 2. Guess the initial values to get the iteration going. Guess

$$\Delta P = 800, m_1' = 2, m_2' = 1, m' = 4$$

Step 3. Calculate y_1, y_2, y_3, and y_4. Substituting the guesses, we get

$$y_1 = 503, y_2 = 130, y_3 = 85, y_4 = 1$$

Step 4. Compute the partial derivatives.

$$\partial y_1/\partial \Delta P = 1, \partial y_1/\partial m_1' = 0, \partial y_1/\partial m_2' = 0, \partial y_1/\partial m' = -12.6m'$$

$$\partial y_2/\partial \Delta P = 1, \partial y_2/\partial m_1' = 25 + 5m_1', \partial y_2/\partial m_2' = 0, \partial y_2/\partial m' = 0$$

$$\partial y_3/\partial \Delta P = 1, \partial y_3/\partial m_1' = 0, \partial y_3/\partial m_2' = 60 + 50m_2', \partial y_3/\partial m' = 0$$

$$\partial y_4/\partial \Delta P = 0, \partial y_4/\partial m_1' = -1, \partial y_4/\partial m_2' = -1, \partial y_4/\partial m' = 1$$

Step 5. Expand y into the two-term Taylor series.

$$y_1\left(\Delta P, m_1', m_2', m'\right) \cong y_1\left(a_1, a_2, a_3, a_4\right) + \left(\partial y_1/\partial \Delta P\right)\left(\Delta P_{new} - a_1\right) + \left(\partial y_2/\partial m_1'\right)\left(m_1' - a_2\right) + \left(\partial y_3/\partial m_2'\right)\left(m_2' - a_3\right) + \left(\partial y_4/\partial m'\right)\left(m' - a_4\right)$$

The set of equations is

$$\begin{bmatrix} \partial y_1/\partial \Delta P & \partial y_1/\partial \dot{m}_1 & \partial y_1/\partial \dot{m}_2 & \partial y_1/\partial \dot{m} \\ \partial y_2/\partial \Delta P & \partial y_2/\partial \dot{m}_1 & \partial y_2/\partial \dot{m}_2 & \partial y_2/\partial \dot{m} \\ \partial y_3/\partial \Delta P & \partial y_3/\partial \dot{m}_1 & \partial y_3/\partial \dot{m}_2 & \partial y_3/\partial \dot{m} \\ \partial y_4/\partial \Delta P & \partial y_4/\partial \dot{m}_1 & \partial y_4/\partial \dot{m}_2 & \partial y_4/\partial \dot{m} \end{bmatrix} \begin{bmatrix} \Delta P - a_1 \\ \dot{m}_1 - a_2 \\ \dot{m}_2 - a_3 \\ \dot{m} - a_4 \end{bmatrix} = \begin{bmatrix} y_1 \\ y_2 \\ y_3 \\ y_4 \end{bmatrix}. \tag{E6.10.9}$$

Step 6. Solve the set of linear equations. In this specific case, we have

$$\begin{bmatrix} 1 & 0 & 0 & -50.4 \\ 1 & 30 & 0 & 0 \\ 1 & 0 & 110 & 0 \\ 0 & -1 & -1 & 1 \end{bmatrix} \begin{bmatrix} \Delta x_1 \\ \Delta x_2 \\ \Delta x_3 \\ \Delta x_4 \end{bmatrix} = \begin{bmatrix} 503 \\ 130 \\ 85 \\ 1 \end{bmatrix}.$$

The solutions are $\Delta x_1 = 258.3$, $\Delta x_2 = -4.278$, $\Delta x_3 = -1.576$, and $\Delta x_4 = -4.854$.

Step 7. Update the values of the variables.

$$\Delta P_{new} = \Delta P_{old} - \Delta x_1 = 800 - 258.3 = 541.7$$

$$m_{1,new}' = m_{1,old}' - \Delta x_2 = 2 - (-4.278) = 6.278$$

$$m_{2,new}' = m_{2,old}' - \Delta x_3 = 1 - (-1.576) = 2.576$$

$$m_{new}' = m_{old}' - \Delta x_4 = 4 - (-4.854) = 8.854$$

Iterate, i.e. go to Step 3; repeat until $\Delta x_i < \Delta x_{i,req}$.

Second iteration

Step 3. Update y_1, y_2, y_3, and y_4.

$$y_1 = -148.4, y_2 = 45.76, y_3 = 62.09, y_4 = 0.000$$

Step 4. Compute the partial derivatives.

$$\partial y_1/\partial \Delta P = 1, \ldots$$

Step 5. Expand y into the two-term Taylor series.

$$y_1\left(\Delta P, m_1', m_2', m'\right) \cong y_1\left(a_1, a_2, a_3, a_4\right) + \left(\partial y_1/\partial \Delta P\right)\left(\Delta P_{new} - a_1\right) + \left(\partial y_2/\partial m_1'\right)\left(m_1' - a_2\right) + \left(\partial y_3/\partial m_2'\right)\left(m_2' - a_3\right) + \left(\partial y_4/\partial m'\right)\left(m' - a_4\right)$$

The set of equations is as described by Eq. (E6.10.9).

Step 6. Solve the set of linear equations. In this specific case, we have

$$\begin{bmatrix} 1 & 0 & 0 & -111.56 \\ 1 & 51.39 & 0 & 0 \\ 1 & 0 & 188.8 & 0 \\ 0 & -1 & -1 & 1 \end{bmatrix} \begin{bmatrix} \Delta x_1 \\ \Delta x_2 \\ \Delta x_3 \\ \Delta x_4 \end{bmatrix} = \begin{bmatrix} -148.4 \\ 45.76 \\ 62.09 \\ 0.000 \end{bmatrix}.$$

The solutions are $\Delta x_1 = -3.303$, $\Delta x_2 = 0.9547$, $\Delta x_3 = 0.3463$, $\Delta x_4 = 1.301$.

Step 7. Update the values of the variables.

$$\Delta P_{new} = \Delta P_{old} - \Delta x_1 = 545.0$$
$$m_{1,new}' = m_{1,old}' - \Delta x_2 = 5.324$$
$$m_{2,new}' = m_{2,old}' - \Delta x_3 = 2.230$$
$$m_{new}' = m_{old}' - \Delta x_4 = 7.553$$

Third iteration

Step 3. Update y_1, y_2, y_3, and y_4.

$$y_1 = -10.66, y_2 = 2.278, y_3 = 2.999, y_4 = 0.000$$

Step 4. Compute the partial derivatives.

$$\partial y_1/\partial \Delta P = 1, \ldots$$

Step 5. Expand y into the two-term Taylor series.

$$y_1\left(\Delta P, m_1', m_2', m'\right)$$
$$\cong y_1\left(a_1, a_2, a_3, a_4\right) + \left(\partial y_1/\partial \Delta P\right)\left(\Delta P_{new} - a_1\right) + \left(\partial y_2/\partial m_1'\right)\left(m_1' - a_2\right)$$
$$+ \left(\partial y_3/\partial m_2'\right)\left(m_2' - a_3\right) + \left(\partial y_4/\partial m'\right)\left(m' - a_4\right)$$

The set of equations is as described by Eq. (E6.10.9).

Step 6. Solve the set of linear equations. In this specific case, we have

$$\begin{bmatrix} 1 & 0 & 0 & -95.17 \\ 1 & 46.62 & 0 & 0 \\ 1 & 0 & 171.5 & 0 \\ 0 & -1 & -1 & 1 \end{bmatrix} \begin{bmatrix} \Delta x_1 \\ \Delta x_2 \\ \Delta x_3 \\ \Delta x_4 \end{bmatrix} = \begin{bmatrix} -10.66 \\ 2.279 \\ 2.999 \\ 0.000 \end{bmatrix}.$$

The solutions are $\Delta x_1 = -1.2089$, $\Delta x_2 = 0.0748$, $\Delta x_3 = 0.0245$, $\Delta x_4 = 0.0993$.

Step 7. Update the values of the variables.

$$\Delta P_{new} = \Delta P_{old} - \Delta x_1 = 546.2$$
$$m_{1,new}' = m_{1,old}' - \Delta x_2 = 5.249$$
$$m_{2,new}' = m_{2,old}' - \Delta x_3 = 2.205$$
$$m_{new}' = m_{old}' - \Delta x_4 = 7.454$$

Fourth iteration

Step 3. Update y_1, y_2, y_3, and y_4.

$$y_1 = -0.0622, y_2 = 0.0140, y_3 = 0.0151, y_4 = 0.000$$

Step 4. Compute the partial derivatives.

$$\partial y_1/\partial \Delta P = 1, \ldots$$

Step 5. Expand y into the two-term Taylor series.

$$y_1(\Delta P, m_1', m_2', m') \cong y_1(a_1, a_2, a_3, a_4) + (\partial y_1/\partial \Delta P)(\Delta P_{new} - a_1) + (\partial y_2/\partial m_1')(m_1' - a_2) + (\partial y_3/\partial m_2')(m_2' - a_3) + (\partial y_4/\partial m')(m' - a_4)$$

The set of equations is as described by Eq (6.9).

Step 6. Solve the set of linear equations. In this specific case, we have

$$\begin{bmatrix} 1 & 0 & 0 & -93.92 \\ 1 & 46.24 & 0 & 0 \\ 1 & 0 & 170.25 & 0 \\ 0 & -1 & -1 & 1 \end{bmatrix} \begin{bmatrix} \Delta x_1 \\ \Delta x_2 \\ \Delta x_3 \\ \Delta x_4 \end{bmatrix} = \begin{bmatrix} -0.0622 \\ 0.0140 \\ 0.0151 \\ 0.000 \end{bmatrix}.$$

The solutions are $\Delta x_1 = -0.0071$, $\Delta x_2 = 0.0005$, $\Delta x_3 = 0.0001$, $\Delta x_4 = 0.0006$. These changes are very small and, therefore, they are possibly good enough.

Step 7. Update the values of the variables.

$\Delta P_{new} = \Delta P_{old} - \Delta x_1 = 546.2$

$$m_{1,new}' = m_{1,old}' - \Delta x_2 = 5.248$$

$$m_{2,new}' = m_{2,old}' - \Delta x_3 = 2.205$$

$$m_{new}' = m_{old}' - \Delta x_4 = 7.453$$

Problems

6.1 Two boilers running on biofuel from three different plants

Two boilers runs on biofuel of different methane fractions produced by three plants, see Figure 6.15. Boiler 1 requires $m_{B1}' = 1$ kg/s of biofuel with a methane mass fraction, $x_{B1} = 0.4$. Boiler 2 runs on biofuel, with $x_{B2} = 0.5$ at $m_{B2}' = 2$ kg/s. The mass fraction of methane produced by Plants 1, 2 and 3 are $x_1 = 0.25$, $x_2 = 0.45$, and $x_3 = 0.55$, respectively. Find the three linear equations from which the three mass flow rates from the three plants can be determined. Determine the three mass flow rates using successive substitution.

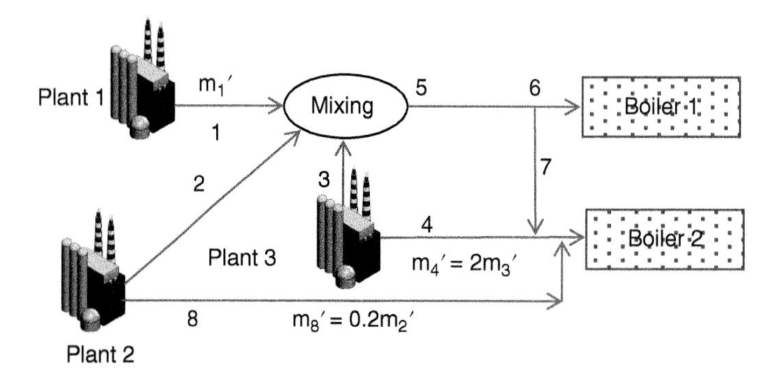

Figure 6.15 Two boilers running on biofuel from three different plants. Source: G. Nagesh and D. Ting.

6.2 How does the Newton-Raphson method work?
Newton-Raphson method improves its guess based on
a) the gradient of the current guess intercepting the $y = 0$ line.
b) the gradient of the current guess intercepting the $x = 0$ line.
c) the second derivative of y with respect to x.
d) setting the second derivative of y with respect to x to zero.
e) taking the integral of y.

6.3 The operation of the Newton-Raphson method for a single-variable problem
The Newton-Raphson method is applied to solve for $y = x^2 - 2$. If the initial guess is $x = 2$, the improved guess will be

x =________________

Hint: The slope, dy/dx can be expressed as $\Delta y/\Delta x$. The value of the slope at the x value, along with the Δy needed to bring y to zero, can be used to deduce the new x, i.e. $x_{new} = x - \Delta x$.

6.4 Success substitution versus the Newton-Raphson method
The operating point of a fan-and-duct system is to be determined. The equations for the two components are
$P_{static} = 70 + 10\,\forall'^{1.7}$
and

$\forall' = 20 - P_{static}{}^2 / 100000,$

where P_{static} = static pressure in Pa and $\forall'$ = airflow rate in m^3/s.
Part I Use successive substitution to solve for the operating point, choosing as trial values $\forall' = 10\,m^3/s$ or $P_{static} = 200$ Pa.
Part II Use the Newton-Raphson method to solve for the operating point, choosing as trial values $\forall' = 10\,m^3/s$ or $P_{static} = 200$ Pa.

6.5 Flashing in cryogenic liquefaction
Flashing is utilized in some cryogenic liquefaction systems to reduce the temperature of the liquid stream. This is achieved by flashing off some of the liquid into vapor via a throttling valve and a heat exchanger. Consider the flashing system shown in Figure 6.16. Find the mass flow rate of liquid leaving the heat exchanger and its temperature using the Newton-Raphson method.

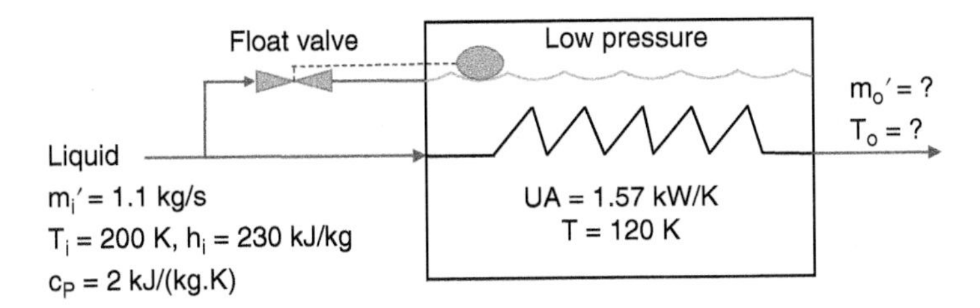

Figure 6.16 A flashing system . Source: Y. Nagaraja, edited by D. Ting).

6.6 Gear pump and centrifugal pump in series

A gear pump and a centrifugal pump work in series to deliver water along a long pipe. The pressure drop-flow rate relations are:

Gear pump: $\Delta P = 40 - 5\forall'$,
Centrifugal pump: $\Delta P = 6 + 2\forall' - 0.5\forall'^2$, and
Pipe: $\Delta P = 0.1\forall'^2$,
where ΔP = pressure rise (or drop in the pipe) in kPa and $\forall'$ = volume flow rate in m^3/s. Plot ΔP versus $\forall'$ performance characteristic curves for all three components. Estimate the approximate solution without executing the simulation of the system. Explain the physical implications of the solution.

References

Burmeister, L.C. (1998). *Elements of Thermal-Fluid System Design*. Upper Saddle River, NJ: Prentice-Hall.
Dhar, P.L. (2016). *Thermal System Design and Simulation*. Cambridge: Academic Press.
Jaluria, Y. (1998). *Design and Optimization of Thermal Systems*. New York: McGraw-Hill.
Penoncello, S.G. (2018). *Thermal Energy Systems: Design and Analysis*, 2nd ed. Boca Raton, FL: CRC Press.
Stoecker, W.F. (1989). *Design of Thermal Systems*, 3rd ed. New York: McGraw-Hill.
Suryanarayana, N.V. and Arici, Ö. (2003). *Design and Simulation of Thermal Systems*. New York: McGraw-Hill.

7

Formulating the Problem for Optimization

The formulation of a problem is often more essential than its solution, which may be merely a matter of mathematical or experimental skills.

– Albert Einstein

Chapter Objectives

- Understand what an objective function is.
- Differentiate equality versus inequality constraints.
- Formulate unconstrained and constrained problems with a single objective function.

Nomenclature

A	area; A_{flame} is the reacting flame front area, A_o is surface area of a smooth sphere enclosing the burned gas
C	cost; C_{cool} is the cooling cost, C_{heat} is the heating cost, C_{light} is the lighting cost, $C_{operate}$ is operating cost, C_{pipe} is pipe cost
c_P	heat capacity at constant pressure
D	diameter
Da	Damköhler number
f	friction coefficient
H	height
h	heat transfer coefficient, or, pressure head; h_{conv} is the convection heat transfer coefficient, $h_{conv,C}$ is the convection heat transfer coefficient from air to the cold surface, $h_{conv,H}$ is the convection heat transfer coefficient from the hot surface to the air, h_{rad} is the radiation heat transfer coefficient, h_1 is the head of Line 1 in m^2/s^2, h_2 is the head of Line 2 in m^2/s^2
K	Empirical coefficient or constant; $K_{operate}$ is the operating cost coefficient, K_{pipe} is the cost per unit pipe material, $K_{u'}$ indicates the magnitude of the influence of the vortex intensity, K_1 is coefficient for Line 1 in $m^2/(kg{\cdot}s)$, K_2 is coefficient for Line 2 in $m^2/(kg{\cdot}s)$
k_{air}	the conductivity of air
L	length

m	mass; m' is mass flow rate, m_{air}' is the mass flow rate of air, m_b' is the mass burning rate, m_1' is the mass flow rate in Line 1
N	number, a constant
P	pressure
PCM	phase change material
Q	heat; Q' is the heat transfer rate, Q_{cond}' is conduction heat transfer rate, Q_{conv}' is convection heat transfer rate, Q_{rad}' is the radiation heat transfer rate
S	speed; S_{flame} is flame propagation speed, S_L is the laminar flame speed
T	temperature; T_{air} is the temperature of the air, T_{avg} is the average temperature, T_C is the cold temperature, T_H is the hot temperature, T_o is the operating temperature
t	thickness
U	velocity
u'	tangential velocity of intensity of a vortex
X, x	independent variable, or, distance, or, gap width
Y, y	objective function

Greek and Other Symbols

α	a positive real number
β	a positive real number
Δ	change
δ	laminar flame front thickness
η	efficiency
λ	size or diameter of a vortex
ρ	density; ρ_u is the density of the unburned gas
σ	Stefan-Boltzmann constant
τ	time or time scale; τ_{chem} is chemical time scale, τ_{flow} is flow time scale
Φ	equality constraint which is equal to zero
ϕ	equality constraint
Ψ	inequality constraint which has zero as the upper or lower limit
ψ	inequality constraint
ε	emissivity; ε_C is the emissivity of the cold surface, ε_H is the emissivity of the hot surface
$\forall$	volume; $\forall'$ = volume flow rate

7.1 Introduction

Engineers are problem-solvers. For typical problems, the first step is to obtain a workable solution. Within the context of design and optimization of thermofluid systems, this workable solution is referred to as a workable or feasible design. Sometimes this is all that is needed, or the appropriate service for the amount of money for the job. More often, however, a better solution is sought, as the reward warrants the extra effort. This is particularly the case for larger problems which involve significant monetary investment and/or are expected to generate long-term returns. In this event, optimization is invoked for achieving *the best design which grants the greatest reward*. The reward is also called the *figure of merit*, or more commonly, the *objective function*, which is to be optimized.

Let us consider an everyday analogy to clarify the distinction between engineering a workable design versus an optimum one. Passing the class is enough for an uninspired student. On the other hand, always doing one's best is the motto of the high achiever. If you are yet to be inspired about optimization, let Albert Einstein remind you that "We have to do the best we can. This is our sacred human responsibility." What if the outcome does not speak for your effort? Do not worry. "If you're doing your best, you won't have any time to worry about failure," according to H. Jackson Brown Jr. As veraciously, Helen Keller attested that "No effort that we take to attain something beautiful is ever lost." Indeed, the upstream striving against the downward current itself is most savor-able. Roy T. Bennett rightly put it, "Life is about accepting the challenges along the way, choosing to keep moving forward, and savoring the journey."

What should a good engineer strive to optimize? The objective function can be quality, efficiency, cost, profit, upkeep, reliability, etc. When efficiency, profitability, or reliability is the objective function, the optimization process aims at maximizing it. On the other hand, if cost or upkeep is the objective function, it is a minimization exercise. Therefore, we see that cost, upkeep, waste, environmental damage, etc. are considered "negative reward," which we strive to minimize. In plain English, minimizing the negative outcome such as waste is equivalent to maximizing "negative waste." As such, minimizing the negative of an objective function is equivalent to maximizing the objective function. Sometimes there is more than one figure of merit. For example, both profit and environmental damage can be equally important. In such cases, proper weighting, and more sophisticated multi-objective optimization techniques are called upon. When feasible and appropriate, two or more objective functions can be logically combined into one overall objective function. We shall limit our scope to single-objective optimization.

There are two levels of optimization. At the more general level, optimization simply involves a comparison of alternate concepts. Once a particular concept is selected, the next phase is to optimize within that concept. For example, compare shipping an item via air, land, or sea. Suppose shipping by air is chosen because the item needs to be shipped over a long distance promptly. Then, the next step is to figure out which company provides the most economical and/or reliable air transportation.

7.2 Objective Function and Constraints

As we can infer from the above, an *objective function* is a quantity representing the feature of prime interest. It is the quantity to be minimized or maximized in an optimization process. For example, if the objective of the optimization process is to maximize the heat transfer rate of a heat exchanger, then the objective function, heat transfer rate,

$$y = f(x_1, x_2, \ldots x_n) \tag{7.1}$$

where the *independent variables*, $x_1, x_2, \ldots$ are, for example, area, material, fluid flow rate, temperature difference, etc. The system is usually subjected to various constraints such as the total size of the heat exchanger, the weight of the system, the operating temperature; these can be *equality* or *inequality* constraints.

Constraints

Almost all practical systems are subjected to various constraints such as maximum operating temperature and pressure. If a particular operating temperature is required, for example, then, we have $\phi_i = T_o$, the operating temperature ϕ_i must be equal to or fixed around T_o. As such, these are equality constraints, which can be expressed mathematically as

$$\phi_1 = \phi_1(x_1, x_2, \ldots x_n) = N_1$$
$$\phi_2 = \phi_2(x_1, x_2, \ldots x_n) = N_2$$
$$\vdots$$
$$\Phi_m = \phi_m(x_1, x_2, \ldots x_n) = N_m \tag{7.2}$$

Many optimization methods function by recasting these equality constraints such that they are equal to zero, i.e.

$$\phi_m = \phi_m(x_1, x_2, \ldots x_n) - N_m = 0 \tag{7.3}$$

Here, N_i denotes the corresponding numerical value of the equality constraint. It is typically more common to have an upper limit for ϕ_i, i.e. $\psi_i < T_o$. Mathematically, we can write

$$\psi_1 = \psi_1(x_1, x_2, \ldots x_n) \leq / \geq N_1$$
$$\psi_2 = \psi_2(x_1, x_2, \ldots x_n) \leq / \geq N_2$$
$$\vdots$$
$$\psi_m = \psi_m(x_1, x_2, \ldots x_n) \leq / \geq N_m. \tag{7.4}$$

Or, setting the inequality constraint with respect to zero,

$$\Psi_m = \Psi_m(x_1, x_2, \ldots x_n) - N_m \leq / \geq 0 \tag{7.5}$$

It is thus clear that there are both equality and inequality constraints, ϕ_i and ψ_i, respectively. The basic conservation principles which must be satisfied are commonly expressed as equality constraints. To ease the solution approach, sometimes it is more convenient to recast the inequality constraints into equality constraints.

7.3 Formulating a Problem to Optimize

From the materials covered thus far, we understand that the objective function has first to be defined. Then, the parameters of which the objective function is a function of need to be delineated. From these parameters, those which can be controlled or varied by the design engineer and/or the operator are singled out. Also, the limits of operation associated with these variables must be spelled out. These are the constraints; they may be equality or inequality constraints. This is formulation, i.e. formulating the problem in mathematical form, including the equations associated with the constraints, so that we can execute an optimization analysis. To sum things up, formulation is expressing a descriptive design optimization problem in mathematical form.

To reiterate, the system under study is required to be described mathematically. For relatively simple systems, such as a heat exchanger, the mathematical description, i.e. the model, is fairly well established. If efficiency is the objective function, then the formulation of the problem, other than specifying the constraints, necessitates an explicit mathematical description of efficiency of the concerned system in terms of the involved parameters. Once this is accomplished, all that is needed is the right attitude (character) to obtain the correct solution; Herbert Read stated, "A man of personality can formulate ideals, but only a man of character can achieve them." In capitalism, however, profit dictates the survival and/or thriving of an individual or a company. This figure of merit, profit, is typically a function of multiple subsystems. Consequently, the formulation of the objective function requires sound engineering knack.

The process for formulating the problem may be summarized into the following steps.

1) Define the objective function.
2) Express the objective function, Y, in terms of the immediate parameters, $X_1, X_2, \ldots$ affecting its value.
3) Relate the objective function to the involved system.
   If the entire system consists of multiple subsystems, the immediate parameters defining the objective function, $X_1, X_2, \ldots$, necessitate proper propagation from the subsystem models, $X_1 = f_1(x_1, x_2, \ldots), X_2 = f_1(x_i, x_{i+1}, \ldots), \ldots$, to the overall system. In context, $x_1, x_2, \ldots$ are the design variables which can be controlled and/or varied to give an optimal Y value.
4) Find out what the constraints are and express them mathematically.
   Equality constraints may be expressed as $\phi_1(x_1, x_2, \ldots) = N_1$, $\phi_2(x_i, x_{i+1}, \ldots) = N_2, \ldots$, while the equations corresponding to inequality constraints may be $\psi_1(x_1, x_2, \ldots) < N_1, \psi_2(x_i, x_{i+1}, \ldots) \geq N_2, \ldots$
5) Convert inequality constraints into equality ones, whenever applicable.

If the optimization of the objective function is to be achieved via selection of off-the-shelf systems and/or subsystems, curve fitting and other manipulations may be required to provide the pertinent equations, especially for Step 3 above.

We will discuss the various approaches of optimization later. Only basic calculus is needed to solve the formulated problems in all the examples furnished in this chapter. The focus is on illustrating the formulation of the problem at hand, leaving the various solution approaches to the other chapters. Let us look at the formulation and optimization in terms of preventing heat loss of a simple thermofluid system, a hot water storage tank.

Example 7.1 *Formulation and minimization of a cylindrical container surface area*

Given

A cylindrical container can be described by diameter D and height H. The container is used for storing 100 kg of hot water.

Find

The dimensions which give the minimum surface area, to reduce heat loss.

Solution

The surface area of the container is the parameter of interest. The goal is to minimize it, in an effort to reduce heat loss. Therefore, the objective function is the surface area,

$$y = \pi\,(\tfrac{1}{2}\,D)^2 \times 2 + \pi\,D\,H \tag{E7.1.1}$$

The container is to be utilized for holding 100 kg of hot water. This requirement can be considered as the constraint. This constraint can easily be expressed as the volume of the container,

$$\phi = \forall = \pi\,(\tfrac{1}{2}\,D)^2\,H = 100\ \text{kg}/1000\ \text{kg/m}^3 = 0.1\ \text{m}^3 \tag{E7.1.2}$$

At this point, the problem of interest has been formulated. The formulation gives the objective function, Eq (E7.1.1), subjected to a constraint, Eq (E7.1.2). Once formulated, we can proceed to solve the problem, i.e. minimize y.

From Eq (E7.1.2), we can get

$$\pi\,(\tfrac{1}{2}\,D)^2\,H = 0.1\ m^3$$

This can be rearranged into

$$H = 0.1/\left[\pi\,(\tfrac{1}{2}\,D)^2\right] = 0.4/\left[\pi\,D^2\right]$$

Substituting this into Eq (E7.1.1) leads to

$$y = 2\pi\,(\tfrac{1}{2}\,D)^2 + \pi\,D\,\left[0.4/\left(\pi\,D^2\right)\right] = 0.5\,\pi\,D^2 + 0.4/D$$

To obtain the optimum, differentiate this with respect to D, and set it to zero, i.e.

$$\partial y \partial D = \pi\,D - 0.4/D^2 = 0$$

This gives

$$D^* = 0.5031\ m$$

The asterisk, *, denotes at optimum. With this, we can solve for the height,

$$H^* = 0.4/\left[\pi\,D^2\right] = 0.5031\ m$$

The minimum surface area, achieved with $D^* = H^* = 0.50$ m, is $y^* = 1.19\ m^2$.

The calculus approach that has been invoked will be expounded in Chapter 8. A more abstract, but fundamental, figure of merit is entropy. Quixotic it may be, it is an oxymoron to talk about sustaining tomorrow if we do not strive to minimize molecular dance, i.e. our generation of entropy, at least within the green planet on which we reside.

> It sometimes seems as if curbing entropy is our quixotic purpose in this universe.
>
> James Gleick

> Heat energy of uniform temperature [is] the ultimate fate of all energy. The power of sunlight and coal, electric power, water power, winds and tides do the work of the world, and in the end all unite to hasten the merry molecular dance.
>
> Frederick Soddy

Let us go through a classroom example concerning the entropy associated with harnessing wind energy.

Example 7.2 ***Formulating wind turbine entropy generation***

Given

A wind turbine spawns entropy as a function of its yawing speed, x_1, and pitching rate, x_2. Specifically, the entropy generation varies with respect to the square of the yawing speed, where yawing in the counter-clockwise direction in the plane view is positive. The entropy generation also changes with the square of the pitching rate, where counter-clockwise pitching when looking into the turbine (blade) is positive. The total entropy generated is the sum of the contribution from yawing and pitching.

Find

The condition at which the wind turbine generates the least amount of entropy.

Solution

English astronomer, physicist, and mathematician, Sir Arthur Stanley Eddington said, "Entropy is time's arrow." By way of explanation, all we can do is move forward in time and, in doing so, we generate entropy. As we embrace this gloomy fact, we should also note that there is a huge difference between generating a large quantity of unnecessary entropy and letting off the minimal unavoidable amount. The hope associated with the latter can be appreciated from Vaclav Havel's observation, "Just as the constant increase of entropy is the basic law of the universe, so it is the basic law of life to be ever more highly structured and to struggle against entropy." This "struggle against entropy" is the purpose of finding the solution for the wind turbine problem at hand.

Our objective is to minimize entropy generation. As entropy generation, y, is from the square of the yawing speed, x_1, plus the pitching speed, x_2, squared, we can write

$$y = x_1^{\,2} + x_2^{\,2} \tag{E7.2.1}$$

Practically, neither the yawing not the pitching can take place too rapidly. With this, we can set the appropriate limits, and explicitly express them as constraints. Specifically,

$$\psi_1(x_1) = x_1^{\,2} < N_1 \tag{E7.2.2}$$

$$\psi_2(x_2) = x_2^{\,2} < N_2 \tag{E7.2.3}$$

The squares of yaw and pitch have been used to account for equal limits of the positive and negative yawing and pitching. Concerning entropy, we know that the more we struggle, the more entropy we generate. For this reason, we are unlikely to approach a high yawing or pitching speed when we are after the minimization of entropy generation.

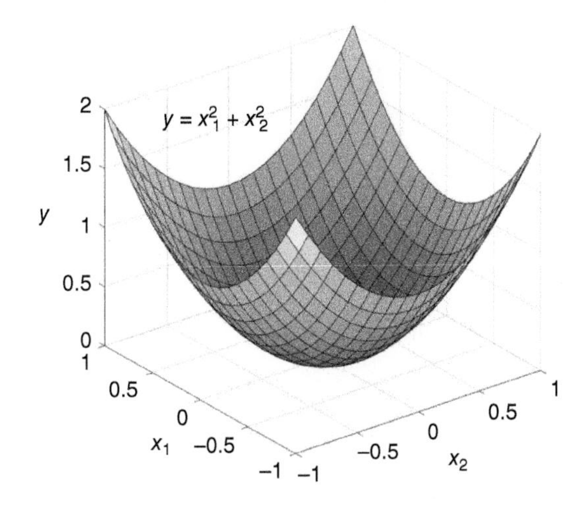

Figure 7.1 Variation of objection function, $y = x_1^{\,2} + x_2^{\,2}$, with respect to x_1 and x_2. Source: Y. Yang.

In a nutshell, the objective function is the entropy generation, as described by Eq (E7.2.1). The goal is to minimize the objective function. For this simple expression, we can invoke calculus to locate the minimum y. Pointedly,

$$\partial y/\partial x_1 = \partial y/\partial x_2 = 0 \tag{E7.2.4}$$

This gives $x_1 = x_2 = 0$, at which condition there is no entropy generation, as shown in Figure 7.1. This implies that the best operation, as far as entropy minimization is concerned, is steady operation, where there is neither yawing nor pitching.

Practically, however, some proper yawing and pitching can significantly increase the amount of wind energy harnessed. For this reason, a more appropriate objective function is the amount of entropy generated per unit harnessed energy.

Sometimes, as is the case for Example 7.2, the best thing to do is to do nothing. A case in point, setting the wind turbine to yaw and pitch around when there is not much to gain can lead to a waste of energy (used for yawing and pitching) and unnecessary wear and tear. As such, Laozi is right in saying, "Doing nothing is better than being busy doing nothing."

Example 7.3 *Formulation of a boiler to maximize water heating*

Given

A hot-water boiler consists of a combustion chamber and a heat exchanger arranged as shown in Figure 7.2. Fuel with a heating value of 4.2×10^7 J/kg is fed into the combustion chamber at a mass flow rate of 0.0025 kg/s. Heat is recovered from the flue gases (for heating the water) by cooling the flue gases to 150°C. The combustion efficiency increases with an increase in the rate of airflow according to

$$\eta\ [\%] = 100 - 0.023/\left(m_{air}' + 0.0025\right)^2 \tag{E7.3.1}$$

where m_{air} is the mass flow rate of air in kg/s. The specific heat of the gas mixture leaving the combustion chamber is 1.05×10^3 J/(kg·K). You wish to maximize the heat transfer rate for heating the water.

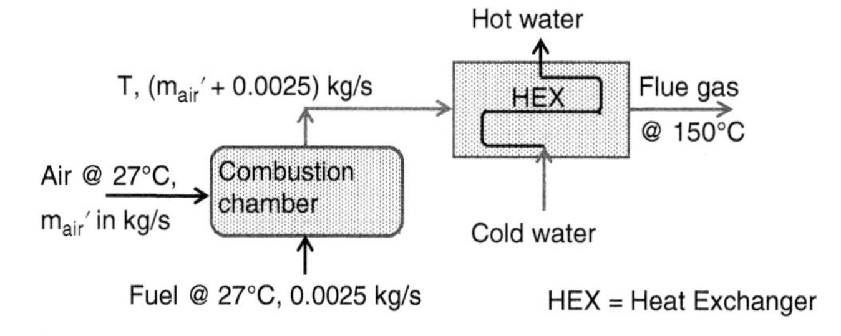

Figure 7.2 A hot-water boiler consisting of a combustion chamber and a heat exchanger. Source: D. Ting.

Find

Set up the objective function explicitly; include the proper units. You do not need to solve the problem.

Solution

From Figure 7.2, we can see that

$$\left(m_{air}' + 0.0025\ \text{kg/s}\right)\ c_P\ \left(T - 27°C\right) = \eta\ \left(0.0025\ \text{kg/s} \times 4.2 \times 10^7\ \text{J/kg}\right) \tag{E7.3.2}$$

where m_{air}' is in kg/s, c_P is in J/kg·C (or J/kg·K), T is in °C (if using K, make sure you replace 27°C with 300.15 K), and η is a percentage.

Substituting for Eq (E7.3.1), $\eta\ [\%] = 100 - 0.023 / (m_{air}' + 0.0025)^2$, we have

$$\left(m_{air}' + 0.0025\right)\ c_P\,(T - 27) = \left[100 - 0.023/\left(m_{air}' + 0.0025\right)^2\right]\ \left(0.0025 \times 4.2 \times 10^7\right) \tag{E7.3.3}$$

This can be rewritten in terms of temperature, T in °C, as a function of m_{air}', i.e.

$$T = \left\{\left[100 - 0.023/(m_{air}' + 0.0025)^2\right] (0.0025 \times 4.2 \times 10^7)\right\} / \left[c_P (m_{air}' + 0.00250)\right] + 27. \quad \text{(E7.3.4)}$$

The objective function is the heat transfer rate,

$$Q' = (m_{air}' + 0.00250)\ c_P (150 - T) \quad \text{(E7.3.5)}$$

Let us look at a purely fluid mechanic example. Transporting fossil fuels through long pipelines can take place over thousands of kilometers. On a shorter scale, water and chemical species are commonly conveyed over tens of meters.

Example 7.4 ***Formulation of a liquid conveying system for minimum cost***

Given

A liquid is to be conveyed via a pipe of diameter D over a distance L at a mass flow rate of m'. The cost of this liquid-conveying system, consisting primarily of the operating cost and pipe cost, is to be minimized.

Find

Set up the objective function and show the conceptual solution in detail.

Solution

The objective function is the total cost, y, which is made up of the operating cost, $C_{operate}$, and the piping cost, C_{pipe},

$$y = C_{operate} + C_{pipe} \quad \text{(E7.4.1)}$$

The operating cost is proportional to the head (pressure) loss multiplied by the mass flow rate,

$$C_{operate} \propto \text{head loss} \times m' \quad \text{(E7.4.2)}$$

Introducing a proportionality constant, $K_{operate}$, to signify the cost associated with a unit of head loss times m′, we have

$$C_{operate} = K_{operate} \times \text{head loss} \times m' \quad \text{(E7.4.3)}$$

Similarly, once the appropriate material for transporting the concerned liquid is chosen, the cost of the pipe is largely related to the amount of pipe material, i.e.

$$C_{pipe} \propto \pi\, D\, t\, L \quad \text{(E7.4.4)}$$

where t is the pipe wall thickness. With K_{pipe} denoting the cost per unit pipe material, we have

$$C_{pipe} = K_{pipe}\ \pi\, D\, t\, L \quad \text{(E7.4.5)}$$

Therefore, we can express the objective function as

$$y = K_{operate} \times \text{head loss} \times m' + K_{pipe}\ \pi\, D\, t\, L \quad \text{(E7.4.6)}$$

The goal is to minimize this objective function.

Recall from Chapter 3, the pressure (head) loss in a pipe with turbulent flow,

$$\Delta P = f\ (L/D)\ \tfrac{1}{2}\, \rho\, U^2 \quad \text{(E7.4.7)}$$

where f is the friction factor, ρ is the liquid density, and U is the velocity. The corresponding laminar flow expression can be invoked if the flow is laminar instead.

Let us take care of the mass flow rate in Eq (E7.4.6), the mass flow rate,

$$m' = \rho \forall' = \rho \left(\pi D^2/4\right) U \quad (E7.4.8)$$

which can be rearranged into

$$U = 4\, m'/\left(\rho \pi D^2\right) \quad (E7.4.9)$$

Thus,

$$U^2 = 16\, m'^2/\left(\rho^2\, \pi^2\, D^4\right) \quad (E7.4.10)$$

Substituting this into Eq. (E7.4.7) gives

$$\Delta P = 8\, f\, m'^2/\left(\rho^2\, \pi^2\, D^5\right) \quad (E7.4.11)$$

Substituting this into Eq. (E7.4.6), we get

$$y = K_{operate}\, 8\, f\, m'^3/\left(\rho^2\, \pi^2\, D^5\right) + K_{pipe}\, \pi\, D\, t\, L \quad (E7.4.12)$$

In practice, the liquid and the condition it is in are typically known and stay more or less fixed and, hence, the liquid density, ρ is constant. The mass flow rate, m', typically does not vary too much or the maximum mass flow rate is generally known. For most long-distance liquid transportation in a pipe, the flow is fully turbulent. Accordingly, the upper end of the friction factor, f, can be deduced from the Moody diagram. To withstand the required pressure, the pipe wall thickness, t, is known *a priori*. Thereupon, the only main design variable is the size of the pipe, specifically, D.

To minimize the cost, we set the differentiation, dy/dD, to zero, i.e.

$$dy/dD = -5\, K_{operate}\, 8\, f\, m'^3/\left(\rho^2\, \pi^2\, D^6\right) + K_{pipe}\, \pi\, t\, L = 0 \quad (E7.4.13)$$

This gives the optimum diameter, which corresponds to the minimum total cost,

$$D^* = \left[40\, K_{operate}\, f\, m'^3/\left(\pi^3\, K_{pipe}\, \rho^2\, \pi^2\, t\, L\right)\right]^{1/6} \quad (E7.4.14)$$

Substituting this into Eq (E.7.4.12) gives the minimum total cost, y_{min}.

We see that both the minimum cost, y*, and the optimal diameter, D*, are insensitive to the cost of power for pumping the liquid, which varies as $K_{operate}{}^{1/6}$ and $\sqrt{m'}$.

In real life, the situation can be somewhat more complex. For example, multiple pumps appropriately stationed across a certain number of meters or kilometers typically post a better solution, in terms of operation and safety.

Example 7.5 ***Optimal double-pane window air gap***

Given

There is, presumptively, some good engineering rationale behind the fact that the air gap of most double-glazed windows is around 1 cm. With your mind impassioned by Engineering Design and Optimization of Thermofluid Systems, you make it your mission to dig out the truth behind this 1 cm gap. From your thermofluid scholarship, you conjecture that this particular air gap leads to minimum heat transfer across the gap.

Find

Formulate the optimization of a window air gap problem.

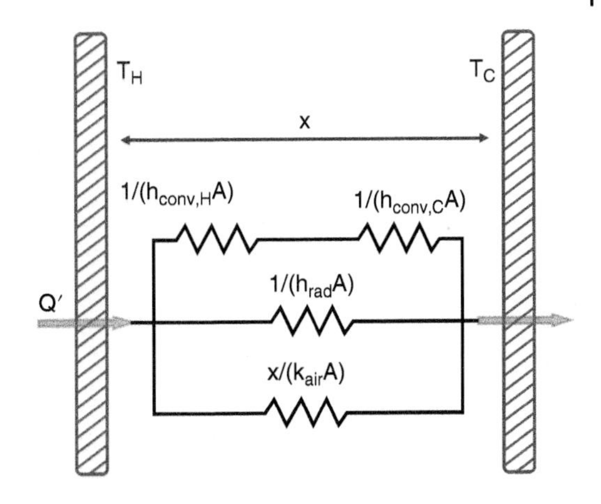

Figure 7.3 Heat transfer through the air gap of a double-glazed window. The three heat transfer resistors corresponding to convection, conduction, and radiation operate in parallel across the gap. Source: D. Ting.

Solution

The objective is to minimize the heat transfer through the air gap. As depicted in Figure 7.3, the rate of heat transfer from the warmer (hot) surface at T_H to the colder surface at T_C is

$$Q' = h_{conv,H}\,A\,\left(T_H - T_{air}\right) + h_{conv,C}\,A\,\left(T_{air} - T_C\right) \\ + h_{rad}\,A\,\left(T_H - T_C\right) \\ + k_{air}\,A\,\left(T_H - T_C\right)/x \tag{E7.5.1}$$

where $h_{conv,H}$ is the convection heat transfer coefficient from the hot surface to the air, A is area, T_{air} is the temperature of the air inside the gap, $h_{conv,C}$ is the convection heat transfer coefficient from air to the cold surface, h_{rad} is the radiation heat transfer coefficient,[1] k_{air} is the conductivity of air, and x is the gap width. Knowing that radiation heat transfer takes place via electromagnetic waves, any moderate changes in the air gap do not affect h_{rad}. For a narrow gap, the two convection heat transfer coefficients may be combined into one such that

$$Q_{conv}' = h_{conv}\,A\,\left(T_H - T_C\right) \tag{E7.5.2}$$

The convection heat transfer coefficient increases as the gap widens, i.e. there is less restriction to hold back the rising hot air and descending cold air. In plain English, Q_{conv}' increases with increasing x, i.e.

$$Q_{conv}' \propto h_{conv} = \alpha\, x^{\beta} \tag{E7.5.3}$$

where α and β are positive real numbers, which can be deduced from experiment or numerical simulation. For the problem at hand, the thermal conductivity of the air, k_{air}, is a constant. It follows that the conduction heat transfer rate is inversely proportional to the gap width, i.e.

$$Q_{cond}' \propto h_{cond} = k_{air}/x \tag{E7.5.4}$$

where h_{cond} is the equivalent conduction heat transfer coefficient.

---

1 The radiation heat transfer rate, $Q_{rad}' = \varepsilon\sigma A(T_H^4 - T_C^4)$, where ε is the emissivity, σ is the Stefan-Boltzmann constant, and T_H and T_C are the hot and cold surfaces, respectively, in absolute units, i.e., degrees Kelvin. When T_H and T_C do not differ too much, such as that between the two inner surfaces of a double-glazed window, and the view factor is close to one, i.e., the two surfaces "see" each other, it is convenient to approximate the radiation heat transfer rate as $Q_{rad}' = h_{rad}A(T_H - T_C)$. It follows that the radiation heat transfer coefficient, $h_{rad} \approx 4\sigma T_{avg}^3 / (1/\varepsilon_H + 1/\varepsilon_C - 1)$, where T_{avg} is the average of T_H and T_C, and ε_H and ε_C are the emissivities of the hot and cold surfaces, respectively.

Therefore, the objective function can be expressed as

$$y = \alpha\, x^{\beta} + k_{air}/x \tag{E7.5.5}$$

To optimize (minimize) y, we take its derivative with respect to x and set it to zero,

$$dy/dx = \alpha\, \beta\, x^{\beta-1} - k_{air}\, x^{-2} = 0 \tag{E7.5.6}$$

This can be rearranged into

$$x^{1-\beta} = k_{air}/\,(\alpha\beta) \tag{E7.5.7}$$

Substituting for k_{air}, α, and β would lead to $x^* \approx 1.2$ cm; see, for example, (Arici and Kan, 2015).

Thermos flasks, especially conventional ones, see Figure 7.4, are designed and optimized to minimize heat transfer similar to, but at a cut above, that of double-paned windows. Our ingenious ancestors realized that by removing the air in the gap, we remove the medium for conduction and convection. To minimize radiation heat transfer, the two inner surfaces are coated with a shiny, silver-colored material which has a very low emissivity.

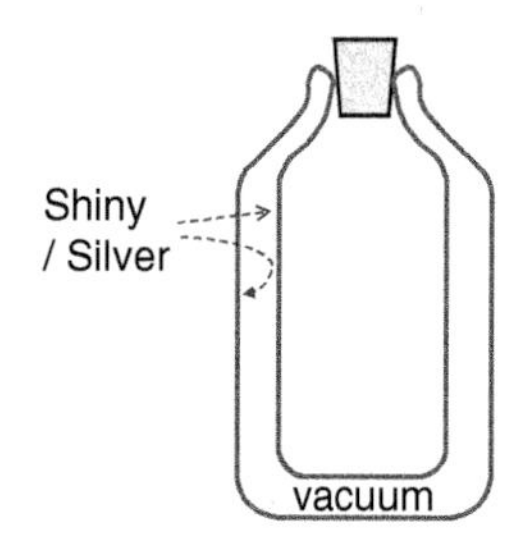

Figure 7.4 Minimizing heat transfer in a thermos flask. The approximate vacuum in the cavity literally eliminates conduction and convection heat transfer. The silver inner surfaces lessen radiation heat transfer by diminishing the emissivity. Source: D. Ting.

Phase change material (PCM) is getting a lot of attention these days. For example, Cunha et al. (2020) eliminated the high cost of encapsulating PCM by directly incorporating PCM into building mortars. Within the scope of their study, it was found that about 20% of PCM led to the best performance in terms of decreasing extreme temperatures and heating and cooling loads. Hu et al. (2020) demonstrated significant savings in heating and cooling by utilizing PCM-enhanced, ventilated windows. Just when we feel great about these recent ingenious advances, we are confronted by the truthfulness of the saying in Ecclesiastes 1:9, "there is nothing new under the sun." For example, 'fruit walls' were part of urban farming in the 1600s (*Low-Tech Magazine*, 2015). These thermally massive fruit walls retain heat during the warm day and slowly release it after the sun sets, keeping the vegetation warm through the night. So, the old has become new again, and one promising way for today's greenhouses to save energy and extend the growing period is to go partially underground (Tiwari and Dhiman, 1986; Stone, 1997; Wang, 2006; Kim et al., 2017), though not as far down into the ground as *fenestraria aurantiaca*, a succulent. Let us attempt to formulate the ideal depth for a greenhouse at a given climatic condition.

Example 7.6 *Formulation of a pit greenhouse for optimum depth*

Given

Greenhouse farmers in places such as southwestern Ontario, Canada, are exploring potent ways to save energy, for heating, in particular. The underground or pit greenhouse is a promising solution. The ground can act as an excellent insulator, and its above-freezing temperature can noticeably extend the growing season. Furthermore, the soil can cool the greenhouse during the hot days, reducing the heat stress imposed on the produce.

Find

Formulate the problem to solve for the ideal depth into the ground.

Solution

If heating and cooling are the only concern, it is likely that a pit deep enough to enclose the entire greenhouse is the way to go. However, sunlight is also an essential ingredient for healthy plant growth. Furthermore, the thermal energy from solar radiation can reduce the required heating during the cold season. Therefore, the cost per yield may be an appropriate objective function. The cost includes both capital cost and operation cost. The primary operation cost is associated with energy which is not freely available from nature, i.e. natural gas, for heating, and electricity from the grid is needed for cooling and lighting. We can approximate the cost per unit of produce as

$$y = C_{heat} + C_{cool} + C_{light} \tag{E7.6.1}$$

where C_{heat} is the heating cost, C_{cool} is the cooling cost, and C_{light} is the lighting cost. The heating cost, the major cost for greenhouse operation in a typical temperate climate, will decrease with depth into the ground, x, i.e.

$$C_{heat} = \alpha_h / x^{\beta_h} \tag{E7.6.2}$$

where α_h and β_h are positive real numbers. The cooling cost is also expected to decrease with x, though not so rapidly compared to the heating cost in a temperate climate region, such as southwestern Ontario. Therefore, we have

$$C_{cool} = \alpha_c / x^{\beta_c} \tag{E7.6.3}$$

where α_c and β_c are positive real numbers. The lighting cost will increase markedly with x, i.e.

$$C_{light} = \alpha_l\, x^{\beta_l} \tag{E7.6.4}$$

where α_l and β_l are positive real numbers. To sum things up, we have

$$y = \alpha_h / x^{\beta_h} + \alpha_c / x^{\beta_c} + \alpha_l\, x^{\beta_l} \tag{E7.6.5}$$

There is also one apparent constraint. In theory, this constraint should automatically be taken care of by the last term in Eq. (E7.6.5). Nevertheless, it is good engineering practice to spell it out explicitly as a constraint, i.e.

$$\psi = x \le H \tag{E7.6.6}$$

where H is the height of the greenhouse. As this constraint is presumably included in the cost of light per unit of produce, we can solve the problem as an unconstrained problem and then check that the obtained solution satisfies this constraint. The optimum occurs at

$$dy/dx = -\alpha_h\, \beta_h\, x^{\beta_h - 1} - \alpha_c\, \beta_c\, x^{\beta_c - 1} + \alpha_l\, \beta_l\, x^{\beta_l - 1} = 0 \tag{E7.6.7}$$

The values for the coefficients, which depend heavily on the location and soil type, can be approximated from systematic modeling analysis performed using software such as TRNSYS and EnergyPlus.

One need not be an activist to strive for greener energy, and fossil fuels are not considered renewable, green, or clean. Nevertheless, today's technologies, which run on fossil fuels, are remarkably cleaner than their predecessors. Moreover, to sustain energy needs, fossil fuels will continue to be a significant part of the energy mix decades into the future (Reader, 2020). It is also not politically correct to assert that the internal combustion engine will be around for a very long time (Reitz et al., 2020). Be that as it may, the future of this workhorse is secured by the continuous greening of its fuel, along with other breakthroughs. For a spark-ignition engine, optimizing the flame propagation process is fundamental. Let us make such an attempt on the elemental level.

Example 7.7 *Enhance a spark-ignited flame propagation in homogeneous vortices*

Given

We have been briefed, in Chapter 5, concerning the underlying physics of a premixed planar flame interacting with a vortex. Let us extend that knowledge into a spherical flame expanding in an engine cylinder where they encounter many turbulent eddies (vortices).

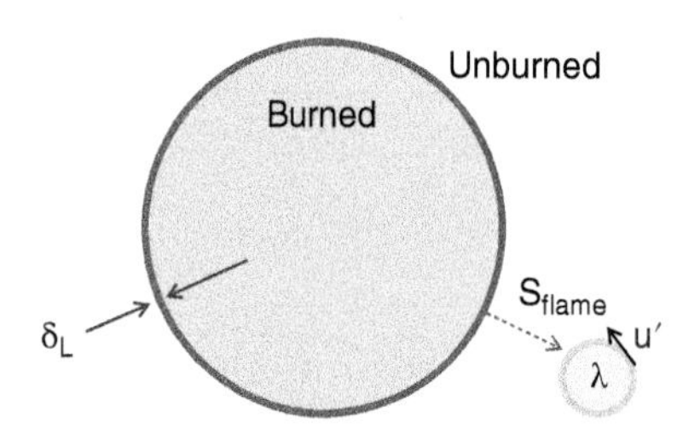

Figure 7.5 A premixed spherical flame approaching a vortex. Source: D. Ting.

Find

Formulate the problem to maximize the combustion rate.

Solution

As in Chapter 5, we assume the fluid dynamics are slow compared to the chemistry of combustion. Specifically, the Damköhler number (Damköhler, 1940), flow-chemical time scale,

$$Da = \tau_{flow}/\tau_{chem} = (\lambda/u') / (\delta_L/S_L) \quad (E7.7.1)$$

is large. From Figure 7.5, we see that λ signifies the diameter of the vortex, u' is the vortex intensity, i.e. its tangential velocity, δ_L is the laminar flame front thickness, and S_L is the laminar flame speed. The flame propagation speed, S_{flame}, which prescribes the rate of combustion, is proportional to the laminar flame speed and the flame surface corrugation (area). In other words, the mass burning rate,

$$m_b' = \rho_u S_L A_{flame} \quad (E7.7.2)$$

where ρ_u is the density of the unburned gas and A_{flame} is the reacting flame front area. The flame surface area increases with the intensity of the vortex, u′, and the number of vortices present, i.e. the reacting surface area, A_{flame}, increases with u' and N, the number of vortices. Therefore,

$$m_b' = \rho_u S_L A_{flame} \approx \rho_u S_L A_o \left(1 + K_{u'} N u'/S_L\right) \quad (E7.7.3)$$

where $K_{u'}$ indicates the magnitude of the influence of the vortex intensity, whose effect is being nullified by the consuming flame, resulting in the relative effect, u'/S_L. The unwrinkled flame area, A_o, corresponds to a smooth sphere enclosing the burned gas. As it stands, Eq (E7.7.3) imposes no upper bound, i.e. the combustion rate increases continuously with increasing number of vortices and/or their intensity. However, we have been enlightened in Chapter 5 that there is no point intensifying a vortex such that it rotates more than once as the flame front passes through it. In plain English, there is little gain in terms of extra reacting area as u'/S_L is pushed beyond approximately one. Similarly, more vortices than what the flame front can interact with do not lead to more gain in reacting surface area. These limits call for two constraints, which may be expressed as

$$u'/S_L \leq 1 \tag{E7.7.4}$$

and

$$N \leq A_o/\left(\pi\lambda^2\right) \tag{E7.7.5}$$

The latter constraint is probably more accurately represented by

$$N \leq A_{flame}/\left(\pi\lambda^2\right) \tag{E7.7.6}$$

because further wrinkling can be wrung into the extra surface created through vortex-induced corrugation.

There are a few points worth mentioning. Turbulent eddying motions are of varying sizes and intensity (Ting, 2016). These vortical structures are tube-like and, therefore, the length also plays a role. More importantly, it takes significant engineering and energy to generate them and they, especially the smaller ones, decay rapidly. On these grounds, we want to generate the right number of vortices just in time, just ahead of the flame front. Enthusiasts are encouraged to refer to Ting (2018).

Problems

7.1 Operating cost of pumping water via two pumps in parallel

Water at standard temperature and pressure is to be delivered at $m' = 2.5$ kg/s via two pumps, Pump 1 and Pump 2, in parallel; see Figure 7.6. The head loss in Line 1, $h_1 = K_1\, m_1'$, where h_1 is in m^2/s^2, constant K_1 is in $m^2/(kg{\cdot}s)$, and the mass flow rate, m_1', is in kg/s. The head loss in Line 2, $h_2 = K_2\, m_2'$, where h_2 is in m^2/s^2, constant K_2 is in $m^2/(kg{\cdot}s)$, and the mass flow rate, m_2', is in kg/s. The operating cost, in terms of pumping power, is to be minimized. Write the objective function and the constraint(s) clearly.

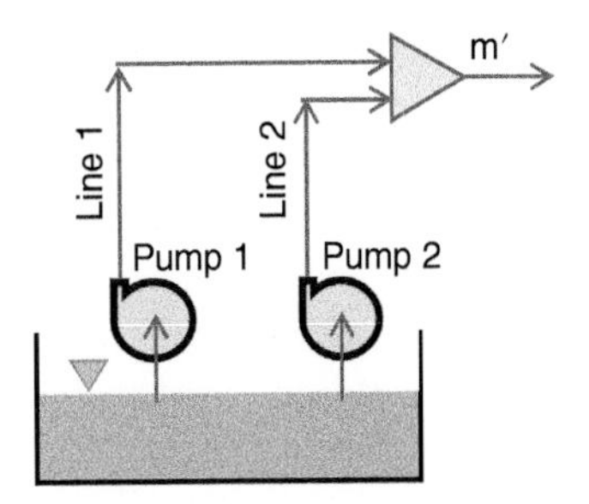

Figure 7.6 Pumping water via two pumps in parallel. Source: D. Ting.

7.2 Minimizing the material used for a canned drink

After acing Engineering Design and Optimization of Thermofluid Systems, two engineering students decide to commercialize a special drink that they have cooked up as their course project. They would like to put 0.1 m^3 of the special drink into each cylindrical can, which is to be made from a 0.1 mm-thick sheet of recycled metal. To ensure environmental friendliness and to reduce material cost, they wish to minimize the material for making the can. Formulate the problem at hand and solve it by determining the dimensions (radius and height) of the can.

7.3 Minimizing flow-induced vibration of a bundle of heat exchanger tubes
Heat exchangers for boilers and other applications typically consist of a bundle of tubes subjected to cross-flow of a fluid. As such, different kinds of flow-induced vibration can occur, leading to failure of the system (Gawande et al., 2011; Khushnood and Nizam, 2017). For that reason, much research has been conducted over the years to better understand the underlying mechanisms. Among others, the vortex-shedding frequency (the Strouhal number) with respect to the natural frequency of the structure, the Reynolds number, and the orientation of the tube array affect the promptness and severity of the flow-induced vibrations. Formulate the problem to minimize the flow-induced vibration of a bundle of 100 2.5-cm-diameter tubes, arranged in 10 rows by 10 columns, occupying a 50 cm x 50 cm cross-sectional area. Assume that the structural-flow frequencies have been taken care of, and the Reynolds number based on the bulk velocity and tube diameter is set around the vicinity of 5000. Note that at the elemental level, the map in Ting et al. (1998) may be helpful.

7.4 Optimize a marching band to resonate a concrete wall
Having been briefed about flow-induced vibration after attempting the previous problem, you surmise that the wall of Jericho, a 2-m-thick and 6–8-m-high mudbrick wall (Sellin and Watzinger, 1973; Kenyon, 1981; Wood, 1990), fell because a marching band hit the resonant frequency of the wall. Specifically, the otherwise-inconsequential imperfections (small cracks) at various places of the impregnable wall must have subtly enlarged as the marching progressed over the week. The synchronized rambunctious shouts most likely pushed the weak points over the threshold, causing the massive barricade to tumble down. Before you object that this problem has nothing to do with thermofluids, recognize that the synchronized squalls are forceful pressure waves, a compressible fluid phenomenon involving the coupling of thermodynamics and fluid mechanics, beating on the fortified architecture. Formulate the problem in terms of synchronization of the entire marching band encircling Jericho. Assume that it is the modulated sound waves, along with coordinated ground vibrations, that created one of the fundamental modes of vibration of the entire roughly circular wall to pulsate out of control. In other words, you are to formulate the problem in terms of the most effective marching and shouting to trigger the wall to resonate and collapse.

7.5 Improve the design of urinals
Serious effort has been invested in understanding the rebound dynamics of a water droplet after impacting a surface. Most of the recent studies along this line, aimed at furthering water-repellence, are for self-cleaning and anti-icing (Yang et al., 2020; Zhang et al., 2020). A headlining application is to harness water droplets for electricity (Xu et al., 2020). One everyday challenge appears to have fallen through the cracks, as far as the general public at large is concerned, i.e. the mitigation of urine from splashing back onto the urinating individual. This problem is particularly common among the newer, water-saving designs. Mitigating this annoyance can conceivably be straightforward, i.e. by maintaining a proper water level in the urinal for minimal rebounding and/or splashing, design an adequately-elevated wall to block off the backlashing pee, and positioning the john at the right distance below the average "human pee faucet." It is worth noting that there are numerous patents detailing auxiliary parts for alleviating the splash. One functioning gizmo is the urinal splash pad. However, the rule of thumb for effective engineering is simplicity. Adding an extra gimmick ultimately increases cost, maintenance, and, above all, cleaning. A generic solution without this com-

plication is to know where on the urinal wall and at what angle to aim the jet of urine (Ferro, 2013). Hurd (pers. comm, June 9, 2020) ascertained that when droplets impact, they create far greater splash than stream impact, see Figure 7.7. The lesson is thereby go full blast or hold your droplets. Unfortunately, this is only applicable for the full-length urinals where the base sits on the floor. For the more common and economical one, with a much shorter length, the outlet is above knee height. For these urinals, optimizing the depth of the water level is a good starting point. Formulate the problem to minimize the splashing of pee.

Stream impact **droplet impact**

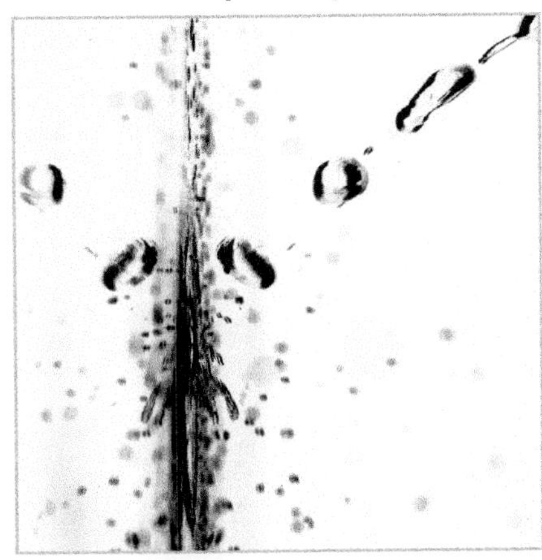

Figure 7.7 Pee (water) stream versus droplets impacting the wall of a full-length urinal. Note that there is next to no splash associated with the stream, but tiny droplets bounce of at all angles after the larger droplets impact the wall. Source: R. Hurd.

7.6 Optimize spray for improving engine combustion
Directly injecting fuel into a spark-ignition gasoline engine has numerous advantages, including cleaner combustion. In general, the smaller the droplets carried by the spray, the faster the liquid fuel vaporizes and, subsequently, the faster and more complete (cleaner) the combustion is. Confined by the small volume when the engine is in the vicinity of top dead center, the engineering of the best combustion becomes particularly challenging. Care should be taken to avoid the spray impinging on the chamber wall or cylinder head. Also, small liquid droplets within close proximity of each other can coalesce. Notwithstanding that, a concentrated region of fuel tends to result in undesirable fuel-rich combustion and non-homogeneous burning throughout the cylinder. The omni-present turbulence can be exploited to mitigate these issues. The two-decades old paper compiled by Sornek et al. (2000) is still an excellent starting point. Formulate the optimal spray which leads to the fastest and cleanest combustion for a typical cylinder with a central spark plug. Keep in mind that, because of the very small reacting surface area associated with the flame growing from a spark kernel, the initial phase of combustion takes the longest time to develop and complete. That being the case, you may wish to pay more attention to nurture the initial flame growth and acceleration (Ting and Checkel, 1997).

References

Arici, M. and Kan, M. (2015). An investigation of flow and conjugate heat transfer in multiple pane windows with respect to gap width, emissivity and gas filling. *Renewable Energy* 75: 249–256.

Arora, J.S. (2010). *Formulating design problems as optimization problems: system engineering - optimization techniques and applications.* In: *Encyclopedia of Aerospace Engineering*, DOI:10.1002/9780470686652.eae494.

Cunha S., Leite, P., and Aguiar J. (2020). Characterization of innovative mortars with direct incorporation of phase change materials. *Journal of Energy Storage*, 30: 101439.

Damköhler, G. (1940).Der Einfluss der Turbulenz auf die Flammengeschwindigkeit in Gasgemischen. Physikalische Chemie 46(11): 601–652. English translation: NACA Technical Memorandum, No. 1112, 1947.

Ferro, S. (2013). Science addresses the problem of pee splashback – Brigham Young University's Splash Lab looks into the dynamics of the male urine stream. Popular Science November 6, 2013. Available at: https://www.popsci.com/article/science/science-addresses-problem-pee-splashback/ accessed May 23, 2020.

Gawande S.H., Keste A.A., Navale, L.G. et al. (2011) Design optimization of shell and tube heat exchanger by vibration analysis. *Modern Mechanical Engineering* 1: 6–11.

Hu, Y., Heiselberg, P.K., and Guo, R. (2020). Ventilation cooling/heating performance of a PCM enhanced ventilated window an experimental study. *Energy and Buildings* 214, 109903.

Kenyon, K.M. (1981). *Excavations at Jericho, 3:110*, London: British School of Archaeology in Jerusalem.

Khushnood S. and Nizam, L.A. (2017). Experimental study on cross-flow induced vibrations in heat exchanger tube bundle. *China Ocean Engineering* 31(1): 91–97.

Kim, S.H., Kwon J.K., Kim H.K., and Joen, J.G. (2017). Heating load analysis of the pit green house. Paper presented at 2017 ASABE Annual International Meeting, 1700734, DOI:10.13031/aim.201700734.

Low-Tech Magazine (2020). Fruit walls: urban farming in the 1600s. Available at: https://www.lowtechmagazine.com/2015/12/fruit-walls-urban-farming.html, accessed May 22, 2020.

Reader, G.T. (2020). Energy, renewables alone? In: Sustaining Resources for Tomorrow (ed. J.A. Stagner and D.S-K. Ting), 1–45. Cham, Switzerland: Springer Nature.

Reitz, R.D. et al. (2020). IJER editorial: The future of the internal combustion engine. *International Journal of Engine Research* 21(1): 3–10.

Sellin, E. and Watzinger, C. (1973). Jericho: die Ergebnisse der Ausgrabungen, Osnabrück: O. Zeller. Reprint of the 1913 edition published by Leipzig: J.C. Hinrichs, 1913.

Sornek, R.J., Dobashi, R., and Hirano, T. (2000). Effect of turbulence on vaporization, mixing, and combustion of liquid-fuel spray. *Combustion and Flame* 120: 479–491.

Stone, G. (1997). *Building a solar-heated pit greenhouse. In: A Storey Country Wisdom Bulletin.* North Adams: Storey Publishing.

Ting, D.S-K. (2016). *Basics of Engineering Turbulence.* New York: Academic Press.

Ting, D.S-K. (2018). *Engineering Combustion Essentials.* Cambridge: Cambridge Scholars Publishing.

Ting, D.S-K. and Checkel, M.D. (1997). The importance of turbulence intensity, eddy size and flame size in spark ignited, premixed flame growth. *Journal of Automobile Engineering* 211: 83–86.

Ting, D.S-K., Wang, D.J., Price, S.J., and Païdoussis, M.P. (1998). An experimental study on the fluid elastic forces for two staggered circular cylinders in cross-flow. *Journal of Fluids and Structures* 12: 259–294.

Tiwari, G.N., and Dhiman, N.K. (1986). Design and optimization of a winter greenhouse for the Leh-type climate. *Energy Conversion and Management* 26(1): 71–78.

Wang, H-X. (2006). Study on heat preservation performance of a pit type sunlight greenhouse on the Loess Plateau area. Journal of Shanxi Agricultural University (Social Science Edition) 2006-S2.

Wood, B.G. (1990). Did the Israelites conquer Jericho? *Biblical Archaeology Review* 16(2): 44–58.

Xu, W., Zheng, H., Liu, Y., et al. (2020). A droplet-based electricity generator with high instantaneous power density. *Nature*, DOI:10.1038/s41586-020-1985-6.

Yang, C. Chao, J., Zhang, J., et al. (2020). Functionalized CFRP surface with water-repellence, self-cleaning and anti-icing properties. *Colloids and Surfaces A* 586: 124278.

Zhang, H., Zhang, X., Yi, X., et al. (2020). Dynamic behaviors of droplets impacting on ultrasonically vibrating surfaces. *Experimental Thermal and Fluid Science* 112: 110019.

8

Calculus Approach

Calculus required continuity, and continuity was supposed to require the infinitely little; but nobody could discover what the infinitely little might be.

– Bertrand Russell

Chapter Objectives

- Understand how the Lagrange Multiplier method works.
- Apply the Lagrange Multiplier method for unconstrained problems.
- Solve constrained and multi-variable problems using the Lagrange Multiplier method.
- Interpret the Lagrange Multiplier as a sensitivity factor.
- Learn to deal with inequality constraints.

Nomenclature

D	diameter
f	function
g	gradient
h	heat transfer coefficient
L	length
S_C	sensitivity coefficient; $S_C = \lambda$
T	temperature
t	tangent
TEG	thermoelectric energy generator
x	a variable
Y, y	a variable, the objective function; Y is the Lagrange expression

Greek and Other Symbols

Δ difference or change
λ Lagrange Multiplier
Φ equality constraint
Ψ inequality constraint
∀ volume; ∀' is volume flow rate
∇ gradient operator

8.1 Introduction

The primary meaning of calculus in Latin is pebble. As such, it is like an abacus, an effective means for counting and performing calculations. Mathematically speaking, calculus is the study of continuous change. Within the context of optimization, the slope of the continuous change is of interest. For this, differential calculus, which is concerned with the rate of change, i.e. the slope of a curve, is instrumental. Consider a dependent parameter, y, that is a function of an independent variable, x; for example, the teaching score of an instructor, y, is a function of the niceness of the instructor, x. A zero slope signifies a maximum, a minimum, or a plateau, as depicted in Figure 8.1. In a three- (or more) dimensional case, we could also have ridges, valleys, and/or saddle points. A saddle point, also called a minimax point, is a point on the surface where the slopes in orthogonal directions are all zero, see Figure 8.2.

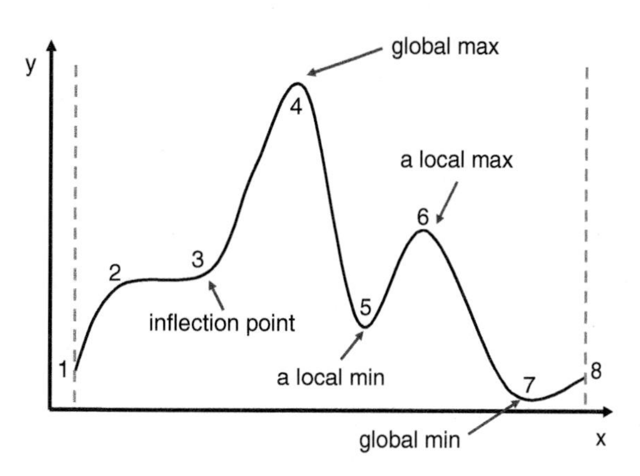

Figure 8.1 Different extrema according to $dy/dx = 0$ for $y = f(x)$. Source: D. Ting.

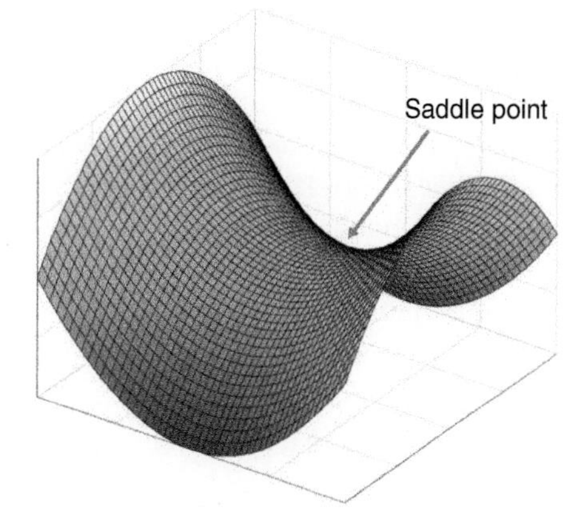

Figure 8.2 A saddle point of $y = f(x_1, x_2)$. The particular surface plotted corresponds to $y = x_1^2 - x_2^2$. Source: Y. Yang.

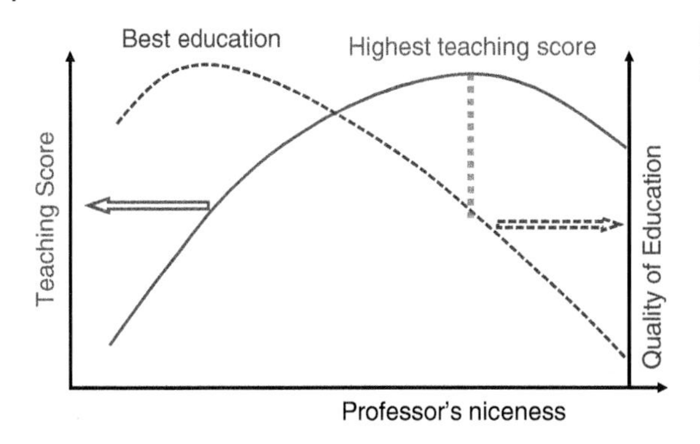

Figure 8.3 Teaching score versus instructor's niceness. Source: D. Ting.

Using the teaching score versus niceness example, a professor is assured to score poorly if he or she has little niceness, i.e. is too strict. Figure 8.3 shows that the student evaluation of teaching score increases with niceness up to a maximum, beyond which further leniency amounts to increasing injustice. For instance, allowing the irresponsible and lethargic students to submit assignments way past the due date is unfair to the conscientious students who submit their work on time. It is more unfair if the solutions have already been posted, allowing the apathetic rascals to simply copy and submit. Thus, the teaching score peaks at excessive niceness at the cost of quality of learning. An educational institution is in luck if its priority is to keep its customers happy, for according to Klionsky (2017), "education is probably the only business where the customer is satisfied with less of the product." Fortunately, he also affirmed that instructors should communicate high expectations, as students tend to put in the effort needed to meet them. Admittedly, one has to be willing to sacrifice teaching scores.

8.2 Lagrange Multiplier

The *Lagrange* (or *Lagrangian*) *Multiplier* is based on the aforementioned continuous differentiable calculus to deduce the zero slope which corresponds to extrema. This method, named after Italian/French mathematician, Joseph-Louis Lagrange (Lagrange, 2018), finds the local maxima and minima of an objective function subject to equality constraints. Its main advantage is in solving the optimization problem, and this can be done without explicitly finding the equation describing the curve or surface of the constraints.

As it is a calculus method, the Lagrange Multiplier works only for differentiable functions and equality constraints. Fortunately, very often, curve fitting can be employed to convert discrete points into a continuous and differentiable function. Also, inequality constraints can oftentimes be converted into equality ones. Moving forward, we synthesize the materials covered particularly in Burmeister (1998), Jaluria (1998), and Stoecker (1989).

For a continuous and differentiable function, $y = f(x)$, such as the teaching score versus instructor's niceness, shown in Figure 8.3, we can locate the extremum by noting the following.

At an extremum, local or global may it be,

$$dy/dx = 0 \tag{8.1}$$

i.e. the slope is zero. If the extremum is a maximum, then,

$$d^2y/dx^2 < 0 \tag{8.2}$$

In other words, the slope changes from a positive one into negative one with increasing x. Imagine you are on a roller coaster ride, such as that shown in Figure 8.1. You will be ascending as you gallop forward, from Points 3 to 4, approaching the summit, Point 5, and then, rapidly plummeting.

If the extremum is a minimum, then you slide downward via a negative slope, Points 6 to 7 in Figure 8.1, until you hit rock bottom, i.e. the minimum, beyond which everything looks upward. Mathematically,

$$d^2y/dx^2 > 0 \tag{8.3}$$

If the second derivative,

$$d^2y/dx^2 = 0 \tag{8.4}$$

we have an inflection point, like Point 3 in Figure 8.1. Figuratively, this can be imagined as ascending or descending a hill and we are relieved to find a plateau (Points 2 to 3 in Figure 8.1) to rest for a while, before continuing the capricious journey.

Let us look at a simple illustration of this.

Example 8.1 ***Reduce grocery costs***

Given

A student decides to fight against the rising cost of grocery. It is known that the amount of food and, hence, the cost, y, needed is a function of the rate of metabolism, which is directly related to the heart rate, x. For a healthy person,

$$y = 16.5\,x^{1.4} + 14.8/x^{2.2} \tag{E8.1.1}$$

where the first term on the right-hand side signifies the activeness and the second term, the depth of meditation.

Find

The minimum grocery cost.

Solution

Take the derivative of the objective function, as described by Eq (E8.1.1), with respect to the independent variable, x, the heart rate. Doing so, we get

$$dy/dx = 23.1\,x^{0.4} - 32.56\,x^{-3.2} = 0 \tag{E8.1.2}$$

This gives the optimum independent variable,

$$x^* = 1.10 \text{ beats/second} = 66 \text{ beats/minute} \tag{E8.1.3}$$

The next question is, does this particular heart rate correspond to a minimum or a maximum grocery cost? To answer this, we take the second derivative,

$$d^2y/dx^2 = 9.24\,x^{-0.6} + 104.2\,x^{-4.2} \tag{E8.1.4}$$

Substituting $x^* = 1.10$ into this gives $d^2y/dx^2 = 78.55$, which is positive or greater than zero. Therefore, the optimum y, y*, corresponds to a (global) minimum.

Note that we are typically interested in the global extremum. It is somewhat problematic for Lagrange Multipliers when maxima or minima are at the boundaries of the domain of interest, like that shown in Figure 8.4. Other challenges faced by the Lagrange Multiplier method shown are saddle points and local extrema. A good understanding of the problem at hand will avoid these potential pitfalls.

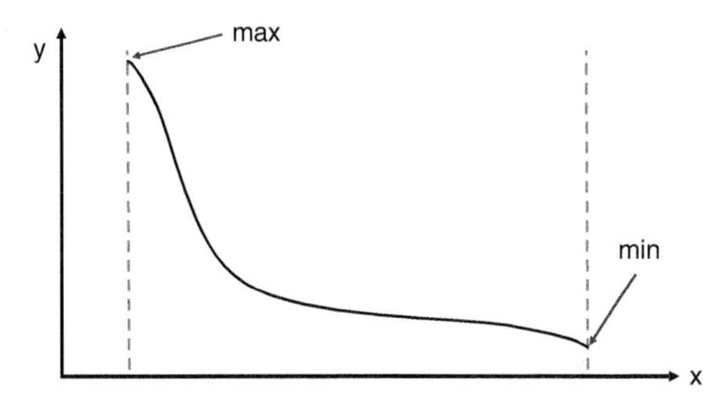

Figure 8.4 Some challenges faced by Lagrange Multiplier method. Source: D. Ting.

8.3 Unconstrained, Multi-Variable, Objective Function

Let us increase the number of independent variables to more than one. For the case with two independent variables involved, i.e. $y = y(x_1, x_2)$, such as that shown in Figure 8.5, we can deduce if the extremum is a minimum or maximum by checking the second derivatives, based on the following criteria:

$$\text{If } d^2y/dx_1^2 > 0 \text{ and } \left(d^2y/dx_1^2\right)\left(d^2y/dx_2^2\right) - \left(d^2y/dx_1dx_2\right) > 0 \tag{8.5}$$

then, it is a minimum.

$$\text{If } d^2y/dx_2^2 < 0 \text{ and } \left(d^2y/dx_1^2\right)\left(d^2y/dx_2^2\right) - \left(d^2y/dx_1dx_2\right) > 0 \tag{8.6}$$

then, it is a maximum.

One can, and probably should, always check to see if the objective function is increasing or decreasing, by moving the independent variable(s) a little away from the deduced optimum.

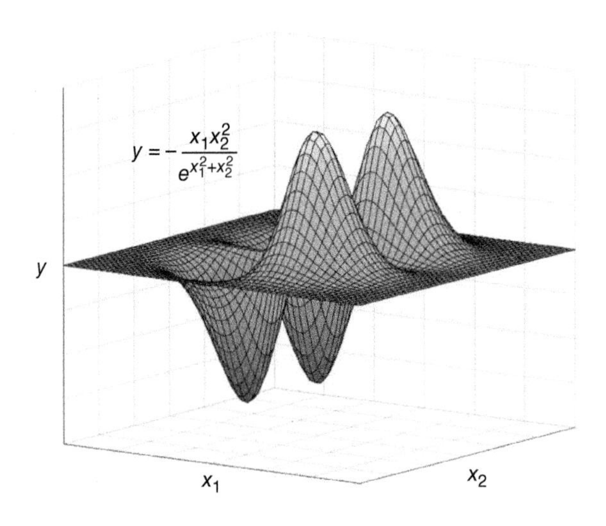

Figure 8.5 Extrema of a typical two-independent-variable objective function, $y = f(x_1, x_2)$. The plotted surface corresponds to $y = -x_1^2\, x_2^2 / \exp(x_1^2+x_2^2)$. Source: Y. Yang.

The general expression for a multi-variable objective function is

$$y = y\left(x_1, x_2, \ldots x_n\right) \tag{8.7}$$

The Lagrange Multiplier equation for this unconstrained optimization is

$$\partial y/\partial x_1 = 0, \partial y/\partial x_2 = 0, \ldots \partial y/\partial x_n = 0 \tag{8.8}$$

In short, there are n expressions that are all equal to zero. These n equations can be solved to obtain the optimum, y*, which takes place at $x_1 = x_1^*, x_2 = x_2^*, \ldots, x_n = x_n^*$.

Example 8.2 ***Minimize our production of disorder***

Given

Among others, Dean Kamen admittedly proclaimed, “I’m a human entropy producer.” In other words, we are but disorder (entropy) generators. Proper knowledge of entropy can enable us to accomplish the same task while producing considerably less disorder. The entropy generation of a compressed air energy storage system, used to mitigate the intermittent nature of renewable energy and the mismatch between supply and demand of a power grid (Ebrahimi et al., 2021), can be expressed as a function of the pressure change, x_1, and the thermal energy (heat) removal or addition, x_2. Specifically, the entropy generation varies with the square of the pressure change, where a positive pressure change corresponds to the charging phase and a negative pressure charge signifies the discharging phase. Similarly, heat extraction and addition are, respectively, positive and negative.

Find

The condition at which the compressed air energy storage system produces the least disorder, using the Lagrange Multiplier method.

Solution

In mathematical form, the entropy production is

$$y = x_1^2 + x_2^2 \tag{E8.2.1}$$

Invoking the Lagrange Multiplier method for this unconstrained optimization,

$$\partial y/\partial x_1 = 2\,x_1 = 0 \tag{E8.2.2}$$

$$\partial y/\partial x_2 = 2\,x_2 = 0 \tag{E8.2.3}$$

The optimum occurs at $x_1^* = 0$ and $x_2^* = 0$, which gives $y^* = 0$. To check if this is a minimum or a maximum, we invoke Eqs. (8.5) and (8.6).

$$d^2y/dx_1^2 = 2 > 0 \tag{E8.2.4}$$

$$\left(d^2y/dx_1^2\right)\,\left(d^2y/dx_2^2\right) - \left(d^2y/dx_1dx_2\right) = (2)\ (2) - 0 > 0 \tag{E8.2.5}$$

Therefore, it is a minimum. This is obvious from Eq (E8.2.1), where x_1 and x_2 of any nonzero values, positive or negative, result in $y > 0$.

The results inform us that the only way to not generate entropy is to do nothing. This, however, is contrary to engineering and to living. Our goal should be in line with what Theodore Roosevelt uttered,

> To waste, to destroy our natural resources, to skin and exhaust the land instead of using it so as to increase its usefulness, will result in undermining in the days of our children the very prosperity which we ought by right to hand down to them amplified and developed.

We are not to waste or destroy, but to improve the usefulness of the resources. The compressed air energy storage system is implemented to enhance the utility of the intermittent renewable energy harnessing system for fluctuating power needs. The charging and discharging phases, where the pressure is increased and decreased, are necessary to realize the usage of free and not-so-organized energy in the Sun's rays, the wind, and/or the waves, into highly-organized energy, electricity. The appropriate removal and addition of heat can significantly improve the performance of the system, reducing the amount of energy wasted. Therefore, the entropy minimization is more judiciously cast in terms of per unit energy harnessed for proper usage. This optimum will not correspond to x_1 and x_2 values of zero. An in-depth version of compressed air energy storage is given in the Appendix.

To explicitly show the objective function of Example 8.2, it is plotted in Figure 8.6. If a figure speaks a thousand words, every word in Figure 8.6 is pointing toward the simple and beautiful minimum, $y^*(x_1^* = 0, x_2^* = 0) = 0$. The three-dimensionality, with the objective function, y, as a function of two independent variables, is more-or-less the limit, as far as clear, illustrative visualization on a piece of paper is concerned. Let us up the game by one level, i.e. look at the case with three independent variables.

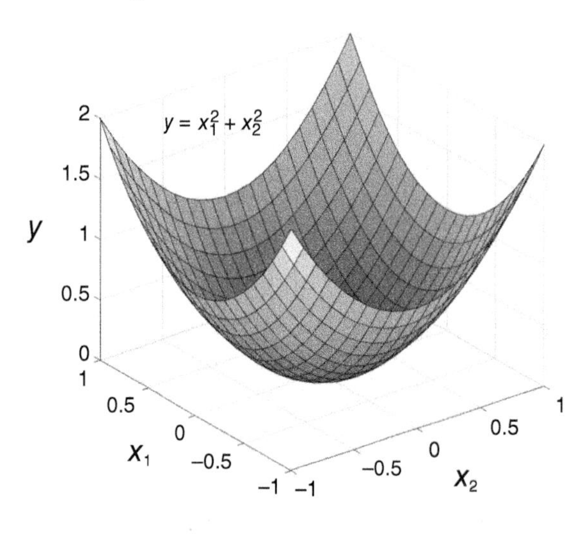

Figure 8.6 Entropy generation as a function of pressure change and heat addition or removal, $y = x_1^2 + x_2^2$. Source: Y. Yang.

Example 8.3 ***Optimize a three-variable objective function***

Given

The drag of a flying bee is related to its body size, x_1, wing flapping frequency, x_2, and wing span, x_3, in the form

$$y = \frac{x_1}{x_2} + \frac{1}{x_1 x_3} + \frac{1}{2}x_2^2 + \frac{x_3}{16} \tag{E8.3.1}$$

Find

The optimum drag.

Solution

Applying the Lagrange Multiplier method, we have

$$\frac{\partial y}{\partial x_1} = \frac{1}{x_2} + \frac{1}{x_1^2 x_3} = 0 \tag{E8.3.2}$$

$$\frac{\partial y}{\partial x_2} = -\frac{x_1}{x_2^2} + x_2 = 0 \tag{E8.3.3}$$

$$\frac{\partial y}{\partial x_3} = -\frac{1}{x_1 x_3^2} + \frac{1}{16} = 0 \tag{E8.3.4}$$

Via substitutions, we get

$$x_1^* = 0.305, x_2^* = 0.673, \text{and } x_3^* = 7.243$$

With these values, the objective function,

$$y^* = 1.585$$

We see that any variation of x_i from x_i^* leads to a larger y. Therefore, the above solution corresponds to the minimum drag.

8.4 Multi-Variable Objective Function with Equality Constraints

Let us add on equality constraints for the multi-variable objective function described by Eq (8.7). We will follow, in particular, Jaluria (1998) and Stoecker (1989). This objective function is subjected to equality constraints,

$$\Phi_1 \left(x_1, x_2, \ldots x_n\right) = 0,$$

$$\vdots$$

and

$$\Phi_m \left(x_1, x_2, \ldots x_n\right) = 0. \tag{8.9}$$

It is noted that, sometimes, some of the equality constraints can be substituted into the objective function to decrease the number of variables and, thus, the number of (constraint) equations. More often, however, the equality constraints are complex, making the substitution challenging. As such, a more general and versatile approach, detailed below, is typically followed.

The objective function and the constraints are combined into a new function, Y, the *Lagrange expression*. The general form of this Lagrange expression is

$$Y \left(x_1, x_2, \ldots x_n, \lambda_1, \lambda_2, \ldots \lambda_m\right)$$
$$= y \left(x_1, x_2, \ldots x_n\right) - \lambda_1 \Phi_1 \left(x_1, x_2, \ldots x_n\right) - \lambda_2 \Phi_2 \left(x_1, x_2, \ldots x_n\right) - \ldots - \lambda_m \Phi_m \left(x_1, x_2, \ldots x_n\right) \tag{8.10}$$

where unknowns, λ_i, are the Lagrange Multipliers. The optimum occurs at

$$\partial Y/\partial x_1 = 0, \partial Y/\partial x_x = 0, \ldots \partial Y/\partial x_n = 0,$$
$$\partial Y/\partial \lambda_1 = 0, \partial Y/\partial \lambda_2 = 0, \ldots \partial Y/\partial \lambda_m = 0 \tag{8.11}$$

Invoking these partial differentiations on Eq (8.10) gives

$$\partial y/\partial x_1 - \lambda_1\, \partial \Phi_1/\partial x_1 - \lambda_2\, \partial \Phi_2/\partial x_1 - \ldots - \lambda_m\, \partial \Phi_m/\partial x_1 = 0,$$
$$\partial y/\partial x_2 - \lambda_1\, \partial \Phi_1/\partial x_2 - \lambda_2\, \partial \Phi_2/\partial x_2 - \ldots - \lambda_m\, \partial \Phi_m/\partial x_2 = 0,$$
$$\vdots$$
$$\partial y/\partial x_n - \lambda_1\, \partial \Phi_1/\partial x_n - \lambda_2\, \partial \Phi_2/\partial x_n - \ldots - \lambda_m\, \partial \Phi_m/\partial x_n = 0 \tag{8.12}$$

We see that there are n equations for the Lagrange expression. These, along with the m constraints, Eq (8.9), grant the n plus m equations required to solve the n+m unknowns, $x_1, x_2, \ldots x_n, \lambda_1, \lambda_2, \ldots \lambda_m$.

Before we proceed with examples to illustrate the working of the Lagrange Multiplier for the multi-variable objective function, let us take a brief look at the single-variable case.

Example 8.4 ***Optimize a teaching score when imposed with a minimum niceness***

Given

A study found that a teaching score varies with the niceness of the professor in the form

$$y = 1 - (x - 0.8)^2 \tag{E8.4.1}$$

where y is the teaching score, and x is the niceness. For good public relations, the administration of a university imposes that the square of the niceness of every professor should be at least 0.25, i.e.

$$x^2 \geq 0.25 \tag{E8.4.2}$$

Find

The optimum teaching score.

Solution

The objective function described by Eq (E8.4.1) is plotted in Figure 8.7. The dotted vertical line indicates the constraint, which has been converted into an equality, i.e.

$$\Phi = x^2 - 0.25 = 0 \tag{E8.4.3}$$

Let us follow the steps as outlined by Eqs. (8.10)–(8.12).

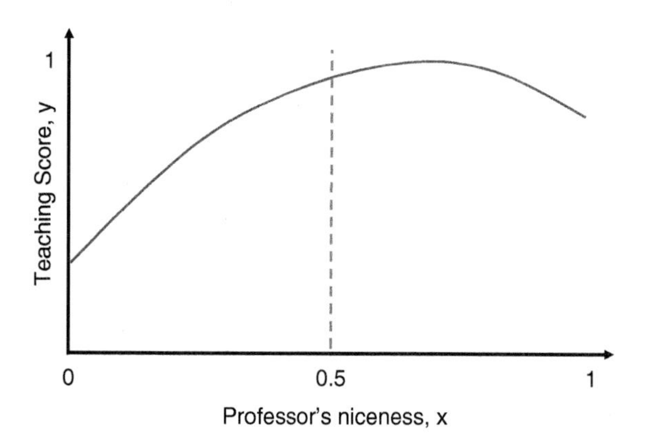

Figure 8.7 Teaching score as a function of niceness, $y = 1 - (x - 0.8)^2$, constrained by $x^2 = 0.25$. Source: D. Ting.

The Lagrange expression,

$$Y(x, \lambda) = y(x) - \lambda\,\Phi(x) = 1 - (x - 0.8)^2 - \lambda\,\left(x^2 - 0.25\right) \tag{E8.4.4}$$

The optimum occurs at

$$\partial Y / \partial x = 0 \tag{E8.4.5}$$

and

$$\partial Y / \partial \lambda = 0 \tag{E8.4.6}$$

These give

$$1.6 - 2x - 2\lambda = 0 \quad \text{(E8.4.7)}$$

and

$$x^2 - 0.25 = 0 \quad \text{(E8.4.8)}$$

From Eq (E8.4.8), we get

$$x = 0.5$$

which is the vertical dotted line in Figure 8.7. Substituting this into Eq (E.8.4.7), we have

$$1.6 - 2(0.5) - 2\lambda = 0$$

which gives

$$\lambda = 0.3$$

The significance of λ will be conveyed shortly. The optimum teaching score when niceness, $x = 0.5$, can be deduced by substituting $x^* = 0.5$ into Eq (E8.4.1). Doing so leads to

$$y^* = 0.91.$$

Note that this is the maximum teaching score when niceness is 0.5. If the teaching quality is allowed to suffer further, i.e. the professor is obliged to be nicer, the teaching score will peak at $y^* = 1$, where the corresponding niceness, $x^* = 0.8$.

Let us consider another example before we appreciate the significance and/or usefulness of the Lagrange Multiplier, λ. We will look at a case where there are two independent variables and one equality constraint.

Example 8.5 ***Optimizing $y(x_1, x_2)$ constrained by $\Phi(x_1, x_2) = 0$***

Given

A shell-and-tube heat exchanger with a total length of 100 m of tubes costs \$900. The cost of the shell is $1100\,D^{2.5}\,L$, where D is the shell diameter in meters and L is the length of the shell in meters. The floor space costs 320 D L. A total of 200 tubes, covering a cross-sectional area of 1 m^2 of the shell, is required.

Find

The D and L which lead to the minimum cost.

Solution

a) This problem can be solved as an unconstrained one by substituting the constraint into the objective function. The constraint is
   200 tubes per m^2 shell [$(\pi D^2/4)$ of m^2 of cross-sectional area] [L in m] = 100 m of tubes
   or

$$50\,\pi D^2\,L = 100 \quad \text{(E8.5.1)}$$

   This can be rewritten as

$$L = 2/\pi D^2 \quad \text{(E8.5.2)}$$

The objective function,

$$y = 900 + 1100D^{2.5}L + 320DL \tag{E8.5.3}$$

Substituting Eq (E8.5.2), the constraint, into Eq (E8.5.3) gives

$$y = 900 + 2200D^{0.5}/\pi + 640/(\pi D) \tag{E8.5.4}$$

Applying the standard calculus method, $\partial y/\partial x = 0$, leads to

$$\partial y/\partial D = 1100D^{-0.5}/\pi - 640/\left(\pi D^2\right) = 0 \tag{E8.5.5}$$

This gives D* = 0.70 m. Substituting this into Eq (E8.5.2) gives L* = 1.3 m. Substituting both D* and L* into Eq (E8.5.3), or substituting D* into Eq (E8.5.4), we get y* = $1,777.

b) Let us solve this problem with the more-versatile Lagrange Multiplier method.
The Lagrange expression,

$$Y\left(x_1, x_2, \lambda\right) = y\left(x_1, x_2\right) - \lambda\,\Phi\left(x_1, x_2\right) \tag{E8.5.6}$$

The optimum occurs at

$$\partial Y/\partial x_1 = 0,\ \partial Y/\partial x_2 = 0, \text{ and } \partial Y/\partial \lambda = 0 \tag{E8.5.7}$$

These are

$$\partial y/\partial x_1 - \lambda\,\partial\Phi/\partial x_1 = 0 \tag{E8.5.8}$$

$$\partial y/\partial x_2 - \lambda\,\partial\Phi/\partial x_2 = 0 \tag{E8.5.9}$$

and

$$\Phi\left(x_1, x_2\right) = 0 \tag{E8.5.10}$$

Equations (E8.5.8) and (E8.5.9) can be expressed in vector (note that a scalar is a tensor of rank 0, a vector, with both magnitude and direction, is a tensor of rank 1) notation, i.e.

$$\nabla y - \lambda\,\nabla\Phi = 0 \tag{E8.5.11}$$

where

$$\nabla\Phi = \frac{\partial\Phi}{\partial x_1}\hat{i} + \frac{\partial\Phi}{\partial x_2}\hat{j} \tag{E8.5.12}$$

Using customary notation, the symbols with a hat, or bolded, are unit vectors in the respective coordinate direction.

For the problem at hand, we have, from Eq (E8.5.1), the constraint,

$$\Phi = 50\,\pi D^2\,L - 100 \tag{E8.5.13}$$

Taking the gradient of the constraint, i.e. invoking Eq. (E8.5.12), with $x_1 = D$ and $x_2 = L$,

$$\nabla\Phi = \left(\partial\Phi/\partial x_1\right)\mathbf{i}_1 + \left(\partial\Phi/\partial x_2\right)\mathbf{i}_2 = 100\,\pi\,D\,L\,\mathbf{i}_1 + 50\,\pi D^2\,\mathbf{i}_2 \tag{E8.5.14}$$

Taking the gradient of the objective function, Eq (E8.5.3), gives

$$\nabla y = \left(\partial y/\partial x_1\right)\mathbf{i}_1 + \left(\partial y/\partial x_2\right)\mathbf{i}_2 = \left(2750\,D^{1.5}\,L + 320\,L\right)\mathbf{i}_1 + 1100\,D^{2.5} + 320\,D)\,\mathbf{i}_2 \tag{E8.5.15}$$

More directly, in terms of Eq. (E8.5.11),

$$\nabla y = \left[2750D^{1.5}L + 320L + \lambda(100\pi DL)\right] \mathbf{i}_1 + \left[1100D^{2.5} + 320D + \lambda\left(50\pi D^2\right)\right] \mathbf{i}_2 = 0. \quad \text{(E8.5.16)}$$

Breaking this into the corresponding two components,

$$\mathbf{i}_1 : 2750D^{1.5}L + 320L - \lambda(100\pi DL) = 0 \quad \text{(E8.5.17)}$$

and

$$\mathbf{i}_2 : 1100D^{2.5} + 320D - \lambda\left(50\pi D^2\right) = 0 \quad \text{(E8.5.18)}$$

The third equation is the constraint, i.e. Eq. (E.8.5.1).

Thus, we have three equations, Eqs. (E8.5.1), (E8.5.17), and (E8.5.18), and three unknowns D*, L*, and λ. From Eq (E8.5.17), we have

$$\lambda = \left(2750D^{1.5} + 320\right) / (100\pi D) \quad \text{(E8.5.17a)}$$

Substituting this into Eq (E8.5.18), we get D* = 0.70 m. Substituting this into Eq (E8.5.1) gives L* = 1.3 m. For these optimal D and L values, the optimal objective function, y* = $1,777. Also, substituting D* = 0.70 m into Eq (E8.5.17a) leads to λ = 8.77.

8.5 Significance of the Lagrange Multiplier Operation

For the unconstrained case, the optimum occurs at $\nabla y = 0$. The gradient vector, ∇y, is normal to the constant-y curve. It represents the direction in which y (the dependent variable, which is the objective function) changes at the fastest rate. Consider the two-variable case, where $y = (x_1, x_2)$. Figure 8.8 is a plot of x_1 versus x_2, where the contours of constant y are depicted. Recall from calculus that

$$dy = \left(\partial y/\partial x_1\right)\, dx_1 + \left(\partial y/\partial x_2\right)\, dx_2 \quad (8.13)$$

Along a y contour line,

$$dy = 0 = \left(\partial y/\partial x_1\right)\, dx_1 + \left(\partial y/\partial x_2\right)\, dx_2 \quad (8.14)$$

Therefore,

$$dx_1 = -dx_2\, \left(\partial y/\partial x_2\right) / \left(\partial y/\partial x_1\right) \quad (8.15)$$

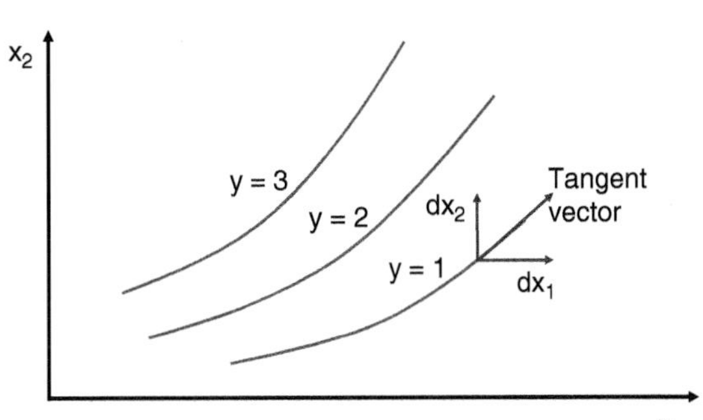

Figure 8.8 Constant contours of $y = f(x_1, x_2)$. Source: D. Ting.

Any arbitrary vector in Figure 8.8 is

$$dx_1\ \mathbf{i}_1 + dx_2\ \mathbf{i}_2 \tag{8.16}$$

Extending the above to any arbitrary unit vector, we have

$$\left(dx_1\ \mathbf{i}_1 + dx_2\ \mathbf{i}_2\right) / \sqrt{\left[\left(dx_1\right)^2 + \left(dx_2\right)^2\right]} \tag{8.17}$$

The particular unit vector, which is tangent to the y = constant line, has dx_1 and dx_2 related according to Eq (8.14). Substituting Eq (8.14) into Eq. (8.17) gives

$$\hat{t} = \frac{-\dfrac{\partial y}{\partial x_2} / \dfrac{\partial y}{\partial x_1}\hat{i}_1 + \hat{i}_2}{\sqrt{\left(\dfrac{\partial y}{\partial x_2} / \dfrac{\partial y}{\partial x_1}\right)^2 + 1}} = \frac{-\dfrac{\partial y}{\partial x_2}\hat{i}_1 + \dfrac{\partial y}{\partial x_1}\hat{i}_2}{\sqrt{\left(\dfrac{\partial y}{\partial x_2}\right)^2 + \left(\dfrac{\partial y}{\partial x_1}\right)^2}} \tag{8.18}$$

This unique (unit) vector is called the *unit tangent vector*, *t*. The procedure above can be repeated for the *unit gradient vector*, g. Doing so results in

$$\hat{g} = \frac{\nabla y}{|\nabla y|} = \frac{-\dfrac{\partial y}{\partial x_1}\hat{i}_1 + \dfrac{\partial y}{\partial x_2}\hat{i}_2}{\sqrt{\left(\dfrac{\partial y}{\partial x_1}\right)^2 + \left(\dfrac{\partial y}{\partial x_2}\right)^2}} \tag{8.19}$$

One can see that the (unit) gradient vector, **g**, is perpendicular to the (unit) tangent vector, **t**. As such, if

$$\mathbf{t} = \delta x_1\ \mathbf{i}_1 + \delta x_2\ \mathbf{i}_2 \tag{8.20}$$

then

$$\mathbf{g} = -\delta x_2\ \mathbf{i}_1 + \delta x_1\ \mathbf{i}_2 \tag{8.21}$$

as depicted in Figure 8.9.

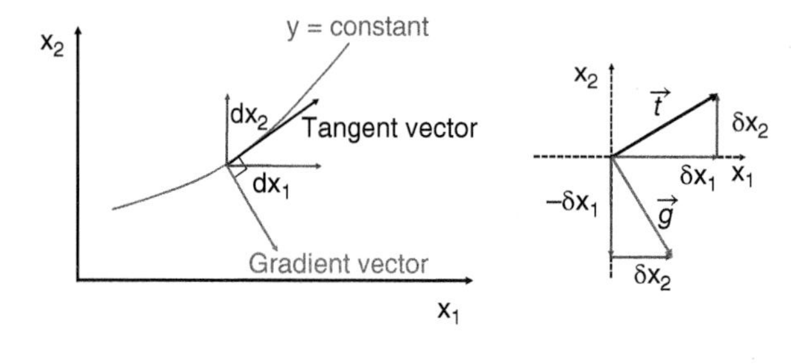

Figure 8.9 Gradient vector-tangent vector relationship. Source: D. Ting.

Analogous outcomes can be derived for objective functions that depend on more independent variables. From $y = f(x_1, x_2)$ to $y = f(x_1, x_2, x_2)$, the constant-contour y lines become constant-contour y surfaces, see Figure 8.10. The gradient vector for the three-variable y is

$$\nabla y = \left(\partial y/\partial x_1\right)\ \mathbf{i}_1 + \left(\partial y/\partial x_2\right)\ \mathbf{i}_2 + \left(\partial y/\partial x_3\right)\ \mathbf{i}_3 \tag{8.22}$$

As a gradient, ∇y, is normal to the constant-y surface that passes through that point. The magnitude of the gradient vector signifies how fast y changes with respect to the x_i. Closely-spaced constant y surfaces imply a large gradient, i.e. the y value changes rapidly with respect to the x_i. Within the context of the optimization of y, the direction of the gradient vector is of primary interest, as it points to the direction to reach the optimal y.

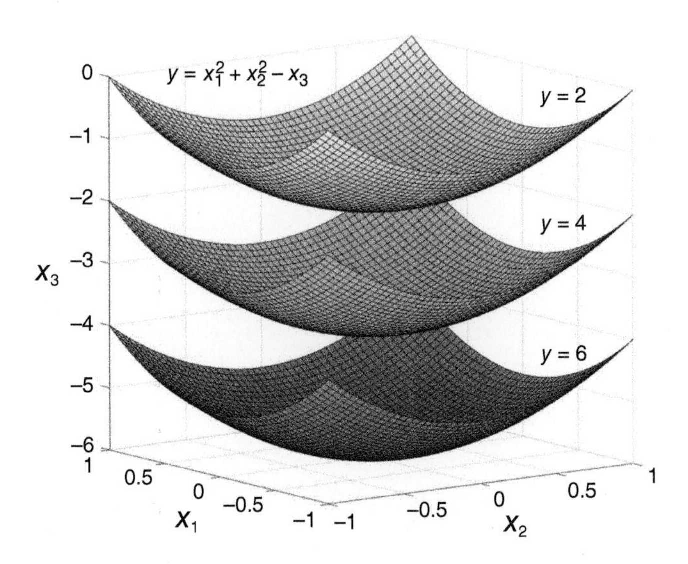

Figure 8.10 Constant-contour-surfaces of $y = x_1^2 + x_2^2 - x_3$. Source: Y. Yang.

Example 8.6 ***Entropy generation in a class, constrained*** $y(x_1, x_2)$

Given

The amount of entropy generated in a class,

$$y = x_1 + x_2 \tag{E8.6.1}$$

where x_1 is the distraction produced by the students and x_2 is the disorganization of the professor. As there is always entropy generation, unless the class consists of dead pupils and instructor, y has to be positive and

$$x_1\, x_2 = 0.25 \tag{E8.6.2}$$

Find

The optimum y.

Solution

Apply the Lagrange Multiplier method. The Lagrange expression,

$$Y\left(x_1, x_2, \lambda\right) = y\left(x_1, x_2\right) - \lambda\, \Phi\left(x_1, x_2\right) \tag{E8.6.3}$$

where

$$\Phi = x_1\, x_2 - 0.25 = 0 \tag{E8.6.4}$$

The optimum occurs at

$$\partial Y/\partial x_1 = 0, \partial Y/\partial x_2 = 0, \text{and } \partial Y/\partial \lambda = 0 \tag{E8.6.5}$$

These are

$$\partial y/\partial x_1 - \lambda\, \partial\phi/\partial x_1 = 0 \tag{E8.6.6}$$

$$\partial y/\partial x_2 - \lambda\, \partial\phi/\partial x_2 = 0 \tag{E8.6.7}$$

and

$$\Phi\left(x_1, x_2\right) = 0 \tag{E8.6.8}$$

Numerically,

$$1 + \lambda\, x_2 = 0, \text{or}, \lambda = -1/x_2 \tag{E8.6.9}$$

$$1 + \lambda\, x_1 = 0, \text{or}, \lambda = -1/x_1 \tag{E8.6.10}$$

and

$$x_1\, x_2 = 0.25 \tag{E8.6.11}$$

From Eqs. (E8.6.9) and (E8.6.10), we see that

$$x_1 = x_2.$$

Substituting this into Eq (E8.6.11), we get

$$x_2^2 = 0.25$$

which gives

$$x_2 = \pm 0.5.$$

If $x_2 = -0.5$, from Eq (E8.6.9), we have $\lambda = 2$ and $y = -1$. This is not possible, as we are all entropy generators, i.e. no human can decrease entropy. If $x_2 = 0.5$, from Eq (E8.6.9), we have $\lambda = -2$ and $y = 1$, i.e. the minimum entropy that the class generates.

Example 8.6 is schematically plotted in Figure 8.11. The declining (to the right) straight lines in the x_2 versus x_1 plane signify constant-y lines. The constraint is illustrated as the curved, dotted line. The point where the constraint intersects with the $y = 1$ line is the minimum. Note that, at this optimum point, the tangent of the $y = 1$ line is perpendicular to the gradient of y, which is also the gradient of the constraint.

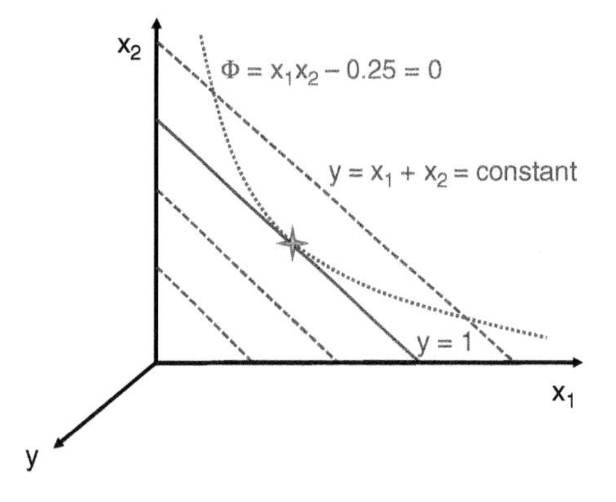

Figure 8.11 Optimizing $y = x_1 + x_2$, constrained by $x_1x_2 = 0.25$, using the Lagrange Multiplier method. Source: D. Ting.

Example 8.7 ***Diminishing intelligence quotient***

Given

The intelligence quotient of an average person diminishes at a rate,

$$y = x_1^2 + x_2^2 \quad \text{(E8.7.1)}$$

where x_1 is the lack of sleep, and x_2 is the lack of brain exercise. The sleep-brain exercise relationship is

$$x_1\, x_2^2 = 10 \quad \text{(E8.7.2)}$$

Find

The optimum y, using the Lagrange Multiplier method.

Solution

The Lagrange expression,

$$Y\left(x_1, x_2, \lambda\right) = y\left(x_1, x_2\right) - \lambda\, \Phi\left(x_1, x_2\right) \quad \text{(E8.7.3)}$$

The optimum occurs at

$$\partial Y/\partial x_1 = 0, \partial Y/\partial x_2 = 0, \text{and } \partial Y/\partial \lambda = 0 \quad \text{(E8.7.4)}$$

These can be expanded as

$$\partial y/\partial x_1 - \lambda\, \partial \Phi/\partial x_1 = 0 \quad \text{(E8.7.5)}$$

$$\partial y/\partial x_2 - \lambda\, \partial \Phi/\partial x_2 = 0 \quad \text{(E8.7.6)}$$

and

$$\Phi\left(x_1, x_2\right) = 0 \quad \text{(E8.7.7)}$$

Equations (E8.7.5) and (E8.7.6) can be expressed in vector notation,

$$\nabla y - \lambda\, \nabla \Phi = 0 \quad \text{(E8.7.8)}$$

where

$$\nabla \Phi = \frac{\partial \Phi}{\partial x_1}\hat{i} + \frac{\partial \Phi}{\partial x_2}\hat{j} \quad \text{(E8.7.9)}$$

The constraint,

$$\Phi = x_1\, x_2^2 - 10 \quad \text{(E8.7.10)}$$

Taking the gradient of the constraint,

$$\nabla \Phi = \left(\partial \Phi/\partial x_1\right)\, \mathbf{i}_1 + \left(\partial \Phi/\partial x_2\right)\, \mathbf{i}_2 = x_2^2\, \mathbf{i}_1 + 2x_1 x_2\, \mathbf{i}_2 \quad \text{(E8.7.11)}$$

keeping in mind that a hatted or bold i signifies a unit vector.

Taking the gradient of the objective function, Eq (E8.7.1), gives

$$\nabla y = \left(\partial y/\partial x_1\right)\, \mathbf{i}_1 + \left(\partial y/\partial x_2\right)\, \mathbf{i}_2 = 2x_1\, \mathbf{i}_1 + 2x_2\, \mathbf{i}_2 \quad \text{(E8.7.12)}$$

Substituting this, and Eq. (E8.7.11), into Eq. (E8.7.8) gives

$$\nabla y = \left[2x_1 - \lambda\left(x_2^2\right)\right]\, \mathbf{i}_1 + \left[2x_2 - \lambda\left(2x_1 x_2\right)\right]\, \mathbf{i}_2 = 0 \quad \text{(E8.7.13)}$$

Breaking this into the corresponding two components,

$$\mathbf{i}_1 : 2x_1 - \lambda\left(x_2^2\right) = 0 \tag{E8.7.14}$$

and

$$\mathbf{i}_2 : 2x_2 - \lambda\left(2x_1x_2\right) = 0 \tag{E8.7.15}$$

The third equation is the constraint, i.e. Eq (E8.7.10).

Thus, we have three equations, Eqs. (E8.7.14), (E8.7.15), and (E8.7.10), and three unknowns, x_1^*, x_2^*, and λ. From Eq (E8.7.14), we have

$$\lambda = 2x_1/x_2^2 \tag{E8.7.14a}$$

Substituting this into Eq. (E8.7.15) gives

$$x_1 = x_2/\sqrt{2} \tag{E8.7.15a}$$

Substituting this into the constraint equation, Eq (E8.7.2), results in

$$x_2^* = 2.418$$

With this, according to Eq (E8.7.15a),

$$x_1^* = 1.710.$$

Substituting these into the objective function, Eq (E8.7.1), leads to

$$y^* = 8.772.$$

We note that this optimum is a minimum. This can be confirmed by simply substituting a slightly different x_1 value. For example, if $x_1 = 1.8$ instead, then, according to Eq (E8.7.15a), $x_2 = 2.546$. These x values lead to y = 9.72, which is larger than y*.

Also, substituting x_1^* and x_2^* into Eq (E8.7.14a) leads to

$$\lambda = 0.585$$

Figure 8.12 shows the x_2 versus x_1 plot with constant-y lines for the optimization of $y(x_1, x_2)$ subjected to a single constraint, such as that described in Example 8.7. The dotted curve describes the constraint. It is clear that Point O is the minimum y along the constraint curve. We can arrive at this optimal point by traveling along the constraint curve until an extremum of y is reached. Incidentally, this extremum is the point where the constraint and the constant-y line are parallel to each other. In other words, the extremum occurs when the tangent vector of the constraint curve lines up with, and points in the same direction as the tangent of the constant-y line. Mathematically, as the gradient vector is normal to a contour line, this is satisfied when

$$\nabla y - \lambda\, \nabla \Phi = 0 \tag{8.23}$$

This can be rearranged into

$$\lambda = -\left(\nabla y/\nabla \Phi\right)^* \tag{8.24}$$

where the asterisk denotes "at the optimum." As such, the Lagrange Multiplier, λ, is used to account for the differing magnitudes of the two gradient vectors and the possible opposing directions the two vectors point.

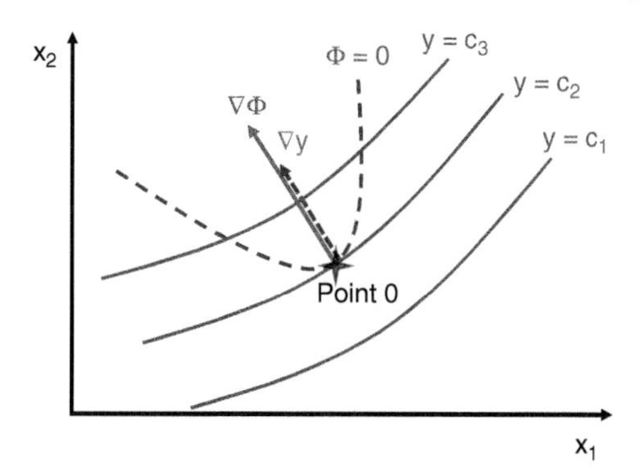

Figure 8.12 Objective function, $y = f(x_1, x_2)$, constrained by $\Phi(x_1, x_2) = 0$. Source: D. Ting.

8.6 The Lagrange Multiplier as a Sensitivity Coefficient

We see that the Lagrange Multiplier, λ, takes care of any difference in the magnitude of the gradient vector of the objective function and that (gradient vector) associated with the constraint. This fascinating Lagrange Multiplier serves one additional purpose, i.e. it indicates the change in the objective function value with respect to the relaxation of the constraint. Let us use a two-variable, single-constraint example to illustrate this.

Example 8.8 ***Lagrange Multiplier as a sensitivity coefficient***

Given

The objective function, the cost, in thousands of dollars, of a solar desalination project

$$y = 7\,x_1 + \tfrac{1}{2}\,x_2^2 \tag{E8.8.1}$$

where x_1 is the number of collector units and x_2 is the radius of the storage tank in meters. The space (length in m) available is limited such that

$$x_1 + x_2 = 9 \tag{E8.8.2}$$

Find

The optimum y, using the Lagrange Multiplier method. What if we relax the space limitation to $x_1 + x_2 = 9.1$?

Solution

The Lagrange expression,

$$Y\left(x_1, x_2, \lambda\right) = y\left(x_1, x_2\right) - \lambda\,\Phi\left(x_1, x_2\right) \tag{E8.8.3}$$

The optimum occurs at

$$\partial Y/\partial x_1 = 0, \partial Y/\partial x_x = 0, \text{and } \partial Y/\partial \lambda = 0 \tag{E8.8.4}$$

These are

$$\partial y/\partial x_1 - \lambda\,\partial\Phi/\partial x_1 = 0 \tag{E8.8.5}$$

$$\partial y/\partial x_2 - \lambda\,\partial\Phi/\partial x_2 = 0 \tag{E8.8.6}$$

and

$$\Phi\left(x_1, x_2\right) = 0 \quad \text{(E8.8.7)}$$

Equations (E8.8.5) and (E8.8.6) can be expressed in vector notation,

$$\nabla y - \lambda\, \nabla\Phi = 0 \quad \text{(E8.8.8)}$$

where

$$\nabla\Phi = \frac{\partial\Phi}{\partial x_1}\hat{i} + \frac{\partial\Phi}{\partial x_2}\hat{j} \quad \text{(E8.8.9)}$$

The constraint,

$$\Phi = x_1 + x_2 - 9 \quad \text{(E8.8.10)}$$

Taking the gradient of the constraint,

$$\nabla\Phi = \left(\partial\Phi/\partial x_1\right)\, \mathbf{i}_1 + \left(\partial\Phi/\partial x_2\right)\, \mathbf{i}_2 = 1\,\mathbf{i}_1 + 1\,\mathbf{i}_2 \quad \text{(E8.8.11)}$$

Taking the gradient of the objective function, Eq. (E8.8.1) gives

$$\nabla y = \left(\partial y/\partial x_1\right)\, \mathbf{i}_1 + \left(\partial y/\partial x_2\right)\, \mathbf{i}_2 = 7\,\mathbf{i}_1 + x_2\,\mathbf{i}_2 \quad \text{(E8.8.12)}$$

Substituting these two equations into Eq (E8.8.8), we get

$$[7 - \lambda(1)]\ \mathbf{i}_1 + \left[x_2 - \lambda(1)\right]\ \mathbf{i}_2 = 0 \quad \text{(E8.8.13)}$$

Breaking this into the corresponding two components,

$$\mathbf{i}_1: \qquad 7 - \lambda = 0 \quad \text{(E8.8.14)}$$

and

$$\mathbf{i}_2: \qquad x_2 - \lambda = 0 \quad \text{(E8.8.15)}$$

From Eq (E8.8.14), we have

$$\lambda = 7. \quad \text{(E8.8.14a)}$$

Substituting this into Eq (E8.8.15) gives

$$x_2 = 7. \quad \text{(E8.8.15a)}$$

That is, $x_2^* = 7$. Substituting this into the constraint equation, Eq (E8.8.2), gives $x_1^* = 2$. Substituting these into the objective function, Eq (E8.8.1), leads to $y^* = 38.5$.

The sensitivity coefficient, from Eq (E8.8.8) and Eq (E8.8.14a),

$$S_C = \lambda = \nabla y/\nabla\phi = \partial y/\partial\Phi = 7 \quad \text{(E8.8.16)}$$

Therefore, an extra 0.1 m of space (length) will increase the cost to 38.5 + 7(0.1) = 39.2. In other words, the cost increases from \$38.5k to \$39.2k, when increasing the space length from 9 to 9.1 m.

Any relaxation of the constraint, such as that illustrated in Example 8.8, can be accounted for by going through the entire calculation again with the new, more relaxed constraint. With the Lagrange Multiplier as the *sensitivity coefficient*, this tedious re-calculation is not needed. Designating the Lagrange Multiplier as the sensitivity coefficient is most appropriate; it illustrates the deviation from the optimal value when the constraint is slightly altered. This relaxation of a constraint is often practically feasible; for example, you can oftentimes convince your boss to up the budget slightly for a much better system.

As the sensitivity coefficient depicts how much the objective function value varies with a small adjustment in the constraint, each constraint has a unique sensitivity coefficient. For the optimization of an objective function confined by m constraints, we have m sensitivity coefficients. In other words, the sensitivity coefficient,

$$S_C = \lambda = \nabla y/\nabla\Phi = \partial y/\partial\Phi, \tag{8.25}$$

or, the sensitivity coefficients,

$$\begin{aligned} S_{C1} &= \lambda_1 = \nabla y/\nabla\Phi_1 = \partial y/\partial\Phi_{1,}\ S_{C2} = \lambda_2 \\ &= \nabla y/\nabla\Phi_2 = \partial y/\partial\Phi_{2,} \ldots S_{Cm} = \lambda_m = \nabla y/\nabla\Phi_m = \partial y/\partial\Phi_m. \end{aligned} \tag{8.26}$$

8.7 Dealing with Inequality Constraints

Quite frequently, the constraints have an upper or lower limit, making them inequality constraints. As such, it prevents the direct utilization of the Lagrange Multiplier, which only works when constraints are equality constraints. Fortunately, as we have mentioned earlier, many practical inequality constraints can be turned into equality ones with some common sense and physical understanding of the problem at hand. Operation temperature and pressure and/or material strength typically set an upper limit to the thermofluid system to be designed. On the other hand, the minimum required cooling rate and process time, etc. set a lower limit.

A different way to deal with inequality constraints is to simply proceed with the optimization process as if the problem is unconstrained. Upon reaching the optimum, the upper or lower limits imposed by the inequality constraints are checked. If no constraint is violated, the optimal solution is, thus, the final answer to the problem at hand. If one or more constraints are violated, then the optimization process is repeated with some reasonable values for the involved limits. As such, the inequality constraints are treated as equality constraints, allowing the execution of the Lagrange Multiplier. Once the corresponding optimum is achieved, the procedure can be repeated for other reasonable values for the limits. This iterative approach will converge timely, when proper rationale is employed in choosing the values for the limits. Another approach to deal with inequality constraints is conveyed next.

Consider the multi-independent-variable objective function $y(x_1, x_2, \ldots x_n)$ subjected to an inequality constraint,

$$\Psi\left(x_1, x_2, \ldots x_n\right) \leq 0 \tag{8.27}$$

We can convert the inequality constraint into a equality one by adding a positive quantity $x_{n+1}{}^2$, where $x_{n+1}{}^2$ is called a *slack variable* (Burmeister, 1998). That is, the constraint can be expressed as

$$\Phi\left(x_1, x_2, \ldots x_n, x_{n+1}\right) = \Psi\left(x_1, x_2, \ldots x_n\right) + x_{n+1}{}^2 = 0 \tag{8.28}$$

Consequently, we have the Lagrange expression

$$Y\left(x_1, x_2, \ldots x_n, x_{n+1}, \lambda\right) = y\left(x_1, x_2, \ldots x_n\right) - \lambda\,\Phi\left(x_1, x_2, \ldots x_n, x_{n+1}\right) \tag{8.29}$$

The optimum occurs at

$$\partial Y/\partial x_1 = 0, \ldots \partial Y/\partial x_{n+1} = 0, \ldots \partial Y/\partial\lambda = 0 \tag{8.30}$$

These partial differentiations can be cast as

$$\partial y/\partial x_1 - \lambda\, \partial \Phi/\partial x_1 = 0,$$
$$\partial y/\partial x_2 - \lambda\, \partial \Phi/\partial x_2 = 0,$$
$$\vdots$$
$$\partial y/\partial x_n - \lambda\, \partial \Phi/\partial x_n = 0,$$
$$\partial y/\partial x_{n+1} - 2\, \lambda\, \partial \Phi/\partial x_{n+1} = 0 \tag{8.31}$$

These n+1 equations, along with the equality constraint equation with the slack variable, Eq (8.28), sum up to n+2 equations needed for the n+2 unknowns, n+1 x's and λ. The "2" in front of the Lagrange Multiplier in Eq (8.31) is due to the square associated with the slack variable, i.e. x_{n+1}^2.

We can see that

$$\partial y/\partial x_{n+1} = 0 \tag{8.32}$$

because y is not a function of x_{n+1}. Substituting this into Eq (8.31), we have

$$2\, \lambda\, \partial \Phi/\partial x_{n+1} = 0 \tag{8.33}$$

Substituting for Φ from Eq (8.28), this becomes

$$2\, \lambda\, \partial \Phi/\partial x_{n+1} = 2\lambda\, \partial \left[\Psi\left(x_1, x_2, \ldots x_n\right) + x_{n+1}^{2}\right] \partial x_{n+1} = 0 + 4\lambda x_{n+1} = 0 \tag{8.34}$$

As such, either λ and/or x_{n+1} are/is zero. If the Lagrange Multiplier, λ, is zero, this implies that the constraint is inactive, i.e. the neighborhood under consideration is way within the constrained region. If the (square root of the) slack variable x_{n+1} is zero, it means that the constraint is in effect, i.e. the optimization operation proceeds along the constraint.

Also, from Eq (8.30) and Eq (8.29), $\partial Y/\partial\lambda = 0$, i.e.

$$\partial Y/\partial \lambda = \partial \left[y\left(x_1, x_2, \ldots x_n\right) - \lambda \Phi\left(x_1, x_2, \ldots x_n, x_{n+1}\right)\right] / \partial \lambda = 0 \tag{8.35}$$

But $\Phi = \Psi(x_1, x_2, \ldots x_n) + x_{n+1}^2$, Eq (8.28), and thus,

$$\partial Y/\partial \lambda = 0 - \partial\left[\lambda \Phi\right]/\partial \lambda = \partial\left\{\lambda[\Psi\left(x_1, x_2, \ldots x_n\right) + x_{n+1}^{2}\right\}/\partial \lambda = 0 \tag{8.35a}$$

This gives

$$\Psi\left(x_1, x_2, \ldots x_n\right) + x_{n+1}^{2} = 0 \tag{8.35b}$$

Thus, if x_{n+1} is zero, $\Psi = 0$, and the constraint is in effect, i.e. satisfied.

Problems

8.1 Understand the Lagrange Multiplier method

True/False statements concerning the Lagrange Multiplier method.

Part 1 The objective function must be unimodal.	True	False
Part 2 The objective function has to be continuous and differentiable.	True	False
Part 3 It can be applied for multi-variable problems.	True	False
Part 4 It can be applied to both constrained and unconstrained problems.	True	False

8.2 Minimizing exergy destruction

The exergy destruction of a thermofluid system is a function of the temperature and flow rate of the working fluid. Specifically, the non-dimensional exergy destruction, $y = 2x_1^2x_2 + 3x_1/x_2^2 + 2/x_1$, where x_1 is the dimension-less temperature and x_2 is the dimension-less flow rate. Minimize the exergy destruction. Note that this question is posed as a two-variable constrained problem. It is, however, much easier to substitute the constraint into the objective function, and solve it as a single-variable unconstrained problem.

8.3 A two-variable unconstrained problem

The cost per unit mass of material processed in an extrusion facility,

$$y = 4\,T^2\forall' + 2\,T/\forall'^2 + 5/T$$

where T is the dimension-less temperature and ∀' is the dimension-less volume flow rate of the material being processed. Find the minimum cost. Use a simple and illustrative plot to depict the solution.

8.4 A three-variable unconstrained problem

The objective function,

$$y = \frac{x_1}{x_2} + \frac{4}{x_1x_3} + \frac{1}{3}x_2^2 + \frac{x_3}{9}$$

Find the optimum y using the Lagrange Multiplier method. Use simple (surface) plots to show the result.

8.5 Power plant water usage as a two-variable constrained problem

The water usage of a power plant, $y = x_1 + x_2$, where x_1 is the ambient temperature and x_2 is the power demand, both in normalized form. The power demand is inversely proportioned to the ambient temperature, i.e. $x_2 = 1/x_1$. Minimize the water usage.

8.6 Minimizing heat loss from a thermoelectric energy generator: solve it as a two-variable constrained problem versus a single-variable unconstrained problem

The heat loss from a thermoelectric energy generator (TEG) power source with L by L sides is to be minimized. The heat transfer coefficient, $h = (2 + 10L^{1/2})\Delta T^{1/4}L^{-1}$, where ΔT is the temperature difference between the TEG and the surroundings. To prevent the bond of the TEG from detaching, $L\Delta T = 5.6$. Minimize the heat loss using the Lagrange Multiplier method. Solve it as a constrained problem and then as an unconstrained problem by substituting the constraint into the objective function.

8.7 Maximize a rectangular duct within an equilateral triangle

A rectangular duct of width W and height H is to be placed in an existing 1 m-side equilateral triangular supporting frame. The cross-sectional area of the duct is to be maximized. Use the Lagrange Multiplier for constrained optimization to determine the optimal width and height.

8.8 Utilize the Lagrange Multiplier for a two-variable constrained problem

Optimize the objective function, $y = 400 + 160x_1 + 80x_2$, constrained by $x_1\,x_2 = 10$, where x_1 and x_2 are positive, by using the Lagrange Multiplier method for a constrained problem. Is the optimum a minimum, a maximum, or an inflection point? Without recalculation (use the Lagrange Multiplier), if the constraint is 10.5 instead, i.e. increase by 5%, what is the optimum y?

8.9 The cost of a heat exchanger as a two-variable constrained optimization problem
A shell-and-tube heat exchanger with a total length of 110 m of tubes costs \$980. The cost of the shell is 700 D^2 L, where D is the shell diameter in meters and L is the length of the shell in meters. The floor space costs 250 D L. A total of 120 tubes, covering a cross-sectional area of 1 m^2 of the shell, is required. Minimize the capital cost by invoking the Lagrange Multiplier for a constrained problem. Use the Lagrange Multiplier to deduce the cost, if 110, instead of 120, tubes are enough.

8.10 Heat loss of a cylindrical drum
A cylindrical container, a drum, is to be designed for a steam boiler unit. The area/volume ratio is to be minimized to reduce heat losses.
Part 1 If the drum volume is not to exceed 2 m^3, express the objective function and the constraint(s) in terms of the diameter and length of the drum.
Part 2 Use the Lagrange Multiplier for constrained optimization to solve for the optimum diameter and length of the drum.
Part 3 If the maximum volume is 2.1 m^3, find the optimum area/volume ratio based on the solution for Part 2 above (that is, without re-solving the problem).

References

Burmeister, L.C. (1998). *Elements of Thermal-Fluid System Design*. Upper Saddle River, NJ: Prentice-Hall.

Ebrahimi, M., Ting, D.S-K., Carriveau, R., and McGillis, A. (2021). Hydrostatically compensated energy storage technology. In *Green Energy and Infrastructure: Securing a Sustainable Future* (ed. J.A. Stagner and D.S-K. Ting). Boca Raton, FL: CRC Press.

Jaluria, Y. (1998). *Design and Optimization of Thermal Systems*. New York: McGraw-Hill.

Klionsky, D.J. (2017). Education is the only business where the customer is satisfied with less of the product. *Journal of Microbiology & Biology Education* 18(2): 1–2.

Lagrange (n.d.) In *Encyclopedia Britannica*, available at: https://www.britannica.com/biography/Joseph-Louis-Lagrange-comte-de-lEmpire, accessed November 10, 2018.

Stoecker, W.F. (1989). *Design of Thermal Systems*, 3rd ed. New York: McGraw-Hill.

9

Search Methods

Searching and learning is where the miracle process all begins.

—Jim Rohn

Chapter Objectives

- Recognize the common types of search methods.
- Become familiar with the exhausting, idiot-proof, elimination method, the Exhaustive Search.
- Learn to apply the effective elimination methods: the Dichotomous Search, the Fibonacci Search, and the Golden Section Search.
- Differentiate multi-variable optimization from single-variable optimization.
- Appreciate the Lattice Search, the Univariate Search, and the Steepest Ascent/Descent methods for unconstrained, multi-variable problems.
- Master the Penalty Function and Search-Along-A-Constraint (Hemstitching) methods for constrained multi-variable problems.

Nomenclature

A	area
D	diameter
d	distance
F	Fibonacci series; $F_o = F_1 = 1, F_2 = 2, F_3 = 3, F_4 = 5, F_5 = 8$
h	height, heat transfer coefficient
L	limit of the constraint
I	interval of uncertainty; I_f is the final interval of uncertainty, I_i is the initial interval of uncertainty
n	number of points/calculations
P	penalty
Q	heat; Q' is heat transfer rate
r	radius
R	reduction ratio; initial-final uncertainty interval ratio
T	temperature
X, x	independent variable
Y, y	dependent variable, objective function, composite objective function

Greek and Special Symbols

α an independent variable
β an independent variable
Δ a small change
δ a small change
Φ equality constraint, $\Phi = 0$
φ equality constraint, $\varphi = L$
∀ volume

9.1 Introduction

There are basically three types of optimization approaches; see, for example, (Burmeister, 1998; Juleria, 1998; Stoecker, 1989). These, as summarized in Table 9.1, are: (i) elimination methods, (ii) hill-climbing techniques, and (iii) constrained optimization. Under the elimination type, the most common methods are: (i) the Exhaustive Search, (ii) the Dichotomous Search, (iii) the Fibonacci Search, and (iv) the Golden Section Search. Lattice Search, Univariate Search and Steepest Ascent/Descent are familiar hill-climbing procedures. For constrained optimization, the Penalty Function method and Search-Along-a-Constraint are versatile.

Table 9.1 Basic types of optimization approaches.

Elimination methods	Hill-climbing techniques	Constrained optimization
Exhaustive Search	Lattice Search	Penalty Function Method
Dichotomous Search	Univariate Search	Search-Along-a-Constraint
Fibonacci Search	Steepest Ascent/Descent	
Golden Section Search		

Another way to classify the approaches is to base the approach on the number of variables and whether the problem is constrained or not. As categorized in Table 9.2, there are: (i) single-variable, (ii) multi-variable unconstrained, and (iii) multi-variable constrained classes.

Table 9.2 Classification of optimizations techniques.

Single variable	Multi-variable unconstrained	Multi-variable constrained
Exhaustive	Lattice	Penalty Functions
Efficient	Univariate	Search-Along-a-Constraint
i) Dichotomous ii) Fibonacci	Steepest Ascent/Descent	

We will follow the sorting of Table 9.1. Elimination methods will be expounded first, followed by the unconstrained, multi-variable technique. The constrained, multi-variable approach will be discussed last.

9.2 Elimination Methods

The most familiar elimination methods are: (i) Exhaustive Search, (ii) Dichotomous Search, (iii) the Fibonacci Search, and (iv) the Golden Section Search. We will use the generic, single-variable case, y = y(x), to illustrate the workings of these methods. In practice, the range over which the independent variable can be varied is typically known a priori. Accordingly, we have the initial interval of uncertainty, I_i, as depicted in Figure 9.1, where a minimum is sought. The idea is to narrow the interval of uncertainty into an acceptable range, the final interval of uncertainty, I_f, within which the optimum occurs.

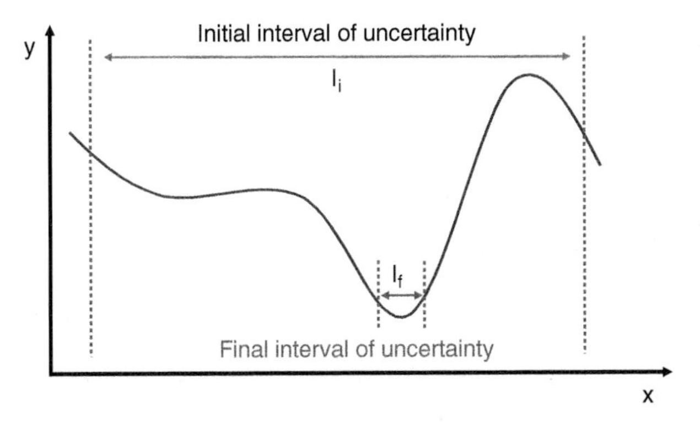

Figure 9.1 Narrowing the solution from the initial interval of uncertainty, I_i, to the final interval of uncertainty, I_f, for the case where a minimum is sought. Source: D. Ting.

9.2.1 Exhaustive Search

This bullet-proof, idiot-friendly method, Exhaustive Search, compares objective function, y, values calculated at uniformly-spaced independent variable x. Since the method calculates the value of y at x, spaced uniformly throughout the entire interval of interest, it is also called *Uniform Exhaustive Search*. Suppose the efficiency of a pump varies with the flow rate, as shown in Figure 9.2, where the range of flow rate of interest is between 0.5 and 5.5 L/s. This defines the initial interval of uncertainty, I_i, within which some optimum is sought. The aim is to find an operating range over which the pump is running around its maximum efficiency. This desired range is the final interval of uncertainty, I_f.

Figure 9.2 Pump efficiency, y, as a function of flow rate, x. Source: D. Ting.

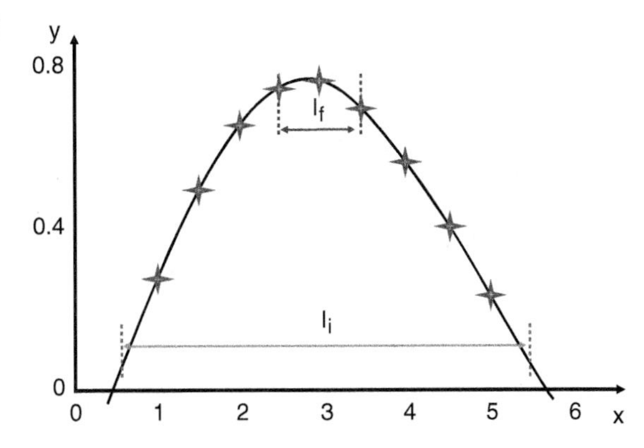

The relation between the final and initial intervals of uncertainty is

$$I_f = 2\ I_i/(n+1) \tag{9.1}$$

where n is the number of calculation points between the initial interval of uncertainty (excluding the boundaries). For an exhaustive search, n is also the number of trial runs needed to reach the specified interval of uncertainty. There are n + 1 uniformly-spaced subdivisions, in the independent variable, x, domain. Because the subdivisions are uniformly spaced, the Exhaustive Search is also called the "Uniform Exhaustive Search." For the pump efficiency example shown in Figure 9.2, the initial interval of uncertainty, I_i, is 5.5–0.5, or 5 L/s. If the desired final interval of uncertainty, I_f, is 1 L/s, then, according to Eq (9.1), the number of points is 9. As can be seen in Figure 9.2, there are three points within the final interval of uncertainty. If we do not know the curve, there is no way of telling if the optimum is to the left or right of $x = 3$, or on $x = 3$. Therefore, we are only certain that the optimum of the objective function, y, is bounded between $x = 2.5$ and 3.5.

We can define the *Reduction Ratio* as the reduction in the interval of uncertainty, that is,

$$R = I_i/I_f = (n+1)/2 \tag{9.2}$$

For the aforementioned pump efficiency example shown in Figure 9.2, n is equal to 9 and, hence, the reduction ratio, $R = 5$. In plain English, the uncertainty interval is reduced by a factor of five after executing nine calculations.

It is clear that this idiot-proof method is not efficient, as it has to run through the entire range of the initial uncertainty interval, i.e. range of possible independent variable values. Nevertheless, it is useful as a first estimate when the basic trend of the objective function is not known. In other words, sparsely-spaced calculations can be performed to have an idea of how y varies with x. Once the overall trend is known, a more robust search method can be invoked to achieve the desired optimum within a fine interval of uncertainty.

Example 9.1 ***Minimize heat transfer area of a hot water tank***

Given

A cylindrical storage tank is to be designed for storing hot water for a solar energy system. The required volume is $2\ m^3$. The surface area is to be minimized to minimize heat loss. Based on the available information, the radius, r, should be between 0.1 m and 2.1 m.

Find

Dimensions of the tank using the Exhaustive Search method, i.e. as an unconstrained single-variable problem. The final uncertainty interval of the radius, I_f, should be no more than 0.2 m.

Solution

The volume of the cylindrical tank, as sketched in Figure 9.3, is

$$\forall = \pi r^2 h = 2\ m^3 \tag{E9.1.1}$$

where h is the height of the tank. This can be rewritten as

$$h = 2/\left(\pi r^2\right) \tag{E9.1.2}$$

The surface area of the tank,

$$A = 2\pi r^2 + 2\pi r h = 2\pi r^2 + 2\pi r\left[2/\left(\pi r^2\right)\right] = 2\pi r^2 + 4/r \tag{E9.1.3}$$

With r between 0.1 and 2.1 m, the initial interval of uncertainty,

$$I_i = 2.1 - 0.1 = 2\ m$$

The desired I_f is 0.2. Invoking Eq (9.1), we have

$$0.2 = 2(2)/(n+1)$$

which gives n = 19. Namely, we need 19 calculations, i.e. at r = 0.2, 0.3, … 1.9, 2.0. The reduction ratio, according to Eq (9.2), is

$$R = (19 + 1)/2 = 10$$

Figure 9.3 A cylindrical solar hot water storage tank, where r is the radius and h is the height. Source: D. Ting.

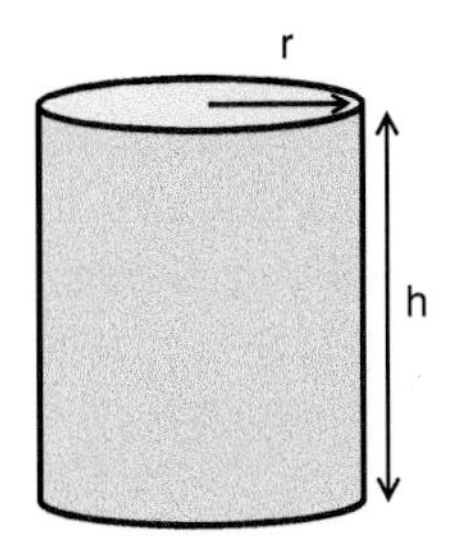

The calculated values are plotted in Figure 9.4. It is clear that the optimum (minimum) y value is found between r equals to 0.6 (y = 8.929) and 0.8 (y = 9.021). In practice, one could choose the midpoint, i.e. r = 0.7 m, where y = 8.793 m^2.

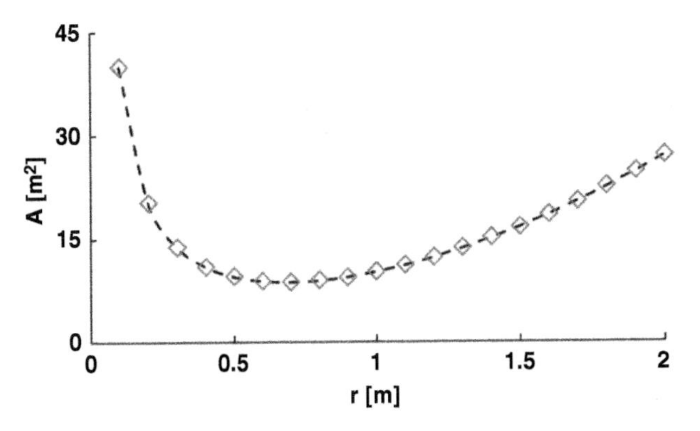

Figure 9.4 Exhaustive search of minimum area cylindrical tank. Source: X. Wang.

For this simple, single-modal, class-illustration problem, we could have employed the calculus method, i.e.

$$dA/dr = 4\pi r - 4/r^2 = 0 \tag{E9.1.4}$$

This gives, $r^* = 0.6828$ m and, thus, $y^* = A^* = 8.788$ m^2. This confirms that our solution obtained via the exhaustive search method is sound.

Note that Example 9.1 is a unimodal case, where there is only one optimum. For this reason, and the fact that the function is continuous, the calculus approach can easily be applied to find the solution. It is important to stress that the Exhaustive Search method is versatile, that is, it is not restricted to unimodal (nor continuous function) problems. In other words, Exhaustive Search can be employed to locate the optimum of multi-modal problems. It enables the comparison of multiple local optima to furnish the global optimum. This is, however, not the case for many other search methods, including those that are discussed below.

9.2.2 Dichotomous Search

A Dichotomous Search is restricted to unimodal functions only. The calculations are carried out in pairs, to see if the objective function is increasing or decreasing to the right or the left of the pair. By comparing the values of the objective function of the pair, one side of the (remaining) calculation domain is eliminated. It is thus clear that, when properly implemented, this method requires many fewer calculations compared to the Exhaustive Search method. Before discussing the bona fide dichotomous search, let us look at the uniform Dichotomous Search.

Uniform Dichotomous Search

Evenly-spaced pairs of runs are put into effect in Uniform Dichotomous Search. The spacing between each pair of points is adequately small, around an order of magnitude less than the desired final interval, so that the obtained solution satisfies the needed resolution without missing any information between the pair, within the spacing. At the same time, this spacing has to be adequate to give differentiable values of the objective function pair, to grant a clear indication of which side of the pair to discard.

The initial interval of uncertainty is divided into $n/2 + 1$ sub-intervals, where n is the number of points or number of runs. The resulting reduction ratio,

$$R = I_i/I_f = (n/2 + 1)/1 = n/2 + 1 \tag{9.3}$$

neglecting ε, the small spacing between the pair of points. Let us revisit Example 9.1, solving it using the uniform dichotomous search this time. From this reduction ratio-number of calculation relation, we observe that this method is only slightly faster than exhaustive search.

Example 9.2 ***Minimize a cylindrical tank surface area using uniform dichotomous search***

Given

A cylindrical storage tank is to be designed for storing hot water for a solar energy system. The required volume is $2\ m^3$. The surface area is to be minimized to minimize heat loss. Based on the available information, the radius, r, should be between 0.1 m and 2.1 m.

Find

Dimensions of the tank using the Uniform Dichotomous Search method, i.e. solve it as an unconstrained single-variable problem. The final uncertainty interval of the radius, I_f, should be no more than 0.2 m.

Solution

From Example 9.1, the surface area of the tank,

$$A = 2\pi r^2 + 4/r$$

With r between 0.1 and 2.1 m, the initial interval of uncertainty,

$$I_i = 2.1 - 0.1 = 2$$

The desired I_f is 0.2. Invoking Eq (9.3), we have

$$R = I_i/I_f = 2/0.2 = 10 = n/2 + 1$$

and thus, n = 18, i.e. 9 pairs. This is only one calculation less than that based on Exhaustive Search. It is thus clear that executing the dichotomous method in this uniform manner defeats its purpose.

The calculations are depicted in Figure 9.5, where the spacing of the pair, δ or Δr, is 0.005 m. The optimum (minimum) y is between r equals to 0.505 (y = 9.523) and 0.7 (y = 8.793).

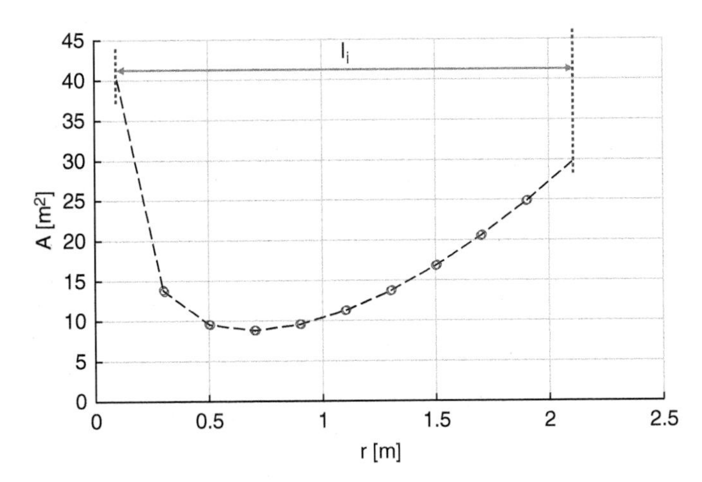

Figure 9.5 Uniform dichotomous search of minimum area cylindrical tank, where a spacing, δ or Δr, of 0.005 m has been employed. Source: Y. Yang.

This final interval of uncertainty, I_f, is 0.195 m, that is, we have met the requirement.

Sequential Dichotomous Search

It is clear that the evenly-spaced dichotomous search method defeats the true efficiency of Dichotomous Search. Strictly speaking, the uniform dichotomous method is not used in practice. Authentic Dichotomous Search uses the information gained from one pair of runs to choose the next pair and, thus, it is also referred to as Sequential Dichotomous Search. The search starts by locating the first pair near the middle of the range. With the first pair of calculations, half of the initial range is eliminated. It should be stressed that this only works for unimodal problems.

The second pair is located at the middle of the reduced range, resulting in further halving. In other words, with each pair, the region of uncertainty is halved. It is this repeated halving that makes this method efficient. Another advantage is that the total number of runs does not need to be fixed or decided a priori. As such, you can proceed with the calculation and stop whenever the remaining uncertainty interval is acceptable. If further refinement beyond the first round of calculations is needed, the refining simply starts from where you have left off, i.e. no recalculation from the beginning is needed.

Due to the sequential halving, the reduction ratio is an exponential function of the number of calculations, n. Expressly,

$$R = I_i/I_f = 2^{n/2} \tag{9.4}$$

Let us appreciate the effectiveness of dichotomous search by solving the same cylindrical tank surface minimization problem.

Example 9.3 ***Cylindrical tank surface area minimization using the Sequential Dichotomous Search method***

Given

A cylindrical storage tank is to be designed for storing hot water for a solar energy system. The required volume is 2 m^3. The surface area is to be minimized to minimize heat loss. Based on the available information, the radius, r, should be between 0.1 m and 2.1 m.

Find

Dimensions of the tank using Sequential Dichotomous Search method, i.e. solve it as an unconstrained single-variable problem. The final uncertainty interval of the radius, I_f, should be no more than 0.2 m.

Solution

From Example 9.1, the surface area of the tank is,

$$A = 2\pi r^2 + 4/r$$

With r between 0.1 and 2.1 m, the initial interval of uncertainty is,

$$I_i = 2.1 - 0.1 = 2$$

The desired I_f is 0.2. According to Eq (9.4),

$$R = I_i/I_f = 2/0.2 = 10 = 2^{n/2}$$

from which we get n = 6.644. The next closest even number is 8, or 4 pairs.

As shown in Figure 9.6, we start the first pair at the middle of the initial interval of uncertainty, where the spacing of the pair, δ or Δr, is 0.01 m. Specifically,

$$y\,(r = 1.05) = 10.74, y\,(r = 1.06) = 10.83$$

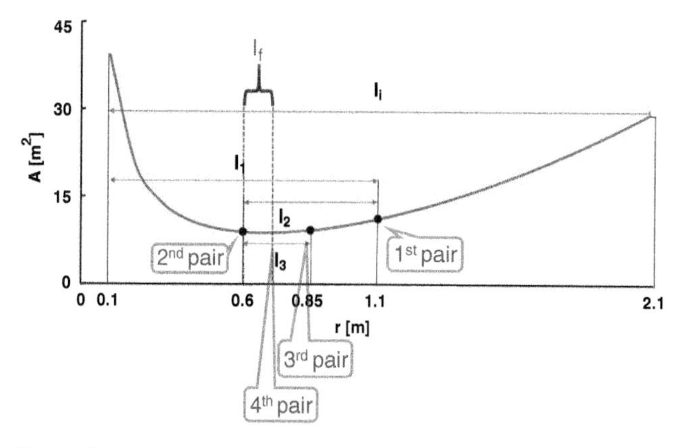

Figure 9.6 Sequential Dichotomous Search of a minimum-area cylindrical tank, where the spacing, δ or Δr, of 0.01 m has been invoked. Source: X. Wang.

For a minimum, we eliminate the right of this first pair. With that, we are left with r = 0.1 to 1.05. The midpoint of this new uncertainty interval is r = 0.475. For hand calculation purposes, it is more convenient to round this up to r = 0.48. The next pair is

$$y(r = 0.48) = 9.781, y(r = 0.49) = 9.672$$

Therefore, we eliminate the left of this second pair. The remaining interval is from r = 0.48 to 1.05, and the midpoint is r = 0.765. With rounding, the third pair is

$$y(r = 0.77) = 8.920, y(r = 0.78) = 8.951$$

So, the right of this pair is eliminated and the new interval is between r = 0.48 and 0.77. With rounding, the fourth pair is

$$y(r = 0.62) = 8.867, y(r = 0.63) = 8.843$$

For a minimum, we eliminate the left of this pair, leaving us with r = 0.63 to 0.77. This interval of uncertainty, 0.14, is less than the required $I_f = 0.2$. Therefore, the optimum (minimum) y is between r equal to 0.63 (y = 8.843) and 0.77 (y = 8.951).

It is important to note the savings, in terms of the number of calculations, in Example 9.3. Compared to exhaustive search, (Sequential) Dichotomous Search saved 11 calculations, from 19. This savings becomes huge in practice, where the reduction ratio is orders of magnitude higher than this simple classroom illustration. For a much larger tank with radius, r, between 100 m and 2,100 m, and a desired I_f of 0.2 m, the reduction ratio, R = 10,000. In this case, the Dichotomous Search involves 28 calculations, whereas the Exhaustive Search demands one less than 20,000 computations!

9.2.3 Fibonacci Search

The Fibonacci series is credited to the twelfth-century mathematician, Leonardo Pisano, from Pisa, Italy (Fibonacci, 2013). Leonardo was the son of Guglielmo Bonaccio, or son of Bonaccio, i.e. "filius Bonacci," which was later misinterpreted as one of the most famous names in mathematics, Fibonacci. Due to this misinterpretation, he became Leonardo Fibonacci. What is even more interesting is that the renounced Fibonacci series was allegedly borrowed from India. Whoever discovered it, it is quite apparent that the series is all over nature, from shells, to pinecones, ocean waves, and galaxies, where the unique numbers and the particular spiral prevail. The Fibonacci series can be described as

$$F_n = F_{n-2} + F_{n-1} \tag{9.5}$$

where $F_o = F_1 = 1$. The first few numbers of the series are tabulated in Table 9.3.

Table 9.3 The Fibonacci series.

n = 0	1	2	3	4	5	6	7	8	9
$F_n = 1$	1	2	3	5	8	13	21	34	55

The formation of the first six numbers is illustrated in Figure 9.7. Starting with a 1 × 1 square at n = 0, and also at n = 1. The side formed by these two squares has a dimension of 2 units. Subsequently, the next square is 2 units × 2 units, i.e. $F_2 = 2$. It is clear from Figure 9.7 that the side made up by this 2 × 2 square and the 1 × 1 unit square has a dimension of 3 units. Consequently, F_3 is portrayed by the 3 × 3 square. Then, stacking the 2 × 2 square on top of the 3 × 3 square leads to the 5 × 5 square. This is followed by an 8 × 8 square spawned from 3 plus 5. Adding 8 and 5 together gives rise to a square of 13 unit sides.

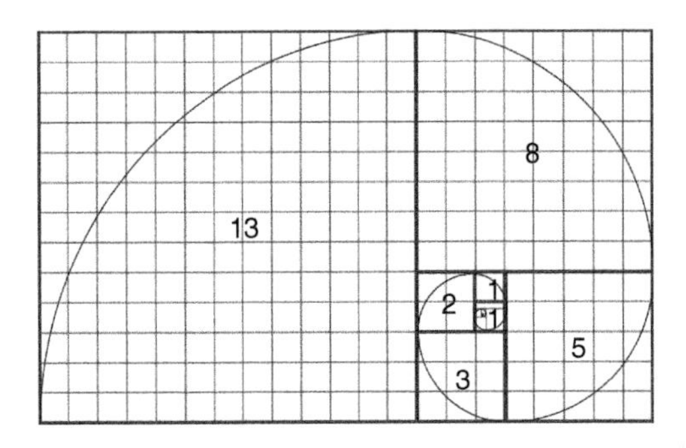

Figure 9.7 Fibonacci series up to n = 6. Source: D. Ting.

The Fibonacci search method follows the following steps.

1) Decide the number of inspections (calculations), n.
2) Calculate the value of the objective function, y, that corresponds to the independent variable, x, at $I_i\ F_{n-2} / F_n$ distance from one (left) end of the uncertainty interval, I_i.
3) Determine the value of the objective function, y, that corresponds to the independent variable, x, at $I_i\ F_{n-2} / F_n$ distance from the other (right) end of the uncertainty interval, I_i.
4) Compare the two y values; depending if a maximum or minimum is sought, eliminate either the region to the right of the right (first) point or to the left of the left (second) point.
5) Repeat Step 2, noting that one of the (two) points at $I_i\ F_{n-2} / F_n$ distance from the new (and shortened) uncertainty range coincides with a previous point, where the y value has been determined. In other words, only one new calculation is needed. Continue until only one point remains to be placed. The final point is the center of the interval of uncertainty.

The reduction ratio is simply

$$R = F_n \tag{9.6}$$

For example, for a reduction factor of 20, only 7 calculations are needed, as compared to 10 for sequential dichotomous search, as per Eq (9.4), where n = 8.6, and thus, 5 pairs or 10 calculations. Let us solve the tank area minimization problem using Fibonacci Search.

Example 9.4 *Cylindrical tank surface area minimization using the Fibonacci Search*

Given

A cylindrical storage tank is to be designed for storing hot water for a solar energy system. The required volume is 2 m^3. The surface area is to be minimized to minimize heat loss. Based on the available information, the radius, r, should be between 0.1 m and 2.1 m.

Find

Dimensions of the tank using the Fibonacci Search method, i.e. as an unconstrained single-variable problem. The final uncertainty interval of the radius, I_f, should be no more than 0.2 m.

Solution

From Example 9.1, the surface area of the tank,

$$A = 2\pi r^2 + 4/r$$

With r between 0.1 and 2.1 m, the initial interval of uncertainty,

$$I_i = 2.1 - 0.1 = 2$$

The desired I_f is 0.2. According to Eq (9.6),

$$R = I_i/I_f = 10 = F_n$$

From Table 9.3, we have n = 6. In other words, only six calculations are needed. The elimination process, as portrayed in Figure 9.8, proceeds in the following sequence.

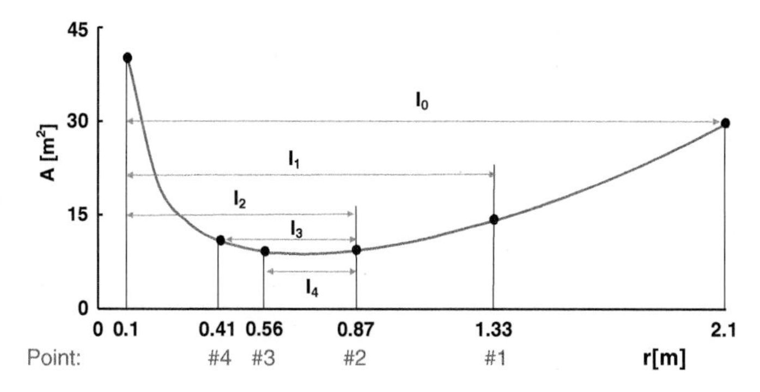

Figure 9.8 Fibonacci search of minimum area cylindrical tank. Source: X. Wang.

The first two calculations are at

$$\begin{aligned} d_1 &= F_{n-2}/F_n\, I_i = F_{6-2}/F_6\, I_i \\ &= 5/13\ \ I_i = 10/13 \end{aligned}$$

from the two ends (limits). The value of y at d_1 from the right end,

$$\text{Point 1}: \ y(r = 1.331) = 14.13$$

and at d_1 from the left end,

$$\text{Point 2}: \ y(r = 0.869) = 9.349$$

As we are after the minimum, we eliminate the right side of Point 1, i.e. r = 1.331. This leaves us with I_1 from r = 0.1 to 1.331, i.e. $I_1 = 1.231$.

The third point is at

$$d_2 = F_{n-3}/F_{n-1}\, I_1 = F_{6-3}/F_{6-1}\, I_1 = 3/8\ \ (1.231) = 0.4616$$

from the left end of the refined uncertainty interval. The point from the right end of this refined interval is Point 2. Therefore, the only point to be calculated is Point 3.

$$\text{Point 3}: \ y(r = 0.5616) = 9.105$$

This y value is less than that at Point 2; therefore, we eliminate the right side of Point 2. With that, we are left with I_2 from r = 0.1 to 0.869, i.e. $I_2 = 0.769$.

The fourth point is at

$$d_3 = F_{n-4}/F_{n-2}\, I_2 = F_{6-4}/F_{6-2}\, I_2 = 2/5\ \ (0.769) = 0.3076$$

from the right limit of the new uncertainty interval. Point 4 is at r = 0.4077, i.e.

$$\text{Point 4 : } y(r = 0.4077) = 10.86$$

This y value is larger than that at Point 3; therefore, we eliminate the left side of Point 4. The remaining range is from Point 4 to Point 2. Specifically, the uncertainty interval, I_3, goes from r = 0.4077 to 0.8692, i.e. $I_3 = 0.4615$.

The fifth point is at

$$d_4 = F_{n-5}/F_{n-3}\, I_3 = F_{6-5}/F_{6-3}\, I_3 = 1/3 \quad (0.4615) = 0.1538$$

from the right side of the remaining uncertainty interval. Point 5 is at r = 0.7154, i.e.

$$\text{Point 5 : } y(r = 0.7154) = 8.807$$

This y value is smaller than that at Point 3; therefore, we eliminate the left side of Point 3. The remaining range is from Point 3 to Point 2. Specifically, the uncertainty interval, I_4, goes from r = 0.5615 to 0.8692, i.e. $I_4 = 0.3077$.

The sixth point is at

$$d_5 = F_{n-6}/F_{n-4} I_4 = \tfrac{1}{2}\ I_4$$

from the new uncertainty interval limits. As this is right in the middle of interval I_4, we introduce a small spacing, δ, to push the right limit just to the right of Point 5. Specifically, Point 6 is at r = 0.7154+δ, i.e.

$$\text{Point 6 : } y(r = 0.7154 + 0.01) = 8.820, \text{where } \delta = 0.01.$$

This y value is larger than that at Point 5; therefore, we eliminate the right of Point 6, or Point 5, if we disregard the small δ.

The remaining range is from Point 3 to Point 5. Specifically, the uncertainty interval, I_5, goes from r = 0.5615 to 0.7154, i.e. $I_5 = 0.1538 = I_f$. This final uncertainty interval, I_f, is 13 times less than the initial uncertainty interval, $I_i = 2$.

It is clear that Fibonacci Search is the most efficient search method covered up to this point. In real life, the cylindrical tank can be much larger. If the radius, r, is between 100 m and 2100 m, and the desired I_f is 0.2 m, the corresponding reduction ratio, R, is 10,000. It is fascinating to see that even for this large reduction ration, only 20 calculations need to be executed, as compared to 28 for Dichotomous Search, and 20,000 for Exhaustive Search!

9.2.4 Golden Section Search

Another interesting feature of the Fibonacci series is that the ratio of the previous number to the current number is approximately fixed beyond the eighth number of the series. Specifically,

$$F_{n-1}/F_n \approx 0.618 \tag{9.7}$$

for n > 8. For example, $F_8/F_9 = 34/55 = 0.6182$, $F_9/F_{10} = 55/89 = 0.6180$, and $F_{20}/F_{21} = 10946/17711 = 0.6180$. As such, the Golden Section Search method is derived from the Fibonacci approach. It is marginally less efficient than the Fibonacci Search method. Its strength is in its versatility, i.e. the total number of runs does not need to be decided a priori. This drastically eases the application.

The reduction ratio,

$$R = I_i/I_f = 1/(0.618)^{n-1} \tag{9.8}$$

The resulting convergence speed is $(0.618)^{n-1}$. This reduction is only slightly less than that of the Fibonacci method.

To illustrate the working of Golden Section Search, let us rework the cylindrical tank surface area minimization problem. For this class problem, which necessitates but only a few calculations, the slight drop in efficiency with respect to the Fibonacci method is hardly detectable. More importantly, the small sacrifice in efficiency lost is worth it considering the gain in convenience. We do not need to decide the number of required calculations ahead of time, but simply stop when the interval of uncertainty is satisfactory. Should further refinement be required, the calculations can be resumed from where we left off, and continue until the new accuracy is reached.

Example 9.5 ***Minimize a cylindrical tank surface area using the Golden Section Search***

Given

A cylindrical storage tank is to be designed for storing hot water for a solar energy system. The required volume is 2 m^3. The surface area is to be minimized to minimize heat loss. Based on the available information, the radius, r, should be between 0.1 m and 2.1 m.

Find

Dimensions of the tank using the Golden Section Search method, i.e. as an unconstrained single-variable problem. The final uncertainty interval of the radius, I_f, should be no more than 0.2 m.

Solution

From Example 9.1, the surface area of the tank,

$$A = 2\pi r^2 + 4/r$$

With r between 0.1 and 2.1 m, the initial interval of uncertainty,

$$I_i = 2.1 - 0.1 = 2$$

The desired I_f is 0.2. According to Eq (9.8),

$$R = I_i/I_f = 10 = 1/(0.618)^{n-1}$$

This results in n = 6. For this particular case, the number of needed calculations is the same as that needed when invoking Fibonacci Search in the previous example. This is due to rounding when dealing with such a small number of calculations.

The elimination process is illustrated in Figure 9.9 and proceeds in the following sequence. The initial uncertainty interval is from r = 0.1 to 2.1.

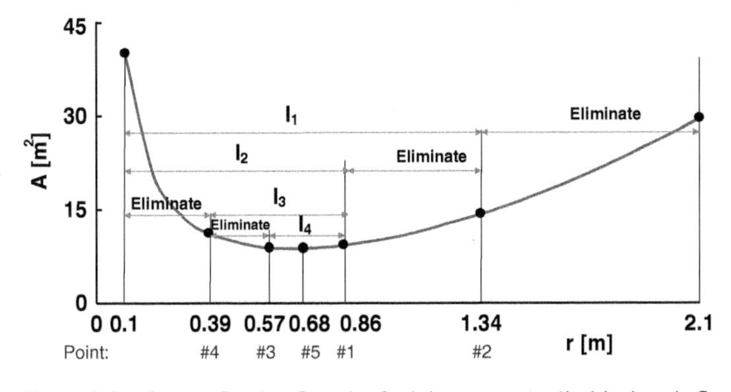

Figure 9.9 Golden Section Search of minimum area cylindrical tank. Source: X. Wang.

The first two calculations are at

$$d_1 = 0.618\ I_i = 1.236$$

from the two ends. Specifically, Point 1 is at r = 2.1 – 1.236 = 0.864, and Point 2 is at r = 1.236 + 0.1 = 1.336. We have

$$\text{Point 1}: \ y(r = 0.864) = 9.320$$

and

$$\text{Point 2}: \ y(r = 1.336) = 14.21.$$

As we are after the minimum, we eliminate the region right of Point 2, i.e. r = 1.336. This leaves us with I_1 from r = 0.1 to 1.336, i.e. $I_1 = 1.236$.

The third point is at

$$d_2 = 0.618\ I_1 = 0.764$$

from the end of the remaining uncertainty interval. This distance from the left end gives us Point 1. Therefore, Point 3 is at d_2 from the new right edge, i.e.

$$\text{Point 3}: \ y(r = 0.5722) = 9.048.$$

This y value is smaller than that at Point 1; therefore, we eliminate the range right of Point 1. With that, we are left with I_2 from r = 0.1 to 0.864, i.e. $I_2 = 0.764$.

The fourth point is at

$$d_3 = 0.618\ I_2 = 0.4722$$

from the limits of the current uncertainty interval. Point 4 is at d_3 from the new right limit, i.e. r = 0.864 - 0.4722 = 0.3918, i.e.

$$\text{Point 4}: \ y(r = 0.3918) = 11.17.$$

This y value is larger than that at Point 3; therefore, we eliminate the extent left of Point 4. The remaining range is thus from Point 4 to Point 1. Specifically, the uncertainty interval, I_3 goes from r = 0.3918 to 0.864, i.e. $I_3 = 0.4722$.

The fifth point is at

$$d_4 = 0.618\ I_3 = 0.2918$$

from the remaining uncertainty interval. Point 5 is d_3 from the left limit, Point 4, i.e. r = 0.3918 + 0.2918 = 0.6836

$$\text{Point 5}: \ y(r = 0.6836) = 8.788$$

This y value is smaller than that at Point 3; therefore, we eliminate the left side of Point 3. The remaining range is from Point 3 to Point 1. Specifically, the uncertainty interval, I_4 goes from r = 0.5722 to 0.864, i.e. $I_4 = 0.2918$.

The sixth point is at

$$d_5 = 0.618\ \ I_4 = 0.1803$$

from the remaining uncertainty interval. Point 6 is at r = 0.7525, i.e.

$$\text{Point 6 : } y(r = 0.7525) = 8.874.$$

This y value is larger than that at Point 5; therefore, we eliminate the right side of Point 6.

The remaining range is from Point 3 to Point 6. Specifically, the uncertainty interval, I_5 goes from r = 0.5722 to 0.7525, i.e. $I_5 = 0.1803$. This uncertainty interval is within the required final uncertainty interval, I_f, of 0.2 m.

9.2.5 Comparison of Elimination Methods

The classroom examples discussed thus far only require a few calculations. The benefits of higher efficiency methods can be better appreciated when a large number of calculations is involved. Such is often the case encountered in practice. Table 9.4 is a comparison of the common elimination methods. It is clear that the savings in the number of required calculations from the Exhaustive Elimination method to the basic sequential Dichotomous Search is huge. Going from Dichotomous to Fibonacci, the gain in efficiency is small and, from Fibonacci to Golden Section, it is nil.

Table 9.4 Comparison of efficiency of common search methods.

Method	Reduction ratio	n for R = 1000	n for R = 10^6
Uniform Exhaustive	(n+1)/2	1999	2×10^6
Sequential Dichotomous	$2^{n/2}$	20	40
Fibonacci	F_n	16	30
Golden Section	$1/(0.618)^{n-1}$	16	30

As far as efficiency is concerned, the Fibonacci method is superior. Other than efficiency, the best method is also a function of ease in setting up the problem. For small and relatively simple systems, ease in setting them up is more important. Accordingly, exhaustive search may be the best and safest choice for small problems, especially because it leaves no region uncovered. The convenience of the Golden Section method in stopping in the middle of a calculation, and resuming when needed, makes it a preferred choice over its counterpart and parent, the Fibonacci method.

9.3 Multi-variable, Unconstrained Optimization

In most engineering systems, including practical thermofluid systems, there are more than one or two parameters at play. With the added independent parameters, most of the methods covered in the previous sections become inapplicable or ineffective. The bullet-proof exhaustive Search method still works, but it becomes exceedingly inept with an increase in the number of variables and/or calculations required. A proven class of search methods analogous to hill-climbing becomes handy. Lattice Search, Univariate Search, and Steepest Ascent/Descent Search are the three most common hill-climbing methods. Let us go through these, but only after we set the Exhaustive Search method as the reference.

9.3.1 Exhaustive Search

As discussed earlier for the single-variable case, this bullet-proof method involves the calculation of the objective function, y, at uniformly-spaced independent variables, $x_1, x_2, \ldots x_m$. The exertion of this search is akin to a good shepherd exhaustively "turning every stone" searching for a lost sheep, as depicted in Figure 9.10. To find the lost sheep via the exhaustive search method, the shepherd has to turn over every rock at every crossing of the dotted x_1 (longitude) and x_2 (latitude) values. The physical elevation may be the objective function; the lost sheep has strayed to the bottom of the pit and could not get itself up, whereas the shepherd starts the exhaustive search from the summit.

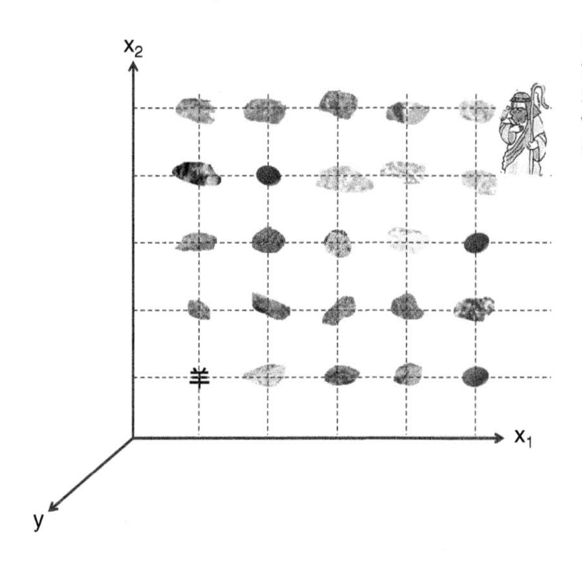

Figure 9.10 Exhaustive Search for y as a function of x_1 and x_2. The objective of the shepherd, top right, is to locate the lost sheep, which is at the bottom left, pictured by the Chinese script for sheep. Source: D. Ting.

Let us see the working of this base method via a mathematical example. Within the context of optimization of thermofluid systems, we will look at minimizing the cost of an air distribution system.

Example 9.6 *Minimize the cost of an air distribution system*

Given

The cost of an air distribution system, y, is a function of the fan capacity, x_1, and the length of the duct, x_2, where

$$y = x_1^2/6 + 4/\left(x_1 x_2\right) + 3x_2 \quad \text{(E9.6.1)}$$

where

$$0 < x_1 < 5 \text{ and } 0 < x_2 < 1$$

Find

The minimum cost system, for $\Delta x_1 = 1$ and $\Delta x_2 = 0.1$ via the Exhaustive Search method.

Solution

For the required resolutions, we have $x_1 = 1, 2, 3, 4$ and $x_2 = 0.1, 0.2, 0.3, 0.4, 0.5, 0.6, 0.7, 0.8, 0.9$. In short, there are $4 \times 10 = 40$ calculations. These points are shown in Figure 9.11. Sample calculations are depicted and also tabulated in Table 9.5. For the utilized resolutions in the two independent variables, the lowest cost corresponds to $y(x_1 = 3, x_2 = 0.7) = 5.50$.

In practice, resolutions of orders of magnitude finer that those used are not uncommon. If we refine the calculations to the extent that 400 x_1 and 1000 x_2 values are searched, the number of calculations will increase to 400,000!

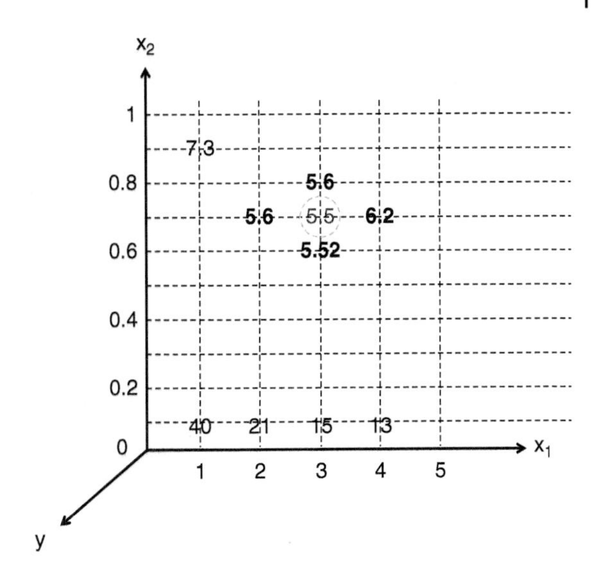

Figure 9.11 Least expensive air distribution system via the Exhaustive Search method. Source: D. Ting.

Table 9.5 Sample calculations of Exhaustive Search for the lowest-cost air distribution system.

x_1	1			2		3				4	
x_2	0.1	...	0.9	0.1	0.7	0.1	0.6	0.7	0.8	0.1	0.7
y	40.47	...	7.31	20.97	5.62	15.13	5.52	5.50	5.57	12.97	6.20

9.3.2 Lattice Search

In the Lattice Search, the objective function value corresponding to an initial point is deduced. This value is then compared to those of the immediate neighboring points. The central point of interest is moved to the neighboring point which is the local optimum. The solution is reached when the central point of interest gives the optimum objective function value, compared to the neighboring points. Let us execute the Lattice Search method for the same air distribution system example.

Example 9.7 ***Air distribution system cost minimization using the Lattice Search***

Given

The cost of an air distribution system, y, is a function of the fan capacity, x_1, and the length of the duct, x_2, where

$$y = x_1^2/6 + 4/\left(x_1 x_2\right) + 3\,x_2$$

where

$$0 < x_1 < 5 \text{ and } 0 < x_2 < 1$$

Find

The minimum cost system, for $\Delta x_1 = 1$ and $\Delta x_2 = 0.1$, via the Lattice Search.

Solution

As illustrated in Figure 9.12, we choose the first point at $x_1 = 2$ and $x_2 = 0.2$, i.e.

Point #1: $y\left(x_1 = 2,\ \ x_2 = 0.2\right) = 11.27$

Let us select the first neighboring point at the top right corner, and go clockwise, i.e.

Point #2: $y(x_1{=}3, x_2{=}0.3) = 6.84$, Point #3: $y(x_1{=}3, x_2{=}0.2) = 8.77$
Point #4: $y(x_1 = 3, x_2 = 0.1) = 15.13$, Point #5: $y(x_1 = 2, x_2 = 0.1) = 20.97$
Point #6: $y(x_1 = 1, x_2 = 0.1) = 40.47$, Point #7: $y(x_1 = 1, x_2 = 0.2) = 20.77$
Point #8: $y(x_1 = 1, x_2 = 0.3) = 14.40$, Point #9: $y(x_1 = 2, x_2 = 0.3) = 8.23$

It is clear that Point #2 has the lowest values among all the 8 neighbors, and that it is smaller than Point #1. Therefore, we move from Point #1 to Point #2.

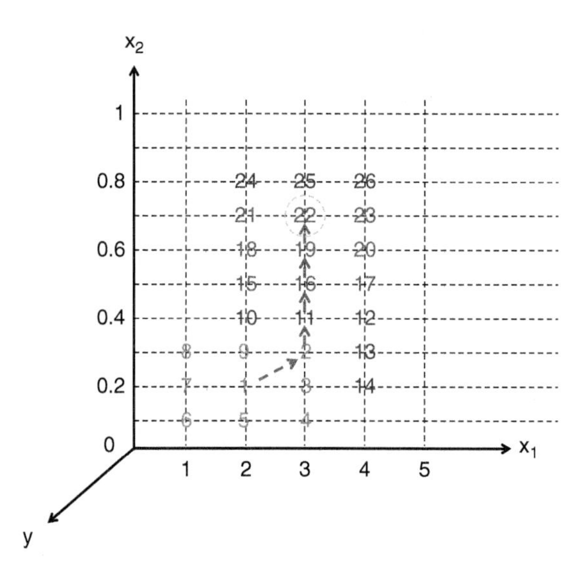

Figure 9.12 Least expensive air distribution system via the Lattice Search Numbers within the grid denote the calculation points (steps). Source: D. Ting.

For Point #2, we start with its first new neighbor at the top left corner, and go clockwise. Specifically,

$$\text{Point \#10: } y\left(x_1 = 2, x_2 = 0.4\right) = 6.87, \text{ Point \#11: } y\left(x_1 = 3, x_2 = 0.4\right) = 6.03$$
$$\text{Point \#12: } y\left(x_1 = 4, x_2 = 0.4\right) = 6.37, \text{ Point \#13: } y\left(x_1 = 4, x_2 = 0.3\right) = 6.90$$
$$\text{Point \#14: } y\left(x_1 = 4, x_2 = 0.2\right) = 8.27$$

We see that Point #11 is smallest and, hence, we move from Point #2 to Point #11.

The three new neighbors for Point #11 are

$$\text{Point \#15: } y\left(x_1 = 2, x_2 = 0.5\right) = 6.17$$
$$\text{Point \#16: } y\left(x_1 = 3, x_2 = 0.5\right) = 5.67$$
$$\text{Point \#17: } y\left(x_1 = 4, x_2 = 0.5\right) = 6.17$$

The next step is to move from Point #11 to the smaller y location, Point #16.

The new neighbors for Point #16 are

$$\text{Point \#18: } y\left(x_1 = 2, x_2 = 0.6\right) = 5.80$$
$$\text{Point \#19: } y\left(x_1 = 3, x_2 = 0.6\right) = 5.52$$
$$\text{Point \#20: } y\left(x_1 = 4, x_2 = 0.6\right) = 6.13$$

As the value at Point #19 is smallest, it becomes our new point of interest.

The new neighbors for Point #19 are

Point #21: $y(x_1 = 2, x_2 = 0.7) = 5.62$
Point #22: $y(x_1 = 3, x_2 = 0.7) = 5.50$
Point #23: $y(x_1 = 4, x_2 = 0.7) = 6.20$

The value at Point #22 is the smallest; hence, it is the prospective optimum point.

With Point #22 as the prospect, we compare its value with those of the new neighbors, i.e.

Point #24: $y(x_1 = 2, x_2 = 0.8) = 5.57$
Point #25: $y(x_1 = 3, x_2 = 0.8) = 5.57$
Point #26: $y(x_1 = 4, x_2 = 0.8) = 6.32$

These new neighbors have larger y values; thus, Point #22 is the minimum.

Compared to the exhaustive search in the previous example, this Lattice Search example requires 14 fewer calculations.

We see that as the Lattice Search method only invokes exhaustive searching of the neighboring central points, it is more efficient than the Exhaustive Search method which searches the entire domain of interest. For the lost sheep example, the good shepherd does not have to turn every stone if he knows that the wayward sheep has wandered down the slippery slope. In Lattice Search, he only has to turn the neighboring stones as he descends downward. Nonetheless, as all neighboring points are explicitly analyzed, as illustrated in Figure 9.12, Lattice Search is still not very efficient. One way to improve the efficiency is to start with a coarse grid and then to zoom into the region where the coarse optimum has been found, using a finer grid. There is, of course, the risk of missing the true (refined) optimum region, if the initial grid is too coarse compared to the gradients of the objective function. For the specific case in point, a coarse grid could result in missing the lost sheep behind a stone located in between two grid points.

9.3.3 Univariate Search

In Univariate Search, the objective function is optimized with respect to one variable at a time, cycling the optimization process through all involved independent variables at every step. Namely, the process involves alternating the search along the variables, one variable at a time. Consider a two-variable problem with the objective function, y, dependent on both α and β, as shown in Figure 9.13. Let us start with Point 1, at the bottom left corner. An athletic example is a climber starting at the foot of a mountain trying to find his way to the apex. At $\alpha = \alpha_1$, we optimize y with respect to β. This gives the optimum y value at $\alpha = \alpha_1$. For the illustration depicted in Figure 9.13, the lines signify contours of the objective function, y. The shrinking areas enclosed by the contours denote increasing elevation, y. As such, Point 2 is the highest point along $\alpha = \alpha_1$. Next, we optimize y along $\beta = \beta_2$. This results in Point 3. Optimizing along $\alpha = \alpha_3$ leads to Point 4. This is followed by Point 5, and finally Point 6, the apex.

We can see that, conceptually, the Univariate Search method appears to require much fewer calculations than the Lattice Search method, which is itself more efficient than the Exhaustive Search method. Let us rework the same air distribution example to see if Univariate Search is indeed superior.

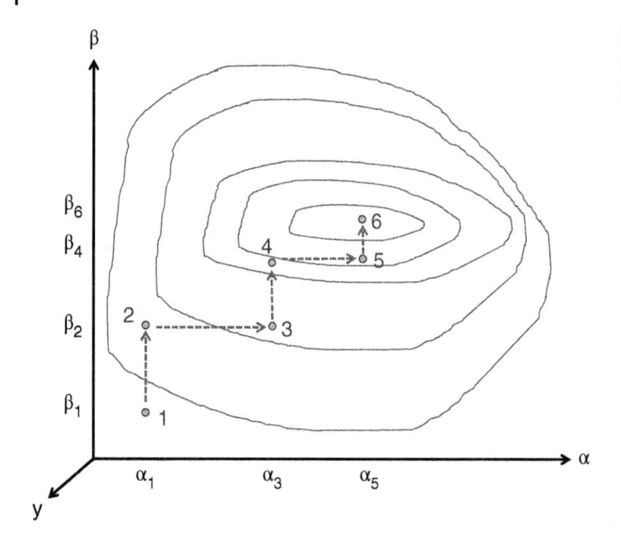

Figure 9.13 The working of Univariate Search for a unimodal, two-variable objective function. Source: D. Ting.

Example 9.8 ***Air distribution system cost minimization using univariate search***

Given

The cost of an air distribution system, y, is a function of the fan capacity, x_1, and the length of the duct, x_2, where

$$y = x_1^2/6 + 4/\left(x_1 x_2\right) + 3x_2$$

where

$$0 < x_1 < 5 \text{ and } 0 < x_2 < 1$$

Find

The minimum cost system, for $\Delta x_1 = 1$ and $\Delta x_2 = 0.1$, via univariate search.

Solution

As illustrated in Figure 9.14, we choose to start at $x_1 = 2$ and optimize y with respect to x_2. Let us use the Exhaustive Search with the given resolutions in x_1 and x_2.

Point #1: $y(x_1 = 2, x_2 = 0.1) = 20.97$
Point #2: $y(x_1 = 2, x_2 = 0.2) = 11.27$
Point #3: $y(x_1 = 2, x_2 = 0.3) = 8.23$
Point #4: $y(x_1 = 2, x_2 = 0.4) = 6.87$
Point #5: $y(x_1 = 2, x_2 = 0.5) = 6.17$
Point #6: $y(x_1 = 2, x_2 = 0.6) = 5.80$
Point #7: $y(x_1 = 2, x_2 = 0.7) = 5.62$
Point #8: $y(x_1 = 2, x_2 = 0.8) = 5.57$
Point #9: $y(x_1 = 2, x_2 = 0.9) = 5.59$.

Point #8 corresponds to the smallest y value for $x_1 = 2$.

With Point #8 having the smallest y value, we set $x_2 = 0.8$ and optimize y with respect to x_1. Namely,

Point #10: $y(x_1 = 1, x_2 = 0.8) = 7.57$

Point #11: $y(x_1 = 3, x_2 = 0.8) = 5.57$

Point #12: $y(x_1 = 4, x_2 = 0.8) = 6.32$

Point #13: $y(x_1 = 5, x_2 = 0.8) = 7.57$

Within the employed resolutions, we have convergence, as the y values at Points #8 and #11 are equally small. Thus, this Univariate Search invokes 13 and 27 fewer calculations than that of the Lattice Search and the Exhaustive Search, respectively.

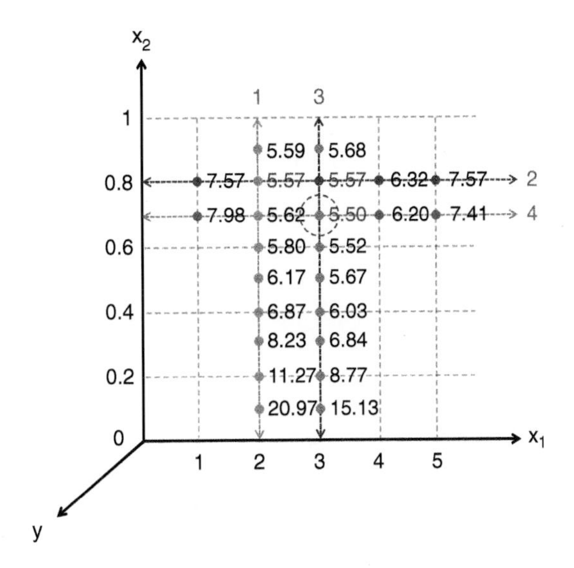

Figure 9.14 Least expensive air distribution system via Univariate Search. Source: Y. Yang.

To reach the same optimum point, $y(x_1 = 3, x_2 = 0.7) = 5.50$, as calculated in the previous example, we need to continue the calculations and to check if there is a point vertically along $x_1 = 3$ which has a y value less than 5.57. This is Line #3, which passes vertically through Point #11 in Figure 9.14. Doing so, after eight additional calculations, we end up with

$$y(x_1 = 3, x_2 = 0.7) = 5.50$$

One more series of computations, horizontally along $x_2 = 0.7$ is needed to close the case. This is shown as Line #4, where four new calculations are necessary to confirm that no points along $x_2 = 0.7$ leads to a y value less than 5.50. Adding these 12 extra computations results in a total of 25 calculations, only one fewer than for the Lattice Search. The use of the Exhaustive Search along the alternating constant-independent-variable paths has contributed to the minuteness in improvement.

9.3.4 Steepest Ascent/Descent Method

Imagine some handsome treasure has been placed at the summit of a mountain. You are one of the many contestants in the finders-keepers game, e.g. The Amazing Race, where every contestant starts at the same spot at the foot of the mountain, see Figure 9.15. Having been enlightened with the various established optimization methods, your brain is on fire to figure out the fastest route to reach the summit. In all likelihood, the shortest passage, i.e. the fastest way is associated with the steepest slope; if you have adequate stamina to sustain you up the slope. In the case where we are climbing from the foot of the mountain to the apex, we move along the steepest ascent.

Figure 9.15 exhibits that the steepest passage changes as you clamber up the elevation. In real life, the course changes continuously, but in calculations involving finite steps or stretches, it takes place as sections of straight lines, as shown in Figure 9.15.

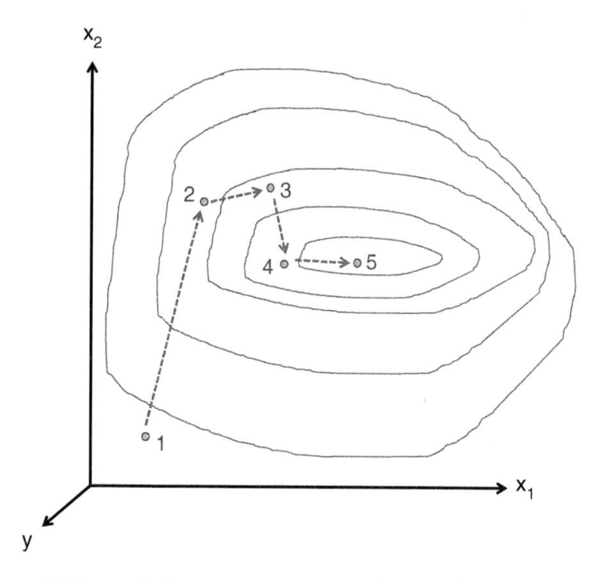

Figure 9.15 The working of the Steepest Ascent/Descent method for a unimodal, two-variable objective function. Source: D. Ting.

If, instead, the contestants were dropped off at the starting point on the summit via a helicopter, and the treasure is buried in the valley, then the steepest descent is the route to take advantage of. Such is the rationale behind the Steepest Ascent/Descent method. The idea is to move toward the optimum over the shortest possible path.

The Steepest Ascent/Descent method directs the calculations along the direction in which the objective function changes at the greatest rate. As such, it requires comparatively fewer iterations. Nonetheless, gradients must be evaluated in order to determine the appropriate direction of motion. Therefore, it is only applicable when the gradients can be obtained accurately and easily. Furthermore, care should be exercised to avoid overshooting a turn and missing the optimum or leading to divergence.

The gradient vector of the objective function, ∇y, is the direction in which the objective function, y, changes at the greatest rate. The gradient vector is

$$\nabla y = \frac{\partial y}{\partial x_1}\hat{i}_i + \frac{\partial y}{\partial x_2}\hat{i}_2 + \cdots \tag{9.9}$$

At an arbitrary point, $y(x_1, x_2, \ldots)$, the largest change in the x_1 direction is $\partial y/\partial x_1$. Similarly, the greatest change in y in the x_2 direction is the tangent, $\partial y/\partial x_2$, etc.

From vector analysis,

$$\frac{\Delta x_1}{\partial y/\partial x_1} = \frac{\Delta x_2}{\partial y/\partial x_2} = \cdots = \frac{\Delta x_n}{\partial y/\partial x_n} \tag{9.10}$$

Therefore, if we choose an infinitely small change in x_1, Δx_1, then we can obtain Δx_2, Δx_3, … accordingly. With these, we can numerically deduce the gradient vector. In the x_i direction, we have

$$\frac{\partial y}{\partial x_i} = \frac{y\left(x_1, x_2, \ldots, x_i + \Delta x_i, \ldots x_n\right) - y\left(x_1, x_2, \ldots x_n\right)}{\Delta x_i} \tag{9.11}$$

For the single-variable case where y is a function of x only, we have,

$$\frac{\partial y}{\partial x} = \frac{y(x + \Delta x) - y(x)}{\Delta x} \tag{9.12}$$

For actual computations involving the steepest ascent/descent method, finite, not infinitely small, Δx_i are employed. As such, the equal signs in the above equations are strictly approximation signs.

The general procedure of the steepest ascent/descent method is:

1) Select a starting point and calculate the objective function, y.
2) Evaluate the gradient at that point and the relationship of the changes of the independent variables, $x_1, x_2, x_3, \ldots$.
3) Decide in which direction (i.e. increasing- or decreasing-y) to move along the gradient.
4) Resolve how far to advance along the gradient (far enough to reduce the number of calculations, but not too far such that the calculation diverges) and move that distance.
5) Test to determine whether the optimum has been achieved. If so, terminate; otherwise return to Step 2 and repeat the process.

Specifically, when executing a two-variable objective function, $y = y(x_1, x_2)$, the calculations can proceed in the following manner.

1) Choose a starting point and calculate y.
2) Select Δx_1.
3) Calculate the derivatives, i.e. gradients.
4) Decide the direction of movement (i.e. positive $\Delta x_1\ \partial y/\partial \Delta x_1$ for maximization and negative $\Delta x_1\ \partial y/\partial \Delta x_1$ for minimization).
5) Deduce Δx_2 from the derivatives and Δx_1; $\Delta x_2 = (\partial y/\partial \Delta x_2)(\Delta x_1/\partial y/\partial \Delta x_1)$.
6) Obtain new values of x_1, x_2, and y.
7) Calculate the derivatives at the new point.
8) Return to Step 4 and iterate until the optimum is reached.

Example 9.9 ***Air distribution system cost minimization using the Steepest Ascent/Descent method***

Given

The cost of an air distribution system, y, is a function of the fan capacity, x_1, and the length of the duct, x_2, where

$$y = x_1^2/6 + 4/\left(x_1 x_2\right) + 3x_2$$

where

$$0 < x_1 < 5 \text{ and } 0 < x_2 < 1$$

Find

The minimum cost system, for $\Delta x_1 = 1$ and $\Delta x_2 = 0.1$, via the Steepest Ascent/Descent method.

Solution

To ease comprehension, the calculation process is shown in Figure 9.16.

Step 1: Choose a starting point.

Let us choose the first point at $x_1 = 2$ and $x_2 = 0.2$, i.e.

Point #1: $y\left(x_1 = 2, x_2 = 0.2\right) = 11.27$

Step 2: Calculate the derivatives.

$$\partial y/\partial x_1 = 2x_1/6 - 4/\left(x_1^2 x_2\right) = -4.333$$

$$\partial y/\partial x_2 = -4/\left(x_1 x_2^2\right) + 3 = -47$$

The negative slopes imply that y is decreasing with increasing x_1 and/or x_2. Therefore, we move in the positive x_1 and/or x_2 directions, as we are after the minimum y.

$$\Delta x_1/\partial y/\partial x_1 = \Delta x_2/\partial y/\partial x_2$$

For $\Delta x_1 = 1$, we have

$$\Delta x_2 = \left(\Delta x_1/\partial y/\partial x_1\right)\ \left(\partial y/\partial x_2\right) = 10.85$$

Step 3: Obtain the new values,

$$x_{1,new} = x_{1,old} + \Delta x_1 = 2 + 1 = 3$$

$$x_{2,new} = x_{2,old} + \Delta x_2 = 0.2 + 10.85 = 11.05$$

Therefore,

$$\text{Point \#2: } y\left(x_1 = 3, x_2 = 11.05\right) = x_1^2/6 + 4/\left(x_1 x_2\right) + 3x_2 = 34.76$$

This new y value is larger than that at Point #1 because of overshoot and/or coarse resolutions. Any coarser resolutions will probably cause the calculation to diverge.

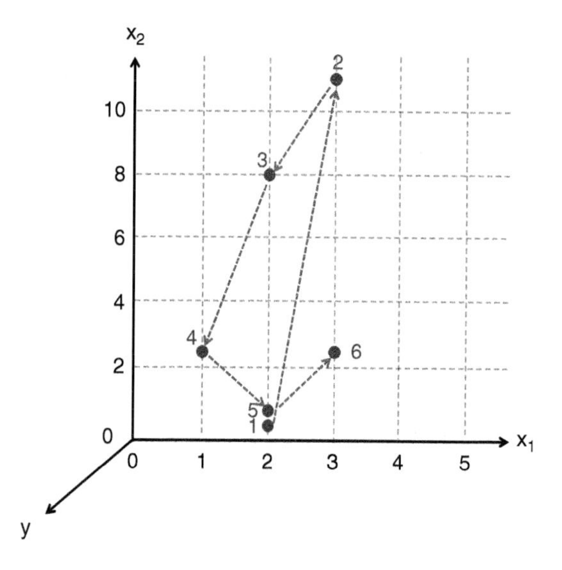

Figure 9.16 Search for the least expensive air distribution system using the Steepest Ascent/Descent method. Source: Y. Yang.

Iteration #2

Step 2: Calculate the derivatives.

$$\partial y/\partial x_1 = 2x_1/6 - 4/\left(x_1^2 x_2\right) = 0.9598$$

$$\partial y/\partial x_2 = -4/\left(x_1\, x_2^2\right) + 3 = 2.989$$

To decrease y, $\Delta x_1 = -1$, we have

$$\Delta x_2 = \left(\Delta x_1/\partial y/\partial x_1\right)\ \left(\partial y/\partial x_2\right) = -3.114$$

Step 3: Obtain the new values,

$$x_{1,new} = x_{1,old} + \Delta x_1 = 3 - 1 = 2$$

$$x_{2,new} = x_{2,old} + \Delta x_2 = 11.05 - 3.114 = 7.932$$

Therefore,

$$\text{Point \#3: } y\left(x_1 = 2, x_2 = 7.932\right) = x_1^2/6 + 4/\left(x_1 x_2\right) + 3x_2 = 24.71$$

This is less than Point #2, but is still much larger than Point #1, presumably due to coarse resolutions.

Iteration #3

Step 2: Calculate the derivatives.

$$\partial y/\partial x_1 = 2x_1/6 - 4/\left(x_1^2 x_2\right) = 0.5406$$
$$\partial y/\partial x_2 = -4/\left(x_1 x_2^2\right) + 3 = 2.968$$

To decrease y, $\Delta x_1 = -1$, we have

$$\Delta x_2 = \left(\Delta x_1/\partial y/\partial x_1\right)\left(\partial y/\partial x_2\right) = -5.491$$

Step 3: Obtain the new values,

$$x_{1,new} = x_{1,old} + \Delta x_1 = 2 - 1 = 1$$
$$x_{2,new} = x_{2,old} + \Delta x_2 = 7.932 - 5.491 = 2.441$$

Therefore,

Point #4: $y\left(x_1 = 1, x_2 = 2.441\right) = x_1^2/6 + 4/\left(x_1 x_2\right) + 3x_2 = 9.129$

This is less than the existing minimum value associated with Point #1. The calculation is converging.

Iteration #4

Step 2: Calculate the derivatives.

$$\partial y/\partial x_1 = 2x_1/6 - 4/\left(x_1^2 x_2\right) = -1.305$$
$$\partial y/\partial x_2 = -4/\left(x_1 x_2^2\right) + 3 = 2.329$$

To decrease y, $\Delta x_1 = 1$, we have

$$\Delta x_2 = \left(\Delta x_1/\partial y/\partial x_1\right)\left(\partial y/\partial x_2\right) = -1.784$$

Step 3: Obtain the new values,

$$x_{1,new} = x_{1,old} + \Delta x_1 = 2$$
$$x_{2,new} = x_{2,old} + \Delta x_2 = 0.6566$$

Therefore,

Point #5: $y\left(x_1 = 2, x_2 = 0.6566\right) = x_1^2/6 + 4/\left(x_1 x_2\right) + 3x_2 = 5.682$

This is smaller than the previous smallest value at Point #4. We are very close to the minimum.

Iteration #5

Step 2: Calculate the derivatives.

$$\partial y/\partial x_1 = 2x_1/6 - 4/\left(x_1^2 x_2\right) = -0.8563$$
$$\partial y/\partial x_2 = -4/\left(x_1 x_2^2\right) + 3 = -1.639$$

To decrease y, $\Delta x_1 = 1$, we have

$$\Delta x_2 = \left(\Delta x_1/\partial y/\partial x_1\right)\left(\partial y/\partial x_2\right) = 1.914$$

Step 3: Obtain the new values,

$$x_{1,new} = x_{1,old} + \Delta x_1 = 3$$

$$x_{2,new} = x_{2,old} + \Delta x_2 = 2.571$$

Therefore,
Point #6: $y(x_1 = 3, x_2 = 2.571) = x_1^2/6 + 4/(x_1 x_2) + 3x_2 = 9.731$
This is larger than Point #5. Due to coarse resolutions, we have overshot the minimum.

Iteration #6
Step 2: Calculate the derivatives.

$$\partial y/\partial x_1 = 2x_1/6 - 4/(x_1^2 x_2) = 0.8271$$

$$\partial y/\partial x_2 = -4/(x_1 x_2^2) + 3 = 2.798$$

To decrease y, $\Delta x_1 = -1$, we have

$$\Delta x_2 = (\Delta x_1/\partial y/\partial x_1)(\partial y/\partial x_2) = -3.383$$

Step 3: Obtain the new values,

$$x_{1,new} = x_{1,old} + \Delta x_1 = 2$$

$$x_{2,new} = x_{2,old} + \Delta x_2 = -0.8123$$

Physically, it is impossible for the length of the duct, x_2, to be negative. This more or less confirms that, due to the coarse resolutions, we have overshot the optimum.

Point #7: $y(x_1 = 2, x_2 = -0.8123) = x_1^2/6 + 4/(x_1 x_2) + 3x_2 = -4.232$

There is no such thing as negative cost, i.e. we have overshot the minimum cost point.

Conclusion
We can conclude that, for the given coarse resolutions, the optimum is Point #4, i.e.

$$y(x_1 = 2, x_2 = 0.6566) = 5.682$$

Note that while significantly fewer calculations are required for the Steepest Ascent/Descent method, the calculations are more involved. The respective gradient and the extent of movement need to be recalculated at every step.

It is clear that, among all the methods covered thus far, the Steepest Ascent/Descent method requires the least number of iterations. In the case of Example 9.9, only six points were calculated. In fact, within the given resolutions, the solution is found at the fifth calculation, where Point #5 confirms that the previous calculation at Point #4 is the solution. As such, the Steepest Ascent/Descent method is superior to the other conferred methods, the Univariate Search, the Lattice Search, and the Exhaustive Search, in terms of number of points to calculate. However, the Steepest Ascent/Descent method requires gradient deductions, which are comparatively more complex. Moreover, it is also more subjective to divergence, especially when the gradients are large.

Measures can be taken to lessen divergence. More positively, the scales of the independent variables can be chosen so that the contours are more spherical, see Figure 9.17. This will hasten the convergence of the calculations (Shah et al., 1964; Wilde, 1964; Jaluria, 1998). In Example 9.9, this could mean that we use $X_2 = 5\,x_2$, instead of using x_2 directly.

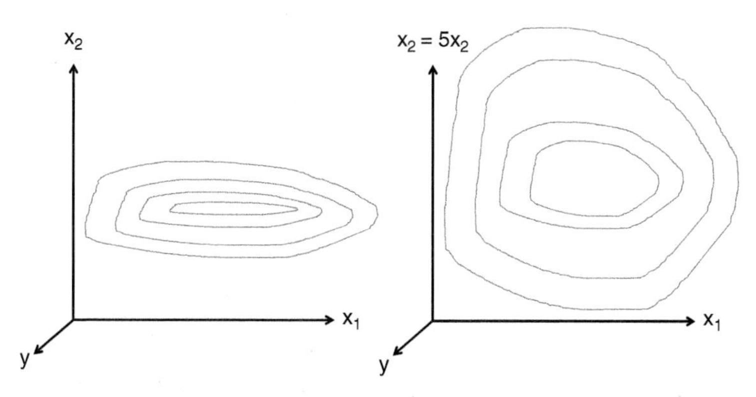

Figure 9.17 Mitigate divergence in the Steepest Ascent/Descent method by making more even the contours of the involved independent parameters. Source: D. Ting.

9.4 Multi-variable, Constrained Optimization

Sometimes, the pertaining constraints are not easily manipulated and incorporated into the objective function. Other times, it is better to keep the constraints separate so that we can soften them and see their respective effects on the results. These effects due to the relaxation of the constraints are revealed through the sensitivity coefficients. A prevailing approach in performing constrained optimization is to prescribe an appropriate penalty when a constraint is not completely satisfied. The other popular approach is to search for the optimum along a constraint.

9.4.1 Penalty Function Method

The idea behind the Penalty Function method is to penalize the violation of the constraint. The more serious the violation, the more severe the penalty. The constraints are encompassed into the objective function, in the form of a composite function. Penalty parameters are utilized to penalize the composite function for violation of the constraints. In general, a moderately light[1] penalty is enforced initially, to ease calculations. With the assigned penalty, the composite function is then optimized using any of the techniques applicable to unconstrained problems. Upon convergence, the penalty is increased and the new composite function is optimized. The calculation is assumed to have converged when there is no significant change in the optimum with further increase in the penalty.

Consider the objective function

$$y = y\left(x_1,\ x_2,\ \ldots x_n\right) \tag{9.13}$$

which is subjected to equality constraints

$$\begin{aligned}\phi_1 &= \phi_1\left(x_1, x_2,\ \ldots\ x_n\right) = L_1\\ \phi_2 &= \phi_2\left(x_1,\ x_2,\ \ldots\ x_n\right) = L_2\\ &\ldots\\ \phi_m &= \phi_m\left(x_1,\ x_2,\ \ldots\ x_n\right) = L_m\end{aligned} \tag{9.14}$$

1 Note that too light a penalty may lead to divergence. This is primarily because there tend to be infinite solutions for some problems when the constraint is completely lifted, i.e., the penalty is set to zero.

or

$$\Phi_1 = \Phi_1(x_1, x_2, \ldots x_n) = \Phi_1(x_1, x_2, \ldots x_n) - L_1 = 0$$
$$\Phi_2 = \Phi_2(x_1, x_2, \ldots x_n) = 0$$
$$\ldots$$
$$\Phi_m = \Phi_m(x_1, x_2, \ldots x_n) = 0$$
$$\ldots \tag{9.15}$$

The corresponding composite (unconstrained) objective function,

$$Y = y + P_1(\Phi_1)^2 + P_2(\Phi_2)^2 + P_3(\Phi_3)^2 + \ldots + P_m(\Phi_m)^2 \tag{9.16}$$

where P_j are the penalties associated with each constraint. The values of penalties, P_j, are positive if the optimum is a minimum. If we are after the maximum, we can rewrite Eq (9.16) as

$$Y = y - P_1(\Phi_1)^2 - P_2(\Phi_2)^2 - P_3(\Phi_3)^2 - \ldots - P_m(\Phi_m)^2 \tag{9.16a}$$

where the values of P_j are positive. The alternative is to stay with Eq (9.16), but assume negative values of penalties, when the optimum is a maximum. It is worth noting that Y is equal to y in Eqs. (9.16) and 9.16a, when either the penalties P_j are all zero or all the constraints are completely satisfied, i.e. all Φ_i are zero.

The following points concerning the penalties are worth noting.

1) When P_j are zero, the constraints are not satisfied.
2) When P_j are small, the constraints are not quite satisfied.
3) When P_j are large, the constraints are more satisfied, but the convergence is slower.

The calculations can proceed in the same manner as the above points are listed. This process is illustrated in Figure 9.18, where the composite objective function is plotted against the first independent variable, x_1. For the minimization example shown, the unconstrained case with P_j equal to zero, leads to the smallest y value. However, this minimum completely violates the constraint, making it an unrealistic solution in real life. For instance, optimizing the environmental friendliness of a residential air conditioner that results in a unit cost of $177,000 is of no practical use. With increasing penalties, P_j, the optimal y value falls progressively shorter of the unconstrained one. In the particular case shown, the corresponding x_1 value decreases.

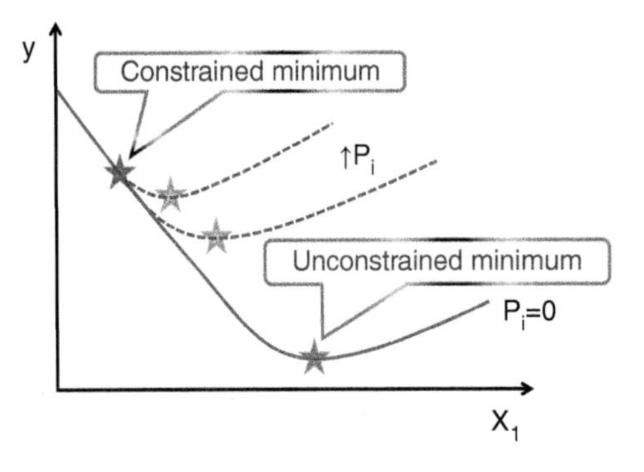

Figure 9.18 The variation of objective function with penalty. Source: D. Ting.

The composite objective function, which includes the penalties multiplied by the square of the equality constraints, is an unconstrained problem. As the procedure is to solve a series of unconstrained composite objective functions with increasing penalties, the technique is known as the

Sequential Unconstrained Optimization Technique. This method can also be applied for objective functions with inequality constraints.

Example 9.10 ***Constrained two-variable system minimization using the Penalty Function method***

Given

In the fall, the rate of the leaves dropping off a tree, y, is found to be a function of the normalized length of the day, x_1, and the normalized ambient temperature, x_2. Specifically,

$$y = 2x_1^2 + 5x_2 \quad \text{(E9.10.1)}$$

In Windsor, Ontario, Canada, the normalized ambient temperature is found to be related to the normalized length of the day in the following form,

$$x_2 = 12/x_1 \quad \text{(E9.10.2)}$$

Find

The minimum leaf-falling rate using the Penalty Function method.

Solution

Basic calculus can be applied to solve this problem. Doing so allows us to know the answer before we proceed and master the penalty function method. Substituting Eq (E9.10.2) into Eq (E9.10.1) gives

$$y = 2x_1^2 + 5\left(12/x_1\right) \quad \text{(E9.10.3)}$$

Differentiating this with respect to x_1 gives

$$dy/dx_1 = 4x_1 - 60/x_1^2 \quad \text{(E9.10.4)}$$

This is zero at the optimum; therefore, we get

$$x_1{}^* = 2.466$$

where asterisk, *, signifies the optimum. Substituting this into Eq (E9.10.1a) gives

$$y^* = 36.49.$$

Knowing the final answer, let us proceed with the penalty function method.

For the penalty function method, the constraint can be recast as

$$\Phi = x_1 x_2 - 12 = 0 \quad \text{(E.9.10.2a)}$$

Encompassing this into the objective function, we form the new composite objective function

$$Y = 2x_1^2 + 5x_2 + P\left(x_1 x_2 - 12\right)^2 \quad \text{(E9.10.5)}$$

Note that we add the penalty function because we are after the minimum leaf-falling rate.

Let us employ the calculus method to minimize the unconstrained function, Y. Differentiating Y in Eq (E9.10.5) with respect to x_1, we get

$$\partial Y/\partial x_1 = 4x_1 + 2\ P\left(x_1 x_2 - 12\right) x_2 = 0 \quad \text{(E9.10.6)}$$

This gives

$$4x_1 = 24\ Px_2 - 2\ Px_1 x_2^2 \quad \text{(E9.10.7)}$$

which can be rearranged into

$$x_1 = 24Px_2/\left(4 + 2Px_2^2\right) \quad \text{(E9.10.8)}$$

Differentiating Y in Eq (E9.10.5) with respect to x_2, we get

$$\partial Y/\partial x_2 = 5 + 2\ P\ \left(x_1 x_2 - 12\right) x_1 = 0 \quad \text{(E9.10.9)}$$

This gives

$$x_2 = \left(24Px_1 - 5\right) / \left(2Px_1^2\right) \quad \text{(E9.10.10)}$$

In short, we have two equations, Eqs. (E9.10.8) and (E9.10.10), with two unknowns, x_1 and x_2, for any given penalty, P. For sample values of P, solving these two equations gives the corresponding optimal values tabulated in Table 9.6. It is clear from Table 9.6 that the constraint is largely satisfied with increasing penalty, P. With P of 10, the optimal Y value is 36.39 (y = 36.27), which is less than 0.3% lower than 36.49 obtained using the conventional calculus method, which corresponds to a P value of infinity. Thus, it is clear that there is no need to use an excessively large penalty to unnecessarily slow down the calculation. Equally important to note is that too small a penalty function can add unnecessary challenges. In this case, extra care is needed to ensure the iterative computations converge and extra calculations are required as we approach the threshold P value, below which the problem diverges.

Table 9.6 Penalty function, two-variable optimization.

P	x_1	x_2	$x_1\ x_2$	Y	y
0	***	***	***	***	***
0.1	**	**	**	**	**
0.3	2.17	3.76	8.16	32.64	28.21
0.5	2.31	4.26	9.83	34.30	31.96
1	2.39	4.58	10.96	35.43	34.34
10	2.46	4.84	11.90	36.39	36.27
100	2.47	4.86	11.99	36.48	36.47
1000	2.47	4.87	12.00	36.49	36.49

***There are infinite solutions.
**The calculation consistently diverges.

9.4.2 Search-Along-a-Constraint (Hemstitching) Method

The idea behind the Search-Along-a-Constraint or Hemstitching method is to stay along, or in the close vicinity of, the constraint and move tangentially in the direction toward the optimum. As the calculations involve finite steps, a movement along the tangent tends to overshoot; therefore, alternate movement normal and toward the constraint is needed. The general procedure, as depicted in Figure 9.19, is:

1) Start with a trial point.
2) Move along a constant-independent-variable line toward, and reach, the constraint.

3) Move tangentially with respect to the constraint, in the direction of increasing or decreasing objective function.
4) Bring the point back to the constraint along a constant-independent-variable line.

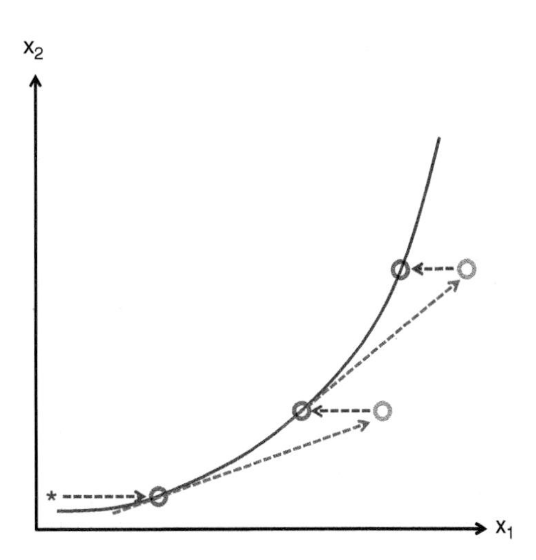

Figure 9.19 Search-Along-a-Constraint (Hemstitching) method. Source: D. Ting.

Let us look at a two-independent-variable objective function constrained by one equality constraint. Specifically, consider $y(x_1, x_2)$ which is subjected to an equality constraint, $\Phi = 0$; see Figure 9.19. An infinitely small step along the tangent of constraint Φ is a move which retains $\Phi = 0$. From chain rule, the change in the constraint during such a small step is

$$\Delta\Phi \approx \partial\Phi/\partial x_1\, \Delta x_1 + \partial\Phi/\partial x_2\, \Delta x_2 = 0 \tag{9.17}$$

From this, we can write,

$$\Delta x_1/\Delta x_2 = -\left(\partial\Phi/\partial x_2\right) / \left(\partial\Phi/\partial x_1\right) \tag{9.18}$$

The change in the objective function, y, with respect to the small movement is

$$\Delta y \approx \frac{\partial y}{\partial x_1}\Delta x_1 + \frac{\partial y}{\partial x_2}\Delta x_2 \tag{9.19}$$

Therefore, from the above two equations, the change in y due to a move tangential to the constraint can be expressed as

$$\Delta y = \left(\frac{\partial y}{\partial x_2} - \frac{\partial y}{\partial x_1}\frac{\partial\Phi/\partial x_2}{\partial\Phi/\partial x_1}\right)\Delta x_2 \tag{9.20}$$

To illustrate the workings of this method, we will resolve Example 9.11 using the Hemstitching method.

Example 9.11 *Constrained two-variable system minimization using the Hemstitching method*

Given

In the fall, the rate of the leaves dropping off a tree, y, is found to be a function of the normalized length of the day, x_1, and the normalized ambient temperature, x_2. Specifically,

$$y = 2x_1^2 + 5x_2 \tag{E9.11.1}$$

In Windsor, Ontario, Canada, the normalized ambient temperature is found to be related to the normalized length of the day in the following form,

$$x_2 = 12/x_1 \tag{E9.11.2}$$

Find

The minimum leaf-falling rate using the Hemstitching method.

Solution

The constraint can be recast as

$$\Phi = x_1 x_2 - 12 = 0 \tag{E9.11.2a}$$

Let us start with the first point at

Point #1: $y\left(x_1 = 2,\ x_2 = 2\right) = 18$

but this point does NOT satisfy the constraint; see Figure 9.20. To satisfy the constraint, we find the corresponding x_2 value for $x_1 = 2$. This is accomplished by substituting $x_1 = 2$ into Eq (E9.11.2a), which gives

$$x_2 = 6$$

In short, the second point is

Point #2: $y\left(x_1 = 2, x_2 = 6\right) = 38$, with $\Phi = x_1 x_2 - 12 = 0$

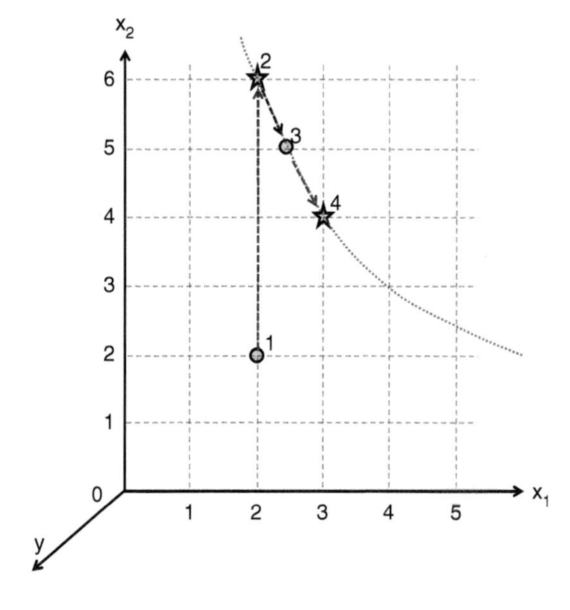

Figure 9.20 Minimization of $y = y(x_1, x_2)$ along a constraint via the Hemstitching method. The lightly dotted curve portrays the equality constraint, $\Phi = x_1\ x_2 - 12 = 0$. Source: D. Ting.

Next, we wish to move along the tangent of the constraint at Point #2, in the direction in which y decreases. For this, we call upon Eq (9.20),

$$\Delta y = \left[\left(\partial y/\partial x_2\right) - \left(\partial y/\partial x_1\right)\left(\partial\Phi/\partial x_2\right) / \left(\partial\Phi/\partial x_1\right)\right]\ \Delta x_2$$

where

$$\partial y/\partial x_2 = 5$$

$$\partial y/\partial x_1 = 4x_1 = 8$$

$$\partial\Phi/\partial x_2 = x_1 = 2$$
$$\partial\Phi/\partial x_1 = x_2 = 6$$

Therefore,

$$\Delta y = \left[5 - 4x_1^2/x_2\right] \Delta x_2 = \left[5 - 4(2)^2/(6)\right] \Delta x_2 = \left[46/6\right] \Delta x_2$$

Let us choose $\Delta x_2 = -1$; the negative sign is for decreasing y. Then,

$$x_2 = x_{2,old} + \Delta x_2 = 5$$

With the chosen Δx_2, we can deduce Δx_1 via Eq (9.18), i.e.

$$\Delta x_1 = -\left[\left(\partial\Phi/\partial x_2\right) / \left(\partial\Phi/\partial x_1\right)\right] \Delta x_2 = -\left(x_1/x_2\right) \Delta x_2 = -(2/5)(-1) = 2/5$$

Thus, we can update x_1, i.e.

$$x_1 = x_{1,old} + \Delta x_1 = 2 + 2/5 = 12/5$$

The third point is

$$\text{Point \#3: } y\left(x_1 = 12/5, \ x_2 = 5\right) = 36.52$$
$$\Phi = x_1 x_2 - 12 = 0.$$

Incidentally, the constraint just happened to be satisfied; see Figure 9.20. This is partly because the step is small and, within the small step, x_1 varies largely linearly with x_2.

Let us continue the movement along the constraint, in the decreasing-y direction. With the new x_1 and x_2 values, we have

$$\Delta y = \left[5 - 4x_1^2/x_2\right] \Delta x_2 = \left[5 - 4(12/5)^2/(5)\right] \Delta x_2 = 0.392 \ \Delta x_2$$

Let us choose $\Delta x_2 = -1$; the negative sign is for decreasing y. Then,

$$x_2 = x_{2,old} + \Delta x_2 = 4$$

With the chosen Δx_2,

$$\Delta x_1 = -\left(x_1/x_2\right) \Delta x_2 = -\left[(12/5)/4\right](-1) = 12/20$$

We thus can update x_1, i.e.

$$x_1 = x_{1,old} + \Delta x_1 = 12/5 + 12/20 = 3$$

Therefore,

$$\text{Point \#4 } : y\left(x_1 = 3, x_2 = 4\right) = 38$$
$$\Phi = x_1 x_2 - 12 = 0$$

Again, the constraint is, by coincidence, exactly satisfied. The y value of Point #4 is larger than that of Point #3 and, hence, we have overshot the optimum y. Let us proceed with one more calculation, and see if we can bring it closer to the optimum.

Proceed to Point #5.

$$\Delta y = \left[5 - 4x_1^2/x_2\right] \Delta x_2 = \left[5 - 4(3)^2/(4)\right] \Delta x_2 = -4\Delta x_2$$

Let us choose $\Delta x_2 = +1$; the positive sign is for decreasing y. Then,

$$x_2 = x_{2,old} + \Delta x_2 = 5$$

With the chosen Δx_2,

$$\Delta x_1 = -\left(x_1/x_2\right)\Delta x_2 = -\left[3/5\right](+1) = -3/5$$

We thus can update x_1, i.e.

$$x_1 = x_{1,old} + \Delta x_1 = 3 - 3/5 = 12/5$$

Therefore, we are back to Point #3. In other words, the optimum is between Points #3 and #4. We need to reduce the resolution, for example, use $\Delta x_2 = \pm 0.5$, for better accuracy. The tradeoff is slower convergence, i.e. many more calculation steps are needed.

In Example 9.11, the calculations tread persistently along the "inversely linear" constraint line in small steps. Consequently, there is no need to alternately bring the point back onto the constraint after every move along the tangent. Let us look at the case where the constraint is not so "linear."

Example 9.12 ***Minimize a two-variable system with a nonlinear constraint using the Hemstitching method***

Given

In the spring, the blooming rate of maple trees, y, depends on the amount of moisture, x_1, and the amount of sunlight, x_2, where all parameters are in the normalized form. Specifically,

$$y = 6 + 4x_1^2 + 5x_2 \tag{E9.12.1}$$

In Calgary, the normalized moisture and sunlight are related in the following manner,

$$x_1^2 x_2^3 = 35 \tag{E9.12.2}$$

Find

The minimum blooming rate using the Hemstitching method.

Solution

The constraint can be recast as

$$\Phi = x_1^2 x_2^3 - 35 = 0 \tag{E9.12.3}$$

This can be expressed as

$$x_1^2 = 35/x_2^3 \tag{E9.12.3a}$$

Let us follow Figure 9.21 with a starting point at

Point #1: $y\left(x_1 = 2,\ x_2 = 2\right) = 32$

However, this point is NOT on the constraint. To satisfy the constraint, we find the corresponding x_1 value for $x_2 = 2$. This is accomplished by substituting $x_2 = 2$ into Eq (E9.12.3), which gives

$$x_1 = 2.092$$

In short, the second point is

Point #2: $y(x_1 = 2.092, x_2 = 2) = 34.80$, with $\Phi = x_1^2 x_2^3 - 35 = 0$, i.e. on the constraint.

Next, we wish to move along the tangent of the constraint at Point #2, in the direction in which y decreases. For this, we rearrange Eq (9.20) into

$$\Delta y = \left[\left(\partial y/\partial x_1\right) - \left(\partial y/\partial x_2\right)\left(\partial \Phi/\partial x_1\right) / \left(\partial \Phi/\partial x_2\right)\right]\ \Delta x_1 \tag{E9.12.4}$$

where

$$\partial y/\partial x_1 = 8x_1, \qquad \partial y/\partial x_2 = 5$$
$$\partial \Phi/\partial x_1 = 2x_1 x_2^3, \quad \partial \Phi/\partial x_2 = 3x_1^2 x_2^2$$

Therefore,

$$\Delta y = \left[8x_1 - (10x_2)/(3x_1)\right] \Delta x_1 = \left[8(2.092) - 10(2)/(3(2.092))\right] \Delta x_1 = 13.55\, \Delta x_2$$

Let us choose $\Delta x_1 = -0.2$ for decreasing y. Then,

$$x_1 = x_{1,old} + \Delta x_1 = 1.892$$

With the chosen Δx_1, we can deduce the corresponding Δx_2 via Eq (9.18) rearranged, i.e.

$$\Delta x_2 = -\left[(\partial\Phi/\partial x_1)/(\partial\Phi/\partial x_2)\right] \Delta x_1 = -\left[2x_2/(3x_1)\right] \Delta x_1 = -\left[2(2)/[3(1.892)](-0.2)\right]$$
$$= 0.1410$$

Hence, we can update x_2, i.e.

$$x_2 = x_{2,old} + \Delta x_2 = 2 + 0.1410 = 2.141$$

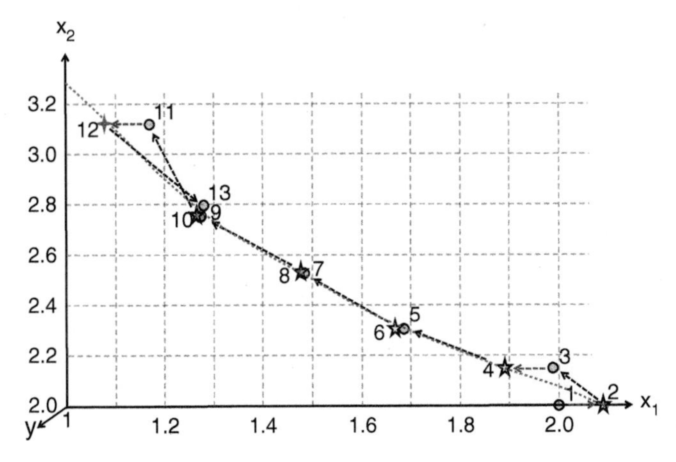

Figure 9.21 Minimization of a two-variable objective function subjected to a highly-nonlinear constraint using the Hemstitching method. The lightly dotted line denotes the equality constraint, $\Phi = x_1^2 x_2^3 - 35 = 0$. Source: D. Ting.

The third point is

$$\text{Point \#3: } y\left(x_1 = 1.892, x_2 = 2.141\right) = 31.02$$
$$\Phi = x_1^2 x_2^3 - 35 = 0.1167$$

While Point #3 is smaller than Point #2, it is a little off from the constraint. Hence, we proceed to bring the prodigal child (point) back onto the path (constraint); see Figure 9.21. To do so, we force the satisfaction of the right path, i.e. the constraint, by finding the corresponding value of x_1 for x_2 = 2.141. From Eq (9.13.3), we have

$$x_1^2 = 35/x_2^3$$

or

$$x_1 = \sqrt{(35/x_2^3)} = \sqrt{[35/(2.141)^3]} = 1.889$$

As such, we have

$$\text{Point \#4: } y\left(x_1 = 1.889, x_2 = 2.141\right) = 30.97$$
$$\Phi = x_1^2 x_2^3 - 35 = 0$$

Let us move along the tangent at Point #4, in search of smaller y.

$$\Delta y = \left[8x_1 - \left(10x_2\right) / \left(3x_1\right)\right] \Delta x_1 = \left[8(1.889) - 10(2.141) / (3(1.889))\right] \Delta x_1 = 3.771 \ \Delta x_2$$

Let us choose $\Delta x_1 = -0.2$ for decreasing y. Then,

$$x_1 = x_{1,old} + \Delta x_1 = 1.689$$

With the chosen Δx_1, we can deduce Δx_2 via Eq (9.18) rearranged, i.e.

$$\Delta x_2 = -\left[\left(\partial\Phi/\partial x_1\right) / \left(\partial\Phi/\partial x_2\right)\right] \Delta x_1 = -\left[2x_2/\left(3x_1\right)\right] \Delta x_1$$
$$= -\left[2(2.141) / [3(1.689)](-0.2)\right] = 0.1691$$

Therefore, we can update x_2, i.e.

$$x_2 = x_{2,old} + \Delta x_2 = 2.141 + 0.1691 = 2.310$$

The fifth point is

$$\text{Point \#5: } y\left(x_1 = 1.691, x_2 = 2.310\right) = 28.95$$
$$\Phi = x_1^2 x_2^3 - 35 = 0.1446$$

Point #5 is smaller than Point #4, but it is a little off from the constraint. Let us bring the wandering sheep back on track,

$$x_1 = \sqrt{\left(35/x_2^3\right)} = \sqrt{(35/(2.310)^3]} = 1.685.$$

Hence, we have, on the constraint,

$$\text{Point \#6: } y\left(x_1 = 1.685, x_2 = 2.310\right) = 28.91$$
$$\Phi = x_1^2 x_2^3 - 35 = 0$$

Let us strive along the tangent at Point #6, in search of an even smaller y.

$$\Delta y = \left[8x_1 - \left(10x_2\right) / \left(3x_1\right)\right] \Delta x_1 = \left[8(1.685) - 10(2.310) / (3(1.685))\right] \Delta x_1 = 8.910 \ \Delta x_2$$

Let us choose $\Delta x_1 = -0.2$ for decreasing y. Then,

$$x_1 = x_{1,old} + \Delta x_1 = 1.485$$

With the chosen Δx_1, we can deduce Δx_2 via Eq (9.18) rearranged, i.e.

$$\Delta x_2 = -\left[\left(\partial\Phi/\partial x_1\right) / \left(\partial\Phi/\partial x_2\right)\right] \Delta x_1 = -\left[2x_2/\left(3x_1\right)\right] \Delta x_1$$
$$= -\left[2(2.310) / [3(1.485)](-0.2)\right] = 0.2074$$

We thus can update x_2, i.e.

$$x_2 = x_{2,old} + \Delta x_2 = 2.310 + 0.2074 = 2.517$$

The next point is

$$\text{Point \#7: } y\left(x_1 = 1.485, x_2 = 2.517\right) = 27.41$$
$$\Phi = x_1^2 x_2^3 - 35 = 0.1840$$

Point #7 is smaller than Point #6, but it is a little off track. Let us bring the stray flock back to the path.

$$x_1 = \sqrt{(35/x_2^3)} = \sqrt{[35/(2.517)^3]} = 1.481$$

Hence, we have, on the constraint,

$$\text{Point \#8: } y\,(x_1 = 1.481, x_2 = 2.517) = 27.36$$
$$\Phi = x_1^2 x_2^3 - 35 = 0$$

Let us continue along the tangent at Point #8 see if there is a smaller y.

$$\Delta y = [8x_1 - (10x_2)\,/\,(3x_1)]\ \Delta x_1 = [8\,(1.481) - 10\,(2.517)\,/\,(3\,(1.481))]\ \Delta x_1 = 6.184\ \ \Delta x_2$$

Let us choose Δx_1 = -0.2 for decreasing y. Then,

$$x_1 = x_{1,old} + \Delta x_1 = 1.281$$

With the chosen Δx_1, we can deduce Δx_2 via Eq (9.18) rearranged, i.e.

$$\Delta x_2 = -\ [(\partial\Phi/\partial x_1)\,/\,(\partial\Phi/\partial x_2)]\ \Delta x_1 = -\ [2x_2/\,(3x_1)]\ \Delta x_1$$
$$= -\ [2\,(2.517)\,/\,[3\,(1.281)]\,(-0.2) = 0.2620]$$

Thus, we can update x_2, i.e.

$$x_2 = x_{2,old} + \Delta x_2 = 2.517 + 0.2620 = 2.779$$

The next point is

$$\text{Point \#9: } y\,(x_1 = 1.281, x_2 = 2.779) = 26.46$$
$$\Phi = x_1^2 x_2^3 - 35 = 0.2422$$

Point #9 is smaller than Point #8, but it is not quite on the path. To return to the constraint, we have

$$x_1 = \sqrt{(35/x_2^3)} = \sqrt{[35/(2.779)^3]} = 1.277$$

Hence, we have, on the constraint,

$$\text{Point \#10: } y\,(x_1 = 1.277, x_2 = 2.779) = 27.36$$
$$\Phi = x_1^2 x_2^3 - 35 = 0$$

Let us pick up the pace on the way home, along the tangent at Point #10.

$$\Delta y = [8x_1 - (10x_2)\,/\,(3x_1)]\ \Delta x_1 = [8\,(1.277) - 10\,(2.779)\,/\,(3\,(1.277))]\ \Delta x_1 = 2.957\ \Delta x_2$$

Let us choose $\Delta x_1 = -0.2$ for decreasing y. Then,

$$x_1 = x_{1,old} + \Delta x_1 = 1.077$$

With the chosen Δx_1, we can deduce Δx_2 via Eq (9.18) rearranged, i.e.

$$\Delta x_2 = -\ [(\partial\Phi/\partial x_1)\,/\,(\partial\Phi/\partial x_2)]\ \Delta x_1 = -\ [2x_2/\,(3x_1)]\ \Delta x_1$$
$$= -\ [2\,(2.779)\,/\,[3\,(1.077)]\,(-0.2) = 0.3442]$$

With that we can update x_2, i.e.

$$x_2 = x_{2,old} + \Delta x_2 = 2.779 + 0.3442 = 3.124$$

The next point is

$$\text{Point \#11: } y\left(x_1 = 1.077, x_2 = 3.124\right) = 26.26$$
$$\Phi = x_1^2 x_2^3 - 35 = 0.3336$$

Point #11 is marginally smaller than Point #10 and it needs to be brought back to the constraint. Therefore,

$$x_1 = \sqrt{\left(35/x_2^3\right)} = \sqrt{\left[35/(3.124)^3\right]} = 1.072$$

Hence, we have, on the constraint,

$$\text{Point \#12: } y\left(x_1 = 1.072, x_2 = 3.124\right) = 26.21$$
$$\Phi = x_1^2 x_2^3 - 35 = 0$$

With the change in y slowing down, we persist along the tangent at Point #12.

$$\Delta y = \left[8x_1 - \left(10x_2\right) / \left(3x_1\right)\right] \Delta x_1 = \left[8\,(1.072) - 10\,(3.124) / (3\,(1.072))\right] \Delta x_1 = -0.2286\,\Delta x_2$$

We note *the sign has changed*; we choose $\Delta x_1 = +0.2$ for decreasing y. Then,

$$x_1 = x_{1,old} + \Delta x_1 = 1.272$$

With the chosen Δx_1, we can deduce Δx_2 via Eq (9.18) rearranged, i.e.

$$\Delta x_2 = -\left[\left(\partial\Phi/\partial x_1\right) / \left(\partial\Phi/\partial x_2\right)\right] \Delta x_1$$
$$= -\left[2x_2 / \left(3x_1\right)\right] \Delta x_1 = -\left[2\,(3.124) / [3\,(1.272)]\,(+0.2)\right] = -0.3275$$

Accordingly, we can update x_2, i.e.

$$x_2 = x_{2,old} + \Delta x_2 = 3.124 - 0.3275 = 2.796$$

The next point is

$$\text{Point \#13: } y\left(x_1 = 1.272, x_2 = 2.796\right) = 26.45$$
$$\Phi = x_1^2 x_2^3 - 35 = 0.3496$$

Point #13 is actually slightly larger than Point #12, indicating that we have converged, i.e. slightly overshot the optimum. Let us bring it back to the constraint and see the value.

$$x_1 = \sqrt{\left(35/x_2^3\right)} = \sqrt{\left[35/(2.796)^3\right]} = 1.265$$

Hence, we have, on the constraint,

$$\text{Point \#14: } y\left(x_1 = 1.265, x_2 = 2.796\right) = 26.38$$
$$\Phi = x_1^2 x_2^3 - 35 = 0$$

This slightly larger y at Point #14, compared to Point #12, confirms that we have passed the optimum. Therefore, within the utilized resolutions, Point #12 is the solution.

Problems

9.1 Minimize the hot water transportation cost

The cost of transporting steam is a function of the pipe diameter, x_1, and the pipe insulation thickness, x_2. Specifically,

$$y = 5x_1 + 20/x_1 + 2.5x_2 + 4/\ln\left[(x_1 + x_2)/x_1\right]$$

Find x_1 and x_2 which give the minimum cost, y.

9.2 Maximize the volume of a box

The volume of a box of height x_1, width x_2 and depth x_2 is to be maximized. The height plus the perimeter of the base is to be no more than 100. What are the corresponding optimal height and width, x_1* and x_2*? Solve the problem as an unconstrained problem and then as a constrained one.

9.3 Maximize a sine function via a Fibonacci Search

The (coefficient of) performance of the cyclic operation of an engine can be described as

$$y = \ln x \sin\left(x^2/25\right)/x$$

where $x^2/25$ is in radians and y is unimodal between $1.5 \leq x \leq 10$. Search for the maximum y using the Fibonacci Search method for an uncertainty of less than 0.3 in x.

9.4 Minimize a two-variable objective function using univariate search

The percentage exergy destruction of a pumping system consisting of two pumps in parallel can be described as

$$y = x_1 + 16/\left(x_1 x_2\right) + \tfrac{1}{2}x_2$$

where x_1 and x_2 are the normalized flow rates through the two pumps. Find the minimum exergy destruction using the univariate search method. Assume the normalized flow rates, x_1 and x_2, have values of positive integers. Start the optimization with $x_2 = 3$.

9.5 Minimizing heat loss from a furnace

The heat loss from a furnace varies with the temperature and the insulation thickness,

$$y = \tfrac{1}{2}x_1^2 + 4/\left(x_1 x_2\right) + 2x_2$$

where x_1 is the temperature and x_2 is the insulation thickness. Start with $y(x_1 = 1, x_2 = 1)$; optimize y via the Univariate Search and Steepest Ascent/Descent methods.

9.6 Optimization of a three-variable unconstrained problem

The cost of extruding a cylindrical plastic part,

$$y = 58x_1/x_2 + 305/\left(x_1 x_3\right) + 3x_1x_2 + 4x_3$$

where x_1 is the diameter of the part, x_2 is the velocity, and x_3 is the temperature. Minimize the cost, starting with $y(x_1 = 1, x_2 = 1, x_3 = 1)$, using $\Delta x = 0.1$, via the (a) Lattice Search, (b) Univariate Search, and (c) Steepest Ascent/Descent methods.

9.7 Minimize a three-variable objective function using the Steepest Ascent/Descent method

The entropy generation of a three-component renewable energy generation system,

$$y = 72x_1/x_2 + 360/\left(x_1 x_3\right) + x_1 x_2 + 2x_3$$

where x_1, x_2, and x_3 represent the capacity of the three components. Determine the minimum entropy generation via the Steepest Ascent/Descent method. Start with $x_1 = 2$ and $x_2 = 7$, with $\Delta x_1 = 1$.

9.8 Minimize the size of an insulated pipe via the Hemstitching method

The size of an insulated pipe for transporting a chilled refrigerant is to be minimized. The size of the insulated pipe can be expressed as,

$$y = D + 2T$$

where D is the diameter of the pipe and T is the insulation thickness. The total annual operating cost consists of the sum of the pumping cost and the heating cost and it is to be no more than $50,000. The pumping cost $= 8/D^5$ and the heating cost $= 1500/T$. Find the objective function and the constraint. Deduce the minimum cost using the Hemstitching method; start with $D = 0.2$ and $T = 0.1$ and employ a ΔD of 0.005.

9.9 Minimize the heat transfer rate from a spherical reactor

For safety reasons, the heat transfer rate from a spherical reactor of diameter D is to be minimized. The heat transfer rate,

$$Q' = hAT$$

where the heat transfer coefficient,

$$h = 2 + 0.5T^{0.2}/D$$

Due to material limitations, $DT = 20$. Find the minimum Q' using the Hemstitching method and compare the solution with that obtained via simple differentiation.

9.10 The effectiveness of common search methods

Which of the following properly rank the common search methods from the most effective one (least number of calculations) to the least effective one?

a) Golden Section, Fibonacci, Dichotomous, Exhaustive
b) Exhaustive, Dichotomous, Fibonacci, Golden Section
c) Fibonacci, Golden Section, Dichotomous, Exhaustive
d) Dichotomous, Golden Section, Fibonacci, Exhaustive
e) Exhaustive, Dichotomous, Fibonacci, Golden Section

9.11 Limited to unimodal objective function

Which of the following search methods can be applied for an objective function which has more than one optimum?

a) Exhaustive
b) Dichotomous
c) Fibonacci
d) Golden Section

References

Burmeister, L.C. (1998). *Elements of Thermal-Fluid System Design*. Upper Saddle River, NJ: Prentice-Hall.

Fibonacci, 2013, https://plus.maths.org/content/life-and-numbers-fibonacci, accessed on September 26, 2019.

Jaluria, Y. (1998). *Design and Optimization of Thermal Systems*. New York: McGraw-Hill.

Shah, B.V., Buehler, R.J., and Kempthorne, O. (1964). Some algorithms for minimizing a function of several variables. *Journal of the Society for Industrial and Applied Mathematics* 12(1): 74–92.

Stoecker, W.F. (1989). *Design of Thermal Systems*, 3rd ed. New York: McGraw-Hill.

Wilde, D.J. (1964). *Optimum Seeking Methods*. Englewood Cliffs, NJ: Prentice-Hall.

10

Geometric Programming

Where there is matter, there is geometry.

– Johannes Kepler

Chapter Objectives

- Be aware of common types of programming optimization methods.
- Learn to apply geometric programming for single- and multi-variable, unconstrained problems.
- Extend the application of geometric programming to constrained, multi-variable problems.

Nomenclature

A	area
A, B	constants, with pertinent numerical values
a, b, c, … , k	constants, with pertinent numerical values
c_p	heat capacity at constant pressure
D	degree of difficulty
F	a function
h	heat transfer coefficient
m	mass
N	number, the total number of terms in the objective function and in the constraints
n	number, the number of independent variables
P	pressure
p	a constant indicating the importance of the constraint
PCM	phase change material
Q	heat; Q' heat transfer rate
R	gas constant

S_C sensitivity coefficient
X, x independent variable
Y, y dependent variable, objective function, composite objective function
w weight

Greek and Special Symbols

γ the specific heat ratio
Δ difference, a small change
λ Lagrange Multiplier
Φ equality constraint which is equal to zero, $\Phi = 0$
ϕ equality constraint, $\phi = k$
∇ mathematical gradient
$\forall$ volume

10.1 Common Types of Programming

According to Jazzwant, "Programming is breaking of one big impossible task into several very small possible tasks." Another way to look at it is via the eyes of Per Brinch Hansen, who said that "Programming is the art of writing essays in crystal clear prose and making them executable." Within the context of this book, programming simply refers to optimization. The most common types of programming are:

1. geometric programming
2. linear programming
3. dynamic programming.

Geometric programming is suited for problems where the objective function and the constraints can be represented as sums of polynomials. Clasen (1963), Duffin et al. (1967), Zener (1971), Beightler and Philips (1976) and Ecker (1980) were some of the earliest who systematically disseminated this method. Linear programming is applicable when the objective function and the constraints are all linear in terms of the involved independent variables. Many operations research problems encountered in industrial engineering fall into this category. An attractive feature of linear programming is its ability to handle large systems of variables, which is commonly the case in traffic networks. The equations describing thermal systems are typically nonlinear in nature. Therefore, linear programming is not normally utilized in optimization of thermofluid systems. Dynamic programming can be exploited to obtain the best path through continuous processes that can be described as a series of distinct stages or steps. For example, it can be used to deduce the optimum configuration of the hot water distribution in a building. It is also pragmatic in optimizing a manufacturing line. The obtained optimum is not a point solution of the objective function, but rather a path or curve of the objective function.

It is clear from the above that linear programming and dynamic programming are ill suited for solving nonlinear thermofluid problems. In thermofluids, polynomials are commonly employed to describe the thermodynamic properties and system characteristics such as that of a pump. In other words, curve fitting of experimental data and numerical results of thermal systems often lead to polynomials and power-law variations. By this very nature, we will focus on geometric programming for the rest of this chapter. Once geometric programming is applied, the optimal solution and the sensitivity of the solution to changes in the constraints can easily be deduced.

10.2 What Is Geometric Programming?

Geometric programming is a nonlinear optimization technique which is applicable when the objective function and the constraints can be expressed as sums of polynomials; see, for example (Stoeker, 1989; Jaluria, 1998). The independent variables in these polynomials may be raised to positive or negative, integer or non-integer exponents. For example, an objective function could be

$$y = a\, x_1^e + b\, x_2^f + c\, x_1^g x_2^h + d \tag{10.1}$$

or

$$y = a\, x_1^e + b\, x_2^f + c\, x_1^g x_2^h + d \text{ with } \phi = x_1^i\, x_2^j = k \tag{10.2}$$

where x_1 and x_2 are the independent variables and ϕ is the equality constraint. The coefficients or constants, a, b, c, ... k, can be positive or negative, integer or non-integer.

When the conditions are appropriate for the application of geometric programming, the calculation process is relatively easy. The values of the independent variables need not be resolved to obtain the optimum objective function. Geometric programming not only generates the optimum objective function, but also reveals the relative contributions of the various terms in the objective function. Namely, the significance of distinct involved aspects is unveiled, along with the optimum value of the objective function.

The extent of the challenge in solving a particular problem using geometric programming is a function of the number of independent variables with respect to the total number of terms. The indicator is called *degree of difficulty*, and it is defined as

$$D = N - (n + 1) \tag{10.3}$$

where N is the total number of terms in the objective function and in the constraints, and n is the number of independent variables. Consider an unconstrained problem with objective function,

$$y = 2\, x_1^3 + 1.4\, x_2^{3.2} + 4.2\, x_1^5\, x_2^7 \tag{10.4}$$

The corresponding degree of difficulty,

$$D = N - (n + 1) = 3 - (2 - 1) = 0 \tag{10.5}$$

A zero degree of difficulty implies that the problem is easy. We shall limit ourselves to $D = 0$ cases only; when D is greater than zero, or it cannot be reduced to zero, the application of geometric programming involves the solutions of nonlinear equations and can be quite complicated.

10.3 Single-Variable, Unconstrained Geometric Programming

Consider the single-independent-variable, unconstrained case with objective function,

$$y = A\, x^a + B\, x^b = y_1 + y_2 \tag{10.6}$$

where the two terms,

$$y_1 = A\,x^a \text{ and } y_2 = B\,x^b \tag{10.7}$$

represent the individual contribution to the overall objective function. For example, the total production cost of a product is equal to the initial cost plus the operating cost. As the system size increases, the initial cost increases, but the operating cost per item decreases. This is illustrated in Figure 10.1. According to geometric programming, the optimum objective function

$$y^* = \left[\left(A\,x^a\right)/w_1\right]^{w_1}\left[\left(B\,x^b\right)/w_2\right]^{w_2} \tag{10.8}$$

where the sum of the weighting factors,

$$w_1 + w_2 = 1 \tag{10.9}$$

and the sum of the independent exponent-weight factor products,

$$aw_1 + bw_2 = 0 \tag{10.10}$$

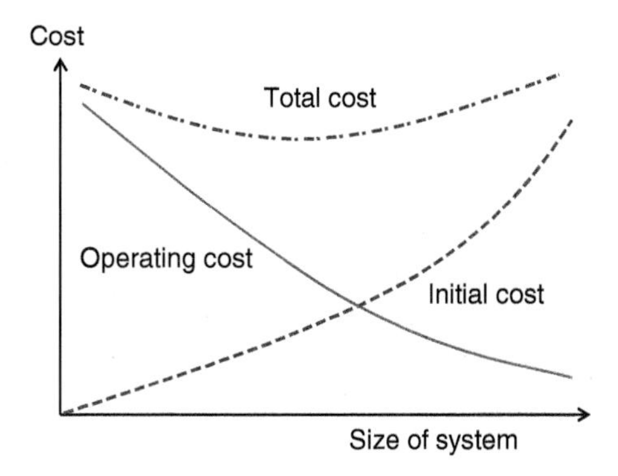

Figure 10.1 Total cost consisting of capital cost and operating cost Source: D. Ting.

Equation (10.8) can be expanded into

$$y^* = \left[\left(A\,x^a\right)/w_1\right]^{w_1}\left[\left(B\,x^b\right)/w_2\right]^{w_2} = \left[(A\,/w_1\right]^{w_1}\left[(B\,/w_2\right]^{w_2} x^{(aw_1+bw_2)} \tag{10.11}$$

To be consistent with the rest of the book, an asterisk, *, denotes "at optimum." According to Eq (10.10), the exponent of x is zero, and hence,

$$y^* = \left[A/w_1\right]^{w_1}\left[B/w_2\right]^{w_2} \tag{10.12}$$

Also, we can see that the weighting factors, which signify the relative contributions of the respective terms to the (total) objective function at the optimum,

$$w_1 = y^*/\left(y_1^* + y_2^*\right) = y_1^*/y^* \tag{10.13}$$

and

$$w_2 = y_2^*/\left(y_1^* + y_2^*\right) = y_2^*/y^* \tag{10.13a}$$

The proof of these formulae is provided as Example 10.1.

Example 10.1 ***Proof regarding the relationships of the weighting factors***

Given

Objective function,

$$y = A\,x^a + B\,x^b = y_1 + y_2 \tag{10.6}$$

where

$$y_1 = A\,x^a \text{ and } y_2 = B\,x^b \tag{10.7}$$

At the optimum, the objective function,

$$y^* = \left[\left(A\,x^a\right)/w_1\right]^{w_1}\left[\left(B\,x^b\right)/w_2\right]^{w_2} \tag{10.8}$$

where the sum of the weighting factors,

$$w_1 + w_2 = 1 \tag{10.9}$$

and the sum of the independent exponent-weight factor products,

$$aw_1 + bw_2 = 0 \tag{10.10}$$

Find

Prove that

$$y^* = \left[(A/w_1\right]^{w_1}\left[(B/w_2\right]^{w_2} \tag{10.12}$$

Solution

Differentiating the objective function,

$$y = A\,x^a + B\,x^b = y_1 + y_2$$

with respect to the independent variable x, we get

$$dy/dx = A\,a\,x^{a-1} + B\,b\,x^{b-1} = 0 \tag{E10.1.1}$$

at the optimum. Multiplying it by x gives

$$A\,a\,x^a + B\,b\,x^b = 0 \tag{E10.1.2}$$

But, from Eq (10.7), $y_1 = Ax^a$ and $y_2 = Bx^b$ and, hence, we can rewrite Eq (E10.1.2) as

$$a\,y_1{}^* + b\,y_2{}^* = 0 \tag{E10.1.3}$$

Introduce a function,

$$F = \left(y_1/w_1\right)^{w_1} + \left(y_2/w_2\right)^{w_2} \tag{E10.1.4}$$

with

$$w_1 + w_2 = 1 \tag{E10.1.5}$$

Taking the natural log, we get

$$\ln F = w_1\left(\ln\,y_1 - \ln w_1\right) + w_2\left(\ln y_2 - \ln w_2\right) \tag{E10.1.6}$$

We can apply the Lagrange Multiplier method to optimize this natural log function, assuming that it has an equality constraint,

$$\Phi = w_1 + w_2 - 1 = 0 \tag{E10.1.7}$$

The Lagrange Multiplier execution is

$$\nabla\,(\ln\ F) + \lambda\nabla\Phi = 0 \tag{E10.1.8}$$

This gives

$$\ln y_1 - \ln w_1 - 1 + \lambda = 0 \tag{E10.1.9}$$

and

$$\ln y_2 - \ln w_2 - 1 + \lambda = 0 \tag{E10.1.10}$$

Subtracting Eq (10.1.10) from Eq (10.1.9), we get

$$(\ln y_1 - \ln w_1) - (\ln y_2 - \ln w_2) = 0 \tag{E10.1.11}$$

Taking the anti-natural-log gives

$$y_1/y_2 = w_1/w_2 \tag{E10.1.12}$$

or

$$w_2 = (y_2/y_1)\ \ w_1 \tag{E10.1.12a}$$

Substituting this into Eq (10.1.5), we get

$$w_1 + [(y_2/y_1)\, w_1] = 1 \tag{E10.1.13}$$

which can be restated as

$$w_1 = y_1/\,(y_1 + y_2) \tag{E10.1.13a}$$

Substituting this into Eq (10.1.12) gives

$$w_2 = y_2/\,(y_1 + y_2) \tag{E10.1.14}$$

From Eq (10.1.3), we have

$$y_1{}^* = -b\, y_2{}^*/a \tag{E10.1.15}$$

and

$$y_2{}^* = -a\, y_1{}^*/b \tag{E10.1.15a}$$

Substituting these into Eqs. (E10.1.13a) and (E10.1.14), respectively, give

$$w_1 = b/\,(a + b) \tag{E10.1.16}$$

and

$$w_2 = a/\,(a - b) \tag{E10.1.17}$$

respectively.

At optimum

$$y^* = F^* = (y_1{}^*/w_1)^{w_1}(y_2{}^*/w_2)^{w_2} \tag{E10.1.4a}$$

Substituting for Eq (10.7) and the exponents for Eqs. (E10.1.16) and (E10.1.17) gives

$$\begin{aligned} y^* = F^* &= (Ax^a/w_1)^{b/(a+b)}(Bx^b/w_2)^{a/(a-b)} = (A/w_1)^{b/(a+b)}(B/w_2)^{a/(a-b)} \\ &= (A/w_1)^{w_1}(B/w_2)^{w_2}. \end{aligned} \tag{E10.1.18}$$

It is interesting to note that

1. The independent variable x has been eliminated.
2. The weighing factors, w_1 and w_2, indicate the relative contributions of the terms y_1 and y_2 at the optimum.

To summarize, the unique features of geometric programming include the partition of the objective function into terms. The respective and relative weights of these terms in altering the value of the objective function in the vicinity of the optimum are the main results. This piece of information is beneficial in practice, advising the engineer which factor to focus on for best operation and to further the design. The optimum objective function is deduced directly from the weights, i.e. not from the corresponding values of the independent variables at optimum. This eases the calculation, for typically more effort is required to solve the independent variables. The optimal independent variables are calculated, lastly and only when desirable, from the optimal objective function value and the weights. Let us go through a single-variable, unconstrained example to illustrate the robustness of geometric programming.

Example 10.2 ***A single-variable, unconstrained problem with two terms***

Given

The cost of air conditioning a building, y, is a function of the required cooling load, x, which is the thermal energy removal rate and/or quantity. Specifically,

$$y = 3.5\,x^{1.4} + 14.8/x^{2.2} \tag{E10.2.1}$$

where y and x are in appropriate non-dimensional form. The first term on the right-hand side signifies the equipment cost, and the second term represents the operating cost. It is typical that a better-performing system costs more initially, but it pays off with time due to its lower operating cost.

Find

The minimum cost via geometric programming.

Solution

We see that Eq (10.2.1) is a polynomial expression and, hence, well posted for geometric programming. The degree of difficulty,

$$D = N - (n + 1) = 2 - (1 + 1) = 0 \tag{E10.2.2}$$

therefore, the problem is simple to solve via geometric programming.

For geometric programming, we have, at optimum,

$$y^* = \left[(A/w_1\right]^{w_1}\left[(B/w_2\right]^{w_2} \tag{E10.2.3}$$

where $w_1 + w_2 = 1$ and $aw_1 + bw_2 = 0$. For the specific problem at hand,

$$y^* = \left[(3.5/w_1\right]^{w_1}\left[(14.8/w_2\right]^{w_2} \tag{E10.2.4}$$

with

$$w_1 + w_2 = 1 \tag{E10.2.5}$$

and

$$1.4w_1 - 2.2w_2 = 0 \qquad \text{(E10.2.6)}$$

From Eqs. (E10.2.5) and (E.10.2.6), we can resolve for

$$w_1 = 0.611, w_2 = 0.389$$

This implies that the equipment cost, the first term, contributes 61% to the total cost of air conditioning, and the operating cost, the second term, forms the remaining 39%. Substituting these into Eq (10.2.4) gives the minimum cost,

$$y^* = 11.96$$

Substituting this into Eq (10.2.1),

$$y = 11.96 = 3.5\, x^{1.4} + 14.8/x^{2.2}$$

we obtain $x^* = 1.692$. To conclude, the minimum cost, $y^* = 11.96$, takes place at a cooling load, x^*, of 1.692.

10.4 Multi-Variable, Unconstrained Geometric Programming

We can extend the application of geometric programming to multiple-variable problems. Let us stay with the unconstrained case for now. Consider an objective function as a function of two variables,

$$y = a\, x_1^d + b\, x_2^e + c\, x_1^f x_2^g = y_1 + y_2 + y_3 \qquad (10.14)$$

The degree of difficulty of this problem is $D = 3 - (2 + 1) = 0$, suggesting that it is easily solvable. The objective function at the optimum,

$$y^* = \left(a\, x_1^d/w_1\right)^{w_1} \left(b\, x_2^e/w_2\right)^{w_2} \left(c\, x_1^f x_2^g/w_3\right)^{w_3} \qquad (10.15)$$

This can be reduced into

$$y^* = \left(a/w_1\right)^{w_1} \left(b/w_2\right)^{w_2} \left(c/w_3\right)^{w_3}, \qquad (10.16)$$

where

$$w_1 + w_2 + w_3 = 1 \qquad (10.17)$$

For the exponents associated with x_1, we have

$$d\, w_1 + f\, w_3 = 0 \qquad (10.18)$$

and, for the exponents associated with x_2,

$$e\, w_2 + g\, w_3 = 0 \qquad (10.19)$$

We will take advantage of a natural thermofluid system, a metabolizing and perspiring living human body, to demonstrate the versatility of geometric programming in handling multi-variable problems.

Example 10.3 ***Optimize sweating rate***

Given

For an average person, the rate of sweating, y, is a function of the individual rate of activity, x_A, the ambient air temperature, x_T, and humidity, x_H. Specifically,

$$y = 70\,x_A x_T + 30\,x_A x_H + 20\,x_T x_H + 500/(x_A x_T x_H) \tag{E10.3.1}$$

Find

The optimum sweating rate via geometric programming.

Solution

Degree of difficulty, D = 4 – (3 – 1) = 0. This implies that the problem is easily solvable via geometric programming.

At the optimum,

$$y^* = (a/w_1)^{w_1}(b/w_2)^{w_2}(c/w_3)^{w_3}(d/w_4)^{w_4} \tag{E10.3.2}$$

i.e. according to Eq (10.3.1),

$$y^* = (70/w_1)^{w_1}(30/w_2)^{w_2}(20/w_3)^{w_3}(500/w_4)^{w_4} \tag{E10.3.3}$$

where the terms on the right-hand side correspond to $x_A x_T$, $x_A x_H$, $x_T x_H$, and $x_A x_T x_H$, respectively.

Also,

$$w_1 + w_2 + w_3 + w_4 = 1 \tag{E10.3.4}$$

and

$$x_A^{w_1+w_2-w_4} = x_A^0 = 1, \text{hence, } w_1 + w_2 - w_4 = 0 \tag{E10.3.5}$$

$$x_B^{w_2+w_2-w_4} = x_B^0 = 1, \text{hence, } w_2 + w_3 - w_4 = 0 \tag{E10.3.6}$$

and

$$x_H^{w_1+w_3-w_4} = x_H^0 = 1$$

hence,

$$w_1 + w_3 - w_4 = 0 \tag{E10.3.7}$$

Subtracting Eq (10.3.6) from Eq (10.3.5) leads to

$$w_1 - w_3 = 0, \text{or, } w_1 = w_3 \tag{E10.3.8}$$

Substituting this into Eq (10.3.7), we get

$$(w_3) + w_3 = w_4, \text{or, } w_4 = 2w_3 = 2w_1 \tag{E10.3.9}$$

Substituting this into Eq (10.3.6) gives

$$w_2 + w_3 - (2w_3) = 0, \text{or, } w_2 = w_3 \tag{E10.3.10}$$

In a word,

$$w_1 = w_2 = w_3 = ½w_4 \tag{E10.3.11}$$

Substituting this into Eq (10.3.4), we get

$$w_1 + (w_1) + (w_1) + 2(w_1) = 1$$

or

$$w_1 = 1/5$$

To sum things up, we have

$$w_1 = w_2 = w_3 = 1/5, \text{and } w_4 = 2/5$$

These results point out that the contribution of the first three terms,$70x_Ax_T$, $30x_Ax_H$ and $20x_Tx_H$, are equally weighted at 20% each. In plain English, the first three terms have an equal effect on the optimal value of the objective function or sweating rate. In contrast, the fourth term, $500/(x_A\ x_T\ x_H)$, has a double effect, i.e. 40%.

Substituting the weights into Eq (10.3.3) we get

$$y^* = (70/0.2)^{0.2}\ (30/0.2)^{0.2}\ (30/0.2)^{0.2}\ (500/0.4)^{0.4} = 304.7$$

This is the minimum normalized sweating rate. The corresponding weights, following Eqs. (E10.3.11) and (E10.3.1) and Eq (10.13), can be expressed in terms of the independent variables when the objective function is at the optimum. Namely,

$$\begin{aligned} w_1 &= 70x_A{}^*x_T{}^*/y^* = w_2 = 30x_A{}^*x_H{}^*/y^* = w_3 = 20x_T{}^*x_H{}^*/y^* \\ &= ½w_4 = ½500/\left(x_A{}^*x_T{}^*x_H{}^*\right)/y^* \end{aligned}$$

This can be rewritten as

$$0.2 = 70\ x_A{}^*\ x_T{}^*/y^* = 30\ x_A{}^*\ x_H{}^*/y^* = 20\ x_T{}^*\ x_H{}^*/y^* = 250/\left(x_A{}^*x_T{}^*x_H{}^*\right)/y^*.$$

Multiply by y*/10, and noting that y* = 304.7, gives

$$0.02y^* = 7\ x_A{}^*x_T{}^* = 3\ x_A{}^*x_H{}^* = 2\ x_T{}^*x_H{}^* = 25/\left(x_A{}^*x_T{}^*x_H{}^*\right)$$

or

$$6.093 = 7\ x_A{}^*x_T{}^* = 3\ x_A{}^*x_H{}^* = 2\ x_T{}^*x_H{}^* = 25/\left(x_A{}^*x_T{}^*x_H{}^*\right) \tag{E10.3.12}$$

From the x* terms, we see

$$x_T{}^* = 3x_H{}^*/7 \tag{E10.3.13}$$

$$x_A{}^* = 2x_H{}^*/7 \tag{E10.3.14}$$

$$x_A{}^* = 2x_T{}^*/3 \tag{E10.3.15}$$

etc. Substitute Eq (10.3.15) into Eq (10.3.12) to obtain

$$6.093 = 7\left(2x_T{}^*/3\right)x_T{}^*, \text{or}, x_T{}^* = 1.143$$

Substitute this into Eq (E10.3.13) and (E10.3.15) to get, respectively,

$$x_H{}^* = 7x_T{}^*/3 = 2.666 \text{ and } x_A{}^* = 2x_T{}^*/3 = 0.7618.$$

In a nutshell, the corresponding normalized rate of activity, normalized air temperature, and normalized humidity at minimum normalized sweating rate are, respectively,

$$x_A{}^* = 0.7618, \quad x_T{}^* = 1.143, \quad x_H{}^* = 2.666.$$

10.5 Constrained Multi-Variable Geometric Programming

Geometric programming is also applicable for equality-constrained problems. Consider the objective function, $y = f(x_1, x_2, x_3, x_4) = y_1 + y_2 + y_3 + y_4$, with constraints $y_4 + y_5 = 1$. According to geometric programming, we can write

$$y^* = (y_1/w_1)^{w_1}(y_2/w_2)^{w_2}(y_3/w_3)^{w_3} \tag{10.20}$$

where

$$w_i = y_i/y^*, w_1 + w_2 + w_3 = 1$$

and

$$(y_4/w_4)^{w_4}(y_5/w_5)^{w_5} = 1 \tag{10.21}$$

where

$$w_4 + w_5 = 1, w_4 = y_4/1 \text{ and } w_5 = y_5/1.$$

From the above, the objective function can then be written as

$$y^* = (y_1/w_1)^{w_1}(y_2/w_2)^{w_2}(y_3/w_3)^{w_3}(y_4/w_4)^{pw_4}(y_5/w_5)^{pw_5} \tag{10.22}$$

where p is a constant indicating the importance of the constraints and the product of the last two terms is equal to one. Applying the Lagrange Multiplier method, we have

$$\nabla\left(y_1 + y_2 + y_5\right) + \lambda\nabla\left(y_4 + y_5\right) = 0 \tag{10.23}$$

recalling that ∇ signifies partial derivative with respect to x_i. Multiplying each term by x_i gives

$$a_{11}w_1 + a_{12}w_2 + a_{13}w_3 + p\,a_{14}w_4 + p\,a_{15}w_5 = 0,$$

$$\vdots$$

$$a_{41}w_1 + a_{42}w_2 + a_{43}w_3 + p\,a_{44}w_4 + p\,a_{45}w_5 = 0, \tag{10.24}$$

where $w_1 + w_2 + w_3 = 1$, and $p(w_4 + w_5) = p$. Note that p is a constant indicating the importance of the constraint. It is related to the sensitivity coefficient and/or the Lagrange Multiplier,

$$S_C = \lambda = -p\,y^* \tag{10.25}$$

which indicates the effect of a small loosening of the constraint in the vicinity of the optimum. Let us revisit Example 10.3, but with a constraint and a slightly modified objective function.

Example 10.4 ***Optimize sweating rate with a constraint***

Given

For an average person, the normalized rate of sweating, y, is a function of the normalized individual rate of activity, x_A, normalized ambient air temperature, x_T, and normalized humidity, x_H. Precisely,

$$y = 70\,x_Ax_T + 30\,x_Ax_H + 20\,x_Tx_H \tag{E10.4.1}$$

To stay within healthy conditions, there has to be some minimal activity, air temperature and humidity, i.e.

$$x_A x_T x_H = 7 \tag{E10.4.2}$$

Simply put, a person's heart has to be pumping, the ambient temperature has to be sufficiently above absolute zero, and all ambient air has finite humidity.

Find

The optimum sweating rate via geometric programming. The corresponding optimum sweating rate, if $x_A\ x_T\ x_H = 7.07$ instead.

Solution

Degree of difficulty, D = 4 − (3 − 1) = 0. This infers that the problem is straightforward to resolve using geometric programming.

Expressing the constraint in the form, $y_4 = 1$, we have

$$y_4 = x_A x_T x_H/7 = 1 \tag{E10.4.2a}$$

At the optimum,

$$y^* = \left(70\ x_A x_T/w_1\right)^{w_1}\left(30\ x_A x_H/w_2\right)^{w_2}\left(20\ x_T x_H/w_3\right)^{w_3}\left(x_A x_T x_H/7/w_4\right)^{pw_4} \tag{E10.4.3}$$

This can be reduced to

$$y^* = \left(70/w_1\right)^{w_1}\left(30/w_2\right)^{w_2}\left(20/w_3\right)^{w_3}\left(1/7/w_4\right) pw_4 \tag{E10.4.3a}$$

where we can see from Eq (10.4.3) that

$$x_A^{w_1+w_2+pw_4} = x_A^0 = 1 \text{ or } w_1 + w_2 + p\,w_4 = 0 \tag{E10.4.4}$$

$$x_T^{w_1+w_3+pw_4} = x_T^0 = 1 \text{ or } w_1 + w_3 + p\,w_4 = 0 \tag{E10.4.5}$$

and

$$x_H^{w_2+w_3+pw_4} = x_H^0 = 1 \text{ or } w_2 + w_3 + p\,w_4 = 0 \tag{E10.4.6}$$

Also,

$$w_1 + w_2 + w_3 = 1 \tag{E10.4.7}$$

and

$$p\,w_4 = p \text{ or } w_4 = 1 \tag{E10.4.8}$$

Substituting these into Eqs. (E10.4.4)–(E10.4.7), we have

$$\begin{bmatrix} 1 & 1 & 0 & 1 \\ 1 & 0 & 1 & 1 \\ 0 & 1 & 1 & 1 \\ 1 & 1 & 1 & 0 \end{bmatrix}\begin{bmatrix} w_1 \\ w_2 \\ w_3 \\ p \end{bmatrix} = \begin{bmatrix} 0 \\ 0 \\ 0 \\ 1 \end{bmatrix}$$

Subtracting Row 2 from Row 1 gives

$$w_2 - w_3 = 0 \text{ or } w_2 = w_3$$

Subtracting Row 3 from Row 1 gives

$$w_1 - w_3 = 0 \text{ or } w_1 = w_3$$

Hence, we have

$$w_1 = w_2 = w_3$$

This asserts that all three terms have equal impact on the sweating rate at the optimum (minimum). Substituting this into Row 4, we have

$$w_1 + w_2 + w_3 = 1 \text{ or } w_1 = w_2 = w_3 = 1/3$$

Substituting this into Row 1 gives

$$1/3 + 1/3 + p\ (1) = 0 \text{ or } p = -2/3$$

Therefore, from Eq (10.4.3a), we have

$$y^* = (70/1/3)^{1/3}(30/1/3)^{1/3}(20/1/3)^{1/3}(1/7/1)^{-2/3} = 381.6$$

In short, the optimum (minimum) normalized sweating rate is 381.6. Besides,

$$w_1 = 70\, x_A{}^*x_T{}^*/y^* = w_2 = 30\ x_A{}^*x_H{}^*/y^* = w_3 = 20\, x_T{}^*x_H{}^*/y^*$$

This can be rewritten as

$$1/3 = 70\ x_A{}^*x_T{}^*/y^* = 30\, x_A{}^*x_H{}^*/y^* = 20\ x_T{}^*x_H{}^*/y^*$$

Multiplying by $y^*/10$ gives

$$y^*/30 = 7\, x_A{}^*x_T{}^* = 3\, x_A{}^*x_H{}^* = 2\, x_T{}^*x_H{}^*. \qquad \text{(E10.4.9)}$$

From the x^* terms, we see

$$x_T{}^* = 3x_H{}^*/7, x_A{}^* = 2x_H{}^*/7, x_A{}^* = 2x_T{}^*/3$$

or

$$x_A{}^* = 2x_H{}^*/7 = 2x_T{}^*/3, \ldots$$

Substituting this and $y^* = 381.6$ into Eq (10.4.9), we have

$$12.72 = 7\ \left(2x_T{}^*/3\right)\ x_T{}^* \text{or } x_T{}^* = 1.651$$

and, thus,

$$x_A{}^* = 2x_T{}^*/3 = 1.101,\ \ x_H{}^* = 7x_T{}^*/3 = 3.852$$

To sum things up, the rate of activeness, the ambient air temperature, and the humidity corresponding to minimum sweating are, respectively,

$$x_A{}^* = 1.651,\ \ x_T{}^* = 1.101,\ \ x_H{}^* = 3.852$$

From Eq (10.25), the sensitivity coefficient,

$$S_C = \lambda = -p\, y^*$$

For the studied case,

$$\left.S_C\right)_a = \lambda_a = -p_a\, y^* = -(-2/3)\,(381.6) = 127.2$$

This is applicable to the constraint as described by Eq (10.4.2a),

$$\phi_a = x_A x_T x_H/7 = 1$$

This is derived from the original equality constraint, Eq (10.4.2),

$$\phi_a = x_A x_T x_H = 7$$

Therefore, the corresponding original sensitivity coefficient,

$$S_C = \lambda = \lambda_a/7 = -\left(p_a/7\right)\ y^* = (127.2/7) = 18.2$$

This indicates the rate of change in y* with respect to a small change in ϕ. For ϕ changing from 7 to 7.07, we have

$$\Delta y^* = 18.2\,(7.07 - 7) = 1.27$$

In other words, the optimum sweating rate, y*, would become 383. Note that this sensitivity coefficient, or Lagrange Multiplier, is strictly applicable at y*, i.e. $\partial y^*/\partial \phi$. To that end, this approximation deteriorates with increasing Δ, especially when the gradient in the vicinity of the optimum is large. For this reason, recalculations are recommended for any significant changes in the constraint.

Let us go over another example so we become comfortable utilizing geometric programming to solve constrained, multi-variable problems. We will examine an everyday problem involving three independent variables and one constraint.

Example 10.5 ***Reduce the cost of rectangular containers***

Given

Right after graduation, you are hired by a leading-edge engineering firm which designs and manufactures phase change material (PCM) for building envelopes to save heating and cooling energy usage. The PCM is shipped in rectangular containers of height x_H, width x_W, and length x_L, without a top; see Figure 10.2. Each container holds 0.5 m^3 of PCM. The material for making the container costs \$7/m^2. Having just mastered Engineering Design and Optimization of Thermofluid Systems, you immediately see the opportunity to save some container material from the existing 1 m high, and 0.5 m by 1 m base container.

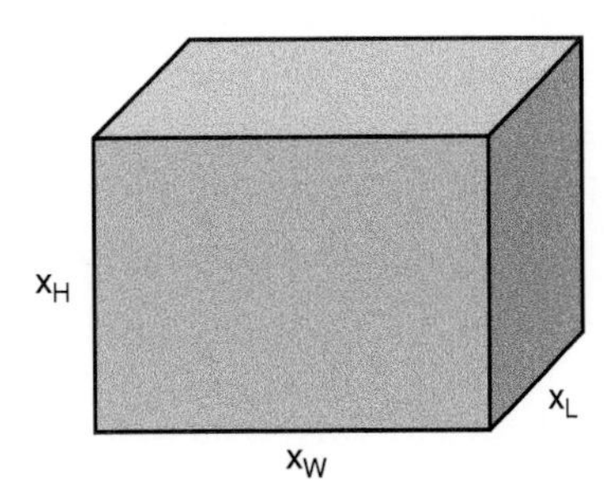

Figure 10.2 A rectangular container of height x_H, width x_W, and length x_L Source: D. Ting.

Find

Use geometric programming to determine the dimensions of the container for minimum container material cost. How much money can you save by using these new dimensions?

Solution

The material cost per container,

$$y = 7\ (x_L\, x_W + 2\, x_L x_H + 2\, x_H x_W) \tag{E10.5.1}$$

The constraint,

$$x_L\, x_H\, x_W = 0.5 \tag{E10.5.2}$$

This can be rearranged into

$$2\, x_L x_H x_W = 1 \tag{E10.5.2a}$$

Degree of difficulty D = 4 − (3 + 1) = 0, i.e. easy.

Let Y = y/7, then, at the optimum,

$$y^* = (x_L x_W/w_1)^{w_1}(2x_L x_H/w_2)^{w_2}(2x_H x_W/w_3)^{w_3}(2x_L x_H x_W/w_4)^{pw_4} \tag{E10.5.3}$$

This can be reduced to

$$y^* = (1/w_1)^{w_1}(2/w_2)^{w_2}(2/w_3)^{w_3}(2/w_4)^{pw_4} \tag{E10.5.4}$$

where, from Eq (10.4.3),

$$x_L^{w_1+w_2+pw_4} = x_L^0 = 1 \text{ or } w_1 + w_2 + p\, w_4 = 0 \tag{E10.5.5}$$

$$x_W^{w_1+w_3+pw_4} = x_W^0 = 1 \text{ or } w_1 + w_3 + p\, w_4 = 0 \tag{E10.5.6}$$

and

$$x_H^{w_2+w_3+pw_4} = x_H^0 = 1 \text{ or } w_2 + w_3 + p\, w_4 = 0 \tag{E10.5.7}$$

Also,

$$w_1 + w_2 + w_3 = 1 \tag{E10.5.8}$$

and

$$p\, w_4 = p \text{ or } w_4 = 1 \tag{E10.5.9}$$

Equations (E10.5.5)–(E10.5.9) can be expressed in matrix form,

$$\begin{bmatrix} 1 & 1 & 0 & 1 \\ 1 & 0 & 1 & 1 \\ 0 & 1 & 1 & 1 \\ 1 & 1 & 1 & 0 \end{bmatrix}\begin{bmatrix} w_1 \\ w_2 \\ w_3 \\ p \end{bmatrix} = \begin{bmatrix} 0 \\ 0 \\ 0 \\ 1 \end{bmatrix}$$

Subtracting Row 2 from Row 1 gives

$$w_2 - w_3 = 0 \text{ or } w_2 = w_3$$

Subtracting Row 3 from Row 1 gives

$$w_1 - w_3 = 0 \text{ or } w_1 = w_3$$

Hence, we have

$$w_1 = w_2 = w_3$$

This implies that all three terms have equal impact on the optimum cost. Substituting this into Row 4, we have

$$w_1 + w_2 + w_3 = 1 \text{ or } w_1 = w_2 = w_3 = 1/3$$

Substituting this into Row 1 gives

$$1/3 + 1/3 + p\ (1) = 0 \text{ or } p = -2/3$$

Therefore, from Eq (10.5.4), we have

$$\begin{aligned} y^* &= (1/w_1)^{w_1}(2/w_2)^{w_2}(2/w_3)^{w_3}(2/w_4)^{pw_4} \\ &= (1/1/3)^{1/3}(2/1/3)^{1/3}(2/1/3)^{1/3}(2/1)^{-2/3} \\ &= 3 \end{aligned}$$

Or

$$y^* = 7y^* = 21$$

In short, the cheapest container for the required purpose costs \$21 each.

At the optimum,

$$w_1\ (= 1/3) = x_L x_W/y^*, w_2\ (= 1/3) = 2x_L x_H/y^*, \text{and } w_3\ (= 1/3) = 2x_H x_W/y^*$$

which give

$$1/3 = x_L x_W/3, 1/3 = 2x_L x_H/3, 1/3 = 2x_H x_W/3$$

These lead to

$$x_H = 1/2, x_W = 1, \text{abd } x_L = 1$$

In other words, containers of 0.5 m height, 1 m width, and 1 m length are the cheapest for holding 0.5 m^3 of PCM.

The area of the existing 1 m high, and 0.5 m by 1 m base container is 3.5 m^2. This costs \$7 × 3.5, or \$24.5 each. Accordingly, the optimized container is \$3.5 cheaper.

The calculation process is the same when there is more than one constraint. Let us go through one such example for completeness. The underlying message is about biomimicry, that is, to learn from and imitate strategies found in nature to advance man-make designs.

Example 10.6 ***Reduce the energy usage while ascending***

Given

Vultures know how to ride on thermals to save energy. The normalized energy usage per unit elevation gain of a vulture for gaining elevation can be described as

$$y = 5\, x_1^2 x_2 + 10/x_3^2$$

where x_1 is the normalized wing span, x_2 is the normalized wing flapping frequency, and x_3 is the normalized buoyant thrust. Bestowed with finite stamina, $x_1 x_2 = 1$; namely, the flapping frequency is inversely proportional to the wing span, i.e. $x_1 = 1/x_2$. On a thermal, the buoyant thrust is proportional to the wing span and inversely proportional to the flapping frequency in the form, $x_3 = 5x_1/x_2$. This suggests that the wing flapping allows rising hot air to escape upward, leading to a downward fall. As such, there are times it is better to do nothing. As A.A. Milne said in *Winnie the Pooh*, "Don't underestimate the value of Doing Nothing."

Find

Minimize the energy usage per elevation gain, y.

Solution

The objective function,

$$y = 5\,x_1^2 x_2 + 10/x_3^2 \tag{E10.6.1}$$

is subject to constraints,

$$x_1\, x_2 = 1 \tag{E10.6.2}$$

and

$$x_3 = 5\,x_1/x_2 \tag{E10.6.3}$$

The second constraint can be rearranged into

$$x_2\, x_3/\left(5\,x_1\right) = 1 \tag{E10.6.3a}$$

Degree of difficulty D = 4 − (3 + 1) = 0, i.e. it is a piece of cake.
At the optimum,

$$y^* = \left(5x_1^2 x_2/w_1\right)^{w_1} + \left[10/\left(x_3^2 w_2\right)\right]^{w_2} \left(x_1 x_2/w_3\right)^{p_1 w_3} \left[x_2\, x_3/\left(5x_1 w_4\right)\right]^{p_2 w_4} \tag{E10.6.4}$$

This can be reduced to

$$y^* = \left(5/w_1\right)^{w_1} \left(10/w_2\right)^{w_2} \left(1/w_3\right)^{p_1 w_3} \left[1/\left(5w_4\right)\right]^{p_2 w_4} \tag{E10.6.4a}$$

where, from Eq (10.4.3),

$$x_1^{2w_1+p_1 w_3-p_2 w_4} = x_1^0 = 1 \text{ or } 2w_1 + p_1\, w_3 - p_2\, w_4 = 0 \tag{E10.6.5}$$

$$x_2^{\,w_1+p_1 w_3+p_2 w_4} = x_2^0 = 1 \text{ or } w_1 + p_1\, w_3 + p_2\, w_4 = 0 \tag{E10.6.6}$$

and

$$x_3^{-2w_2+p_2 w_4} = x_3^0 = 1 \text{ or } -2w_2 + p_2\, w_4 = 0 \tag{E10.6.7}$$

Also,

$$w_1 + w_2 = 1 \tag{E10.6.8}$$

$$p_1\, w_3 = p_1 \text{ or } w_3 = 1 \tag{E10.6.9}$$

and

$$p_2\, w_4 = p_2 \text{ or } w_4 = 1 \tag{E10.6.10}$$

Substituting Eqs. (E10.6.9) and (E10.6.10) into Eqs. (E10.6.5)–(E10.6.7), we get

$$2w_1 + p_1\,(1) - p_2\,(1) = 0 \tag{E10.6.5a}$$

$$w_1 + p_1\,(1) + p_2\,(1) = 0 \tag{E10.6.6a}$$

and

$$-2w_2 + p_2\,(1) = 0 \tag{E10.6.7a}$$

Equation (E10.6.7a) can be rearranged into

$$p_2 = 2w_2 \tag{E10.6.7b}$$

Equation (E10.6.5a) + Eq (10.6.6a) give

$$3w_1 + 2p_1 = 0 \tag{E10.6.8}$$

Subtracting Eq (10.6.6a) from Eq (10.6.5a) leads to

$$w_1 - 2p_2 = 0 \text{ or } p_2 = w_1/2 \tag{E10.6.9}$$

Substituting this into Eq (10.6.7b), we get

$$\left(w_1/2\right) = 2w_2 \text{ or } w_1 = 4w_2 \tag{E10.6.7c}$$

Substituting Eq (10.6.9) into Eq (10.6.8) results in

$$\left(4w_2\right) + w_2 = 1 \text{ or } w_2 = 0.2 \tag{E10.6.8a}$$

Substituting this into Eq (10.6.7c), we have

$$w_1 = 4\,(0.2) = 0.8. \tag{E10.6.7d}$$

Substitute this into Eq (10.6.8) to get

$$3\,(0.8) + 2p_1 = 0$$

or

$$p_1 = -1.2. \tag{E10.6.8b}$$

Substituting Eq (10.6.7d) into Eq (10.6.9), we get

$$p_2 = (0.8)/2 = 0.4 \tag{E10.6.9a}$$

In short, we have

$$w_1 = 0.8, w_2 = 0.2, w_3 = 1, w_4 = 1, p_1 = -1.2, \text{and } p_2 = 0.4$$

We note that the weight of the first term is four times larger than that of the second term. Substituting these into Eq (10.6.4a), we obtain the minimum energy usage per elevation gain,

$$y^* = (5/0.8)^{0.8}(10/0.2)^{0.2}(1/1)^{-1.2(1)}\ \left[(1/5\,(1))\right]^{0.4(1)} = 4.98.$$

10.6 Conclusion

This conventional method still finds many applications, including some rather advanced usages. For example, Swamee et al. (2008) elegantly applied geometric programming to optimize the design of a double-pipe heat exchanger. More recently, Hoburg and Abbeel (2014) invoked geometric programming and demonstrated its potency in the optimization of an aircraft wing design. Ojha and Biswal (2010) disclosed ways to help engineers deal with more realistic applied design problems. At the more advanced level, much has evolved from the concept based on geometric programming, especially in the areas of controls and operations.

Problems

10.1 Minimize the surface area of a cylinder

The surface area of a cylindrical tank with a volume of 25 m^3 is to be minimized. Use geometric programming to determine the height and radius of the tank. Solve it simply as an unconstrained problem.

10.2 Minimize heat input

A hot water piping network has two delivery paths with corresponding mass flow rates, x_1 and x_2. The heat loss decreases with these mass flow rates such that the total required heat input,

$$Q' = 3x_1^2 + 4x_2^2 + 17$$

where 17 illustrates finite heat loss even when the hot water is not flowing. The limitation imposed by the mass flow rates can be expressed as $x_1 x_2 = 20$. Recast the problem as an unconstrained problem and solve it using geometric programming.

10.3 Minimize a three-term, two-variable unconstrained problem

The heat produced by a tri-generation (combined cooling, heat, and power) plant can be expressed as

$$y = 2\ x_1^2 + 3\ x_2^3 - 38\ x_1^{10.5} x_2^{11}$$

where the first term represents the power demand, the second term denotes the heating requirement, and the third term stands for cooling. Independent variables x_1 and x_2 signify the operation of an engine and a reversible heat pump, respectively. Minimize the heat via geometric programming. Which of the three terms ($2x_1^2$, $3x_2^3$, or $38x_1^{10.5}x_2{}^{11}$) has a greater effect on the minimum heat? What are the corresponding weights of influence?

The weight of the first term, $2x_1^2$, is w_1 = ________________

The weight of the second term, $3x_2^3$, is w_2 = ________________

The weight of the third term, $38x_1^{10.5}\ x_2{}^{11}$, is w_3 = ________________

10.4 Optimize a three-variable unconstrained problem

The total thermal energy loss of a solar thermal collector system,

$$y = 2\ x_1 x_2 + 2/\left(x_1 x_3\right) - 12x_2^2 + 3x_3$$

where x_1 is the wind turbulence, x_2 is the temperature of the surroundings such as the roof, and x_3 is wind speed. Minimize the total loss via geometric programming.

10.5 Optimize the cost of wheat drying

A simple but effective way to dry wheat is by blowing it, from the bottom of the bed upward, with heated air. Thermodynamically, the amount of heating is a function of the mass of the wheat, the heat capacity of the wheat, and the temperature rise, i.e. $Q = mc_P\Delta T$. Accordingly, the heating cost has been deduced to be 0.005 ∀ ΔT, where ∀ is the amount (volume) of wheat in m^3/m^2 of bed area, and ΔT, in °C or K, is the required temperature rise. The other cost is associated with the energy input required to operate the blower, and this can be expressed as 5×10^{-9}∀. The drying time in days is $8\times10^7/(\forall^2\Delta T)$, i.e. the larger the flow rate and/or the hotter the air, the shorter the drying duration. Minimize the cost for drying, which is to be achieved in 53 days, via geometric programming as a constrained problem.

10.6 Minimizing heat loss from a chemical reactor

A chemical reaction is administered in a 2-m-long cylindrical stainless steel vessel. The heat transfer coefficient from the outside of the vessel to the ambient is found to be a function of the diameter of the vessel, D, and the temperature difference between the vessel surface and the ambient, ΔT, i.e.

$$h = 0.5/D^{1.5} + 2\Delta T^{0.25}/D^{0.25}$$

The chemical reaction results in $\Delta T\ D^{0.25} = 10$.

Part I Minimize the heat loss by solving it as a two-variable constrained problem via geometric programming.

Part II Which term has a larger effect? What are the corresponding weights of influence?

10.7 Optimize a three-stage compression system

Atmospheric air at 25°C and 0.1 MPa enters the first stage of a three-stage compression system; it exits the third stage at 7 MPa. An inter-cooler is used to cool the air between the stages back to 25°C. The compression work per unit mass of air is

$$\frac{\gamma}{\gamma-1}RT_i\left[1-\left(\frac{P_o}{P_i}\right)^{(\gamma-1)/\gamma}\right]$$

where γ is the specific heat ratio, R is the gas constant, subscript i denotes the initial (atmospheric) conditions, and subscript o signifies the exiting or outlet conditions. Find the intermediate pressures so that the total work is minimized. Solve it using geometric programming; assume the compressions are reversible and adiabatic. Give the values of the minimum work and the two intermediate pressures.

References

Beightler, C.S. and Philips, D.T. (1976). *Applied Geometric Programming*. New York: Wiley.

Clasen, R. (1963). The linear algorithmic problem. Rand Corporation Memorandum RM-37-7-PR, June.

Duffin, R.J., Peterson, E.L., and Zener, C. (1967). *Geometric Programming: Theory and Application*. New York: Wiley.

Ecker, J.G. (1980). Geometric programming: methods, computations and applications. *SIAM Review* 22(3): 338–362.

Hoburg, W. and Abbeel, P. (2014). Geometric programming for aircraft design optimization. *AIAA Journal* 52(11), DOI:10.2514/1.J052732, 2014.

Jaluria, Y. (1998). *Design and Optimization of Thermal Systems*. New York: McGraw-Hill.

Ojha, A.K., and Biswal, K.K. (2010). Posynomial geometric programming problems with multiple parameters. *Journal of Computing* 2(1): 84–90.

Stoecker, W.F. (1989). *Design of Thermal Systems*, 3rd ed. New York: McGraw-Hill.

Swamee, P.K., Aggarwal, N., and Aggarwal, V. (2008). Optimum design of double pipe heat exchanger. *International Journal of Heat and Mass Transfer* 51(9–10): 2260–2266.

Zener, C. (1971). *Engineering Design by Geometric Programming*. New York: Wiley.

Appendix: Sample Design and Optimization Projects

Any intelligent fool can make things bigger and more complex. It takes a touch of genius – and a lot of courage – to move in the opposite direction.

—Albert Einstein

Nomenclature

CAES	compressed air energy storage
c_P	heat capacity at constant pressure; $c_{P,cold}$ is cold stream heat capacity, $c_{P,hot}$ is hot stream heat capacity
Comp	compressor
D	diameter
f	friction factor
g	gravity
H	height
h	head; h_L is the major head loss
HEX	heat exchanger
HVAC	heating, ventilation, and air conditioning
L	length; L_{pipe} is pipe length
m	mass; m' is mass flow rate, m_{air}' is mass flow rate of air, m_{cold}' is cold stream mass flow rate, m_{hot}' is hot stream mass flow rate
P	pressure; P_{amb} is ambient pressure, P_{cavern} is cavern pressure, P_i is inlet pressure, P_o is outlet pressure, P_1 is the pressure at location 1, P_2 is the pressure at location 2
PV	photovoltaic
Q	heat; Q' is heat transfer rate, Q_{act}' is the actual heat transfer rate, Q_{max}' is the theoretical maximum heat transfer rate
R	gas constant
T	temperature; T_{amb} is ambient temperature, T_{cavern} is cavern temperature, $T_{cold,i}$ is incoming cold stream temperature, and $T_{cold,o}$ is outgoing cold stream temperature, $T_{hot,i}$ is incoming hot stream temperature, $T_{hot,o}$ is outgoing hot stream temperature, T_i is inlet temperature, T_o is outlet temperature
TES	thermal energy storage
Turb	turbine

U	velocity, U_1 is the velocity at location 1, U_2 is the velocity at location 2
UWCAES	underwater compressed air energy storage
W	width
X	stream-wise coordinate
Y	cross-stream coordinate
Z	vertical coordinate
z	elevation; z_1 is the elevation at location 1, z_2 is the elevation at location 2

Greek and Other Symbols

α	attack angle
γ	specific heat capacity ratio
Δ	change
η	efficiency; η_{comp} is compressor efficiency, η_{exp} is expander (turbine) efficiency, $\eta_{isentopic}$ is isentropic efficiency
ρ	density
$\forall$	volume

A.1 Introduction

This appendix contains a number of design and optimization examples. A few of these are envisioned, some are existing engineering systems, and a couple of them are new or emerging engineering undertakings. These examples can be customized, depending on the specific need, into small, medium, and large course projects. A small design undertaking can be a conceptual solution which considers only limited aspects of the problem at hand. The engagement of some numerical analysis, including the consideration of limitations or constraints, formulates a medium project. Going beyond the workable design and executing some optimization calculations would make a major course project.

A.2 Cavern-based Compressed Air Energy Storage

Renewable energies such as wind, solar, and waves are intermittent in nature. Oftentimes, when there is abundant "wind" (stands for a typical renewable energy source) to harness, the need may not be there. At other times, the demand is high, but "the wind is not blowing." Because of this discord, curtailment is routinely invoked and much of the free "wind" is forfeited, and environmentally-unfriendly nonrenewable energy has to be called in when the "wind2 is not enough to satisfy the grid. Energy storage can be used to resolve this double-edged problem. Namely, the excess energy can be stored and put to use when the need is there. It is worth noting that the mismatch between supply and demand is not a new challenge. It is simply compounded with the increasing share of intermittent energy sources. For that reason, there is a need for energy storage technologies, such as compressed air energy storage (CAES) is sure (Carriveau et al., 2019; Ebrahimi et al., 2019). CAES is a proven technology whose widespread implementation hinges on further design and optimization. The major components of a CAES system, as shown in Figure A.1, are a storage cavern, a motor compressor, and a turbo-expander-generator. When there is excess energy, the motor runs the compressor, that is, the excess energy is employed to compress air into the storage cavern. When there is power shortage, the compressed air is passed through the turbo-expander (turbine) to generate electricity for the grid.

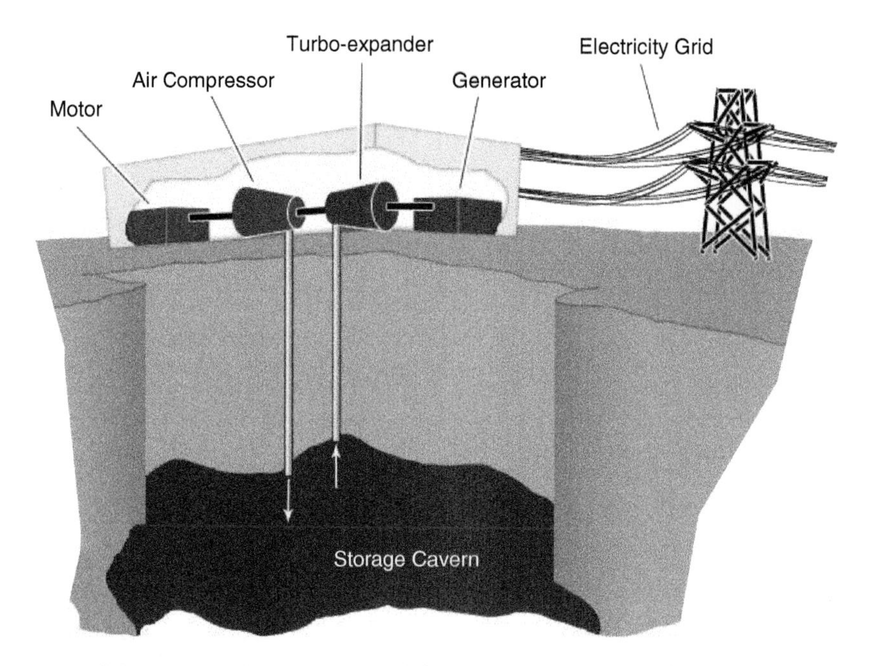

Figure A.1 A conventional compressed air energy storage system. Source: B. Cheung.

The main weakness of a conventional CAES system is its low efficiency. This low efficiency is due to considerable energy loss during the charging phase, when the heat generated by compression is not captured to warm up the compressed air before it is expanded during the discharging phase. Because of this, it makes sense to recover and retain the heat within the operating cycle, as depicted in Figure A.2.

With the appropriate engineering background after mastering Engineering Design and Optimization of Thermofluid Systems, you joined a frontier company to work on designing a cavern-based CAES system; see, for example (Ebrahimi et al., 2020). It is a 100 MW system for supplying electricity during a high-demand period of up to 8 hours at a time. A 50 m wide by 110 long and 50 m high underground cavern, otherwise filled with water, as shown in Figure A.3, is available. When charging, air is compressed into the cavern, pushing the water up through the vertical shaft into a lake. As such, the compressed air is at a constant pressure, which is equal to the hydrostatic water head.

In order not to waste the heat of compression, it is captured via heat exchangers, as depicted in Figure A.4. This garnered heat is used to heat the air before entering each turbine during the discharging mode. To that end, the complete charge-discharge cycle is an adiabatic process. A series of three compressors and three turbines are shown in Figure A.4. These numbers are arbitrary. Having said that, the efficiency tends to increase when the pressure ratio across a compressor or turbine is lower, but so does the cost. To that end, three is possibly more or less the optimal number.

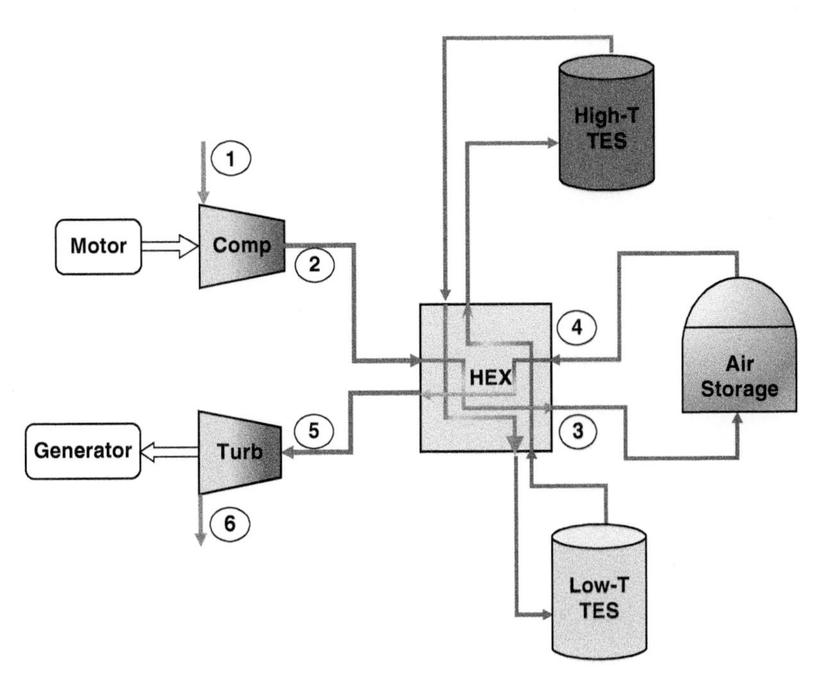

Figure A.2 A CAES system with heat recovery. Source: R.S.S. Dittakavi.

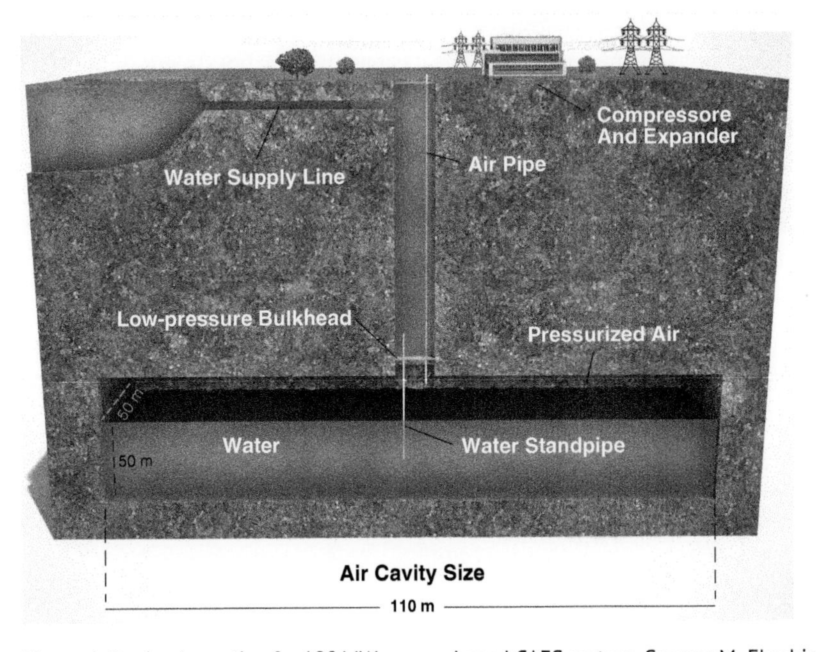

Figure A.3 A schematic of a 100 MW cavern-based CAES system. Source: M. Ebrahimi.

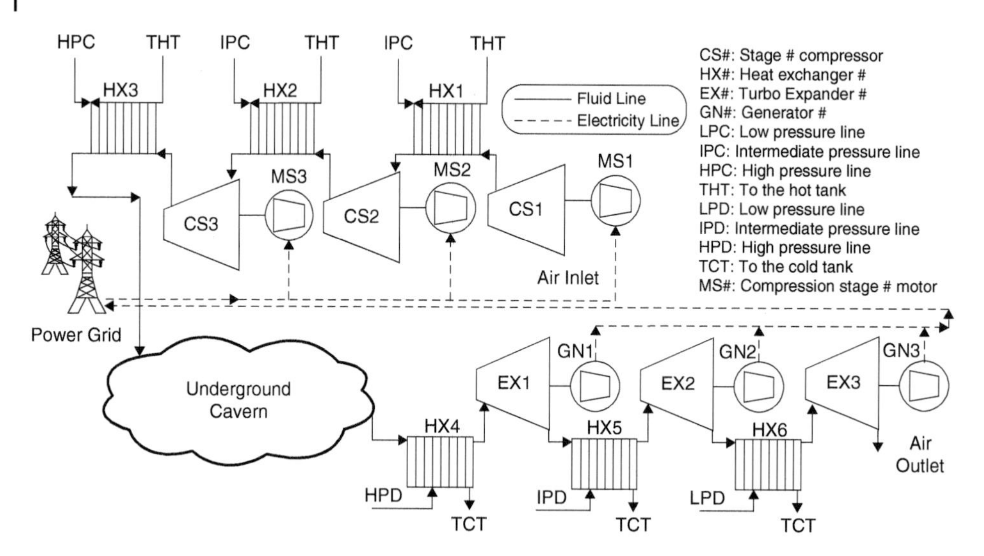

Figure A.4 A schematic showing major components of a cavern-based CAES system. Source: M. Ebrahimi.

The efficiency of the heat exchangers can be defined as

$$\eta = Q_{act}'/Q_{max}' \tag{A.1}$$

where Q_{act}' is the actual heat transfer rate and Q_{max}' is the theoretical maximum heat transfer rate. If we assume negligible heat loss to the environment, the heat transfer rate can be deduced from

$$Q' = m_{hot}'c_{P,hot}\left(T_{hot,i} - T_{hot,o}\right) = m_{cold}'c_{P,cold}\left(T_{cold,o} - T_{cold,i}\right) \tag{A.2}$$

where m_{hot}' is the hot stream mass flow rate, $c_{P,hot}$ is the heat capacity of the hot stream, $T_{hot,i}$ is the temperature of the entering hot stream, $T_{hot,o}$ is the outgoing hot stream temperature, m_{cold}' is the cold stream mass flow rate, $c_{P,cold}$ is the heat capacity of the cold stream, $T_{cold,i}$ is the temperature of the entering cold stream, and $T_{cold,o}$ is the outgoing cold stream temperature.

For isentropic compression and expansion of air as an ideal gas with constant specific heats, the outlet temperature can be determined from

$$T_o = T_i\left(\frac{P_o}{P_i}\right)^{(\gamma-1)/\gamma} \tag{A.3}$$

where T_i is the inlet temperature, P_o is the outlet pressure, P_i is the inlet pressure, and γ is the specific heat capacity ratio. With these parameters, we can compute the power required to drive a compressor, i.e.,

$$\dot{W}_{comp} = \frac{\dot{m}RT_i}{\eta_{comp}}\left[\left(\frac{P_o}{P_i}\right)^{\frac{\gamma-1}{\gamma}} - 1\right]\frac{\gamma}{\gamma-1} \tag{A.4}$$

where R is the gas constant and η_{comp} is the efficiency of the compressor. Similarly, the power produced by a turbine is

$$\dot{W}_{exp} = \frac{\dot{m}RT_i}{\eta_{exp}}\left[1 - \left(\frac{P_o}{P_i}\right)^{\frac{\gamma-1}{\gamma}}\right]\frac{\gamma}{\gamma-1} \tag{A.5}$$

where η_{exp} is the efficiency of the turbine.

Sample charging and discharging conditions are provided in Tables A.1 and A.2, respectively. Exercise your engineering knack and common sense to make and/or adjust appropriate assumptions, as needed, to determine the temperature and pressure at the inlet and outlet of all components. According to the duty (the heat transfer rate) of each heat exchanger, deduce the required mass flow rate of water countering the flowing stream of air. Employ appropriate efficiencies for motors and generators to calculate the corresponding consumed or delivered power. Maximize the system performance in terms of optimal compression ratios for the respective compressors. Relax the assumptions gradually, except the atmospheric conditions, cavern conditions, the 100 MW capacity for a minimum of 5 hours of discharging along with a total (as wind is not always blowing) charging duration of no more than 8 hours, and optimize the system performance. Assume the numbers of compressors, heat exchangers, and expanders (turbines) stay the same. Would changing these numbers further the system performance?

Table A.1 Conditions during the charging phase.

Component	Condition
Atmospheric Inlet	m_{air}' = 180 kg/s, P_{amb} = 101 kPa, T_{amb} = 20°C
Compressor 1 (CS1)	Compression Ratio = 4.36, $\eta_{isentopic}$ = 92.5%
Heat Exchanger 1 (HX1)	Air Pressure Loss = 5 kPa, Duty = 24.2 MW
Compressor 2 (CS2)	Compression Ratio = 3.84, $\eta_{isentopic}$ = 91.5%
Heat Exchanger 2 (HX2)	Air Pressure Loss = 20 kPa, Duty = 24.3 MW
Compressor 3 (CS3)	Compression Ratio = 4.0, $\eta_{isentopic}$ = 90.5%
Heat Exchanger 3 (HX3)	Air Pressure Loss = 20 kPa, Duty = 27.2 MW

Table A.2 Conditions during the discharging phase.

Component	Condition
Cavern	m_{air}' = 280 kg/s, P_{cavern} = 6,500 kPa, T_{cavern} = 25°C
Heat Exchanger 4 (HX4)	Air Pressure Loss = 8 kPa, Duty = 51.8 MW
Turbine 1 (EX1)	Expansion Ratio = 0.38, $\eta_{isentopic}$ = 89%
Heat Exchanger 5 (HX5)	Air Pressure Loss = 10 kPa, Duty = 29.0 MW
Turbine 2 (EX2)	Expansion Ratio = 0.30, $\eta_{isentopic}$ = 89%
Heat Exchanger 6 (HX6)	Air Pressure Loss = 8 kPa, Duty = 35.9 MW
Turbine 3 (EX3)	Expansion Ratio = 0.20, $\eta_{isentopic}$ = 89%

A.3 Underwater Compressed Air Energy Storage

Near-shore, shoreline, and off-shore communities can take advantage of another form of compressed air energy storage, that is, underwater compressed air energy storage (UWCAES) (Cheung et al., 2014a; Wang et al., 2016; Carriveau et al., 2019; Ebrahimi et al., 2019). Figure A.5 is a schematic depicting the key components associated with an UWCAES system. When the need is not there, we operate the charging phase, compressing atmospheric air using the available energy, into the underwater accumulator(s). When the power demand arises, we discharge and expand the compressed air to rotate the turbine(s), feeding the generated power into the starving grid. Both charging and discharging processes take place with the accumulator(s) fixed at the same depth, that is, under a constant pressure condition (Cheung et al., 2014b; Wang et al., 2019). Lowering the accumulators farther into the depth increases the stored energy density, i.e., the stored air is at a higher pressure and, thus, has more potential energy per unit volume of compressed air. On the flip side, higher capacity compressors, longer air conveying pipes, higher pressure loss, etc. arise.

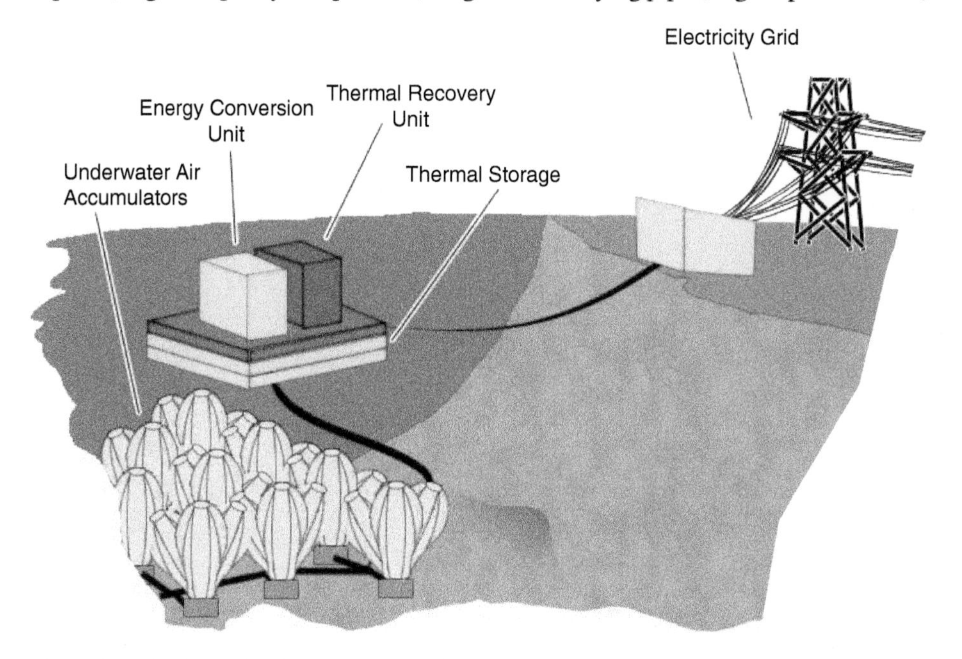

Figure A.5 A schematic of an Underwater Compressed Air Energy Storage System. Source: B. Cheung.

Both compression and expansion of air may be assumed to be a polytropic process, that is,

$$P\forall^{\gamma} = \text{constant} \tag{A.6}$$

where $\forall$ is the volume. Assuming air behaves as an ideal gas, we can write

$$T_o/T_i = \left(P_o/P_i\right)^{[(\gamma-1)/\gamma]} \tag{A.7}$$

where subscripts o and i denote the outlet and inlet of a compressor or turbine. The power required to drive a compressor and produced by a turbine can be calculated via Eqs. A.5 and (A.6), respectively.

Design a UWCAES system for storing compressed air at 80 m underwater. The system is to provide 1 MW of power for up to 4 hours at a time. The atmospheric air varies from -25°C to 35°C, and the atmospheric pressure ranges between 95 and 107 kPa. The charging is to take place at an average rate of 0.5 MW. The compressed air is expected to stay in the accumulator for up to a maximum of 8 hours. Optimize the system for the highest overall system efficiency. Design parameters which you can consider include the heat recovery unit (type of thermal storage; heat exchanger, number, types, working fluid), stages and types of compression (number of compressors with particular compression ratio and efficiency), stages and types of expansion (number of turbines with respective expansion ratio and efficiency). Start with only one compressor and one expander. Focus on recovering the thermal energy produced during the compression process, that is, acquire it, store it, and feed it back to the air during the discharging phase just before it enters the turbine. The specific types of compression and expansion processes are also expected to have a notable impact on the system performance.

A.4 Compressed Air Energy Storage Underground

The classical way of storing energy as compressed air is to utilize an underground cavern such as a depleted mine, as exhibited in Figure A.1. The compressed air energy storage system in Huntorf, Germany, has been in service for about four decades (Crotogino et al., 2001; Cheung et al., 2014b). Another long-servicing CAES is located in McIntosh, Alabama, USA (Davis and Schainker, 2006). Perform a thermodynamic analysis of either of these systems to obtain its first law efficiency. Following that, choose a couple of key independent variables to optimize the system with respect to them. Replacing the combustor is an environmentally-responsive way to improve the system. If time permits, compare this CAES operation with UWCAES for the same capacity and charge and discharge durations.

A.5 Geothermal Heat Exchanger

Coaxial heat exchangers, see Figure A.6, promise to outperform their U-bend counterparts. For this reason, the study of coaxial heat exchangers has received serious attention (Holmberg et al., 2016; Gordon et al., 2018, Dai et al., 2019; Iry and Rafee, 2019; Liu et al., 2019; Luo et al., 2019; Pan et al., 2019; Zhang et al., 2019; Cazorla-Marín et al., 2020, Hu et al., 2020). Potential reasons behind the better performance include: (i) a larger heat transfer surface area for a given borehole size, (ii) the fluid in the inner pipe is significantly more insulated from the ground and, thus, exchanges minimum heat with the ground, (iii) possibly less pressure (head) loss because of the larger (compared to its conventional U-bend counterpart) and potentially shorter (to deliver the same amount of heat from or to the ground) conduit. Note that residence time and flow rate may have some counter-acting effects. Also, an increase in flow turbulence is generally good for heat transfer, but it comes with increased pressure loss and, hence, pumping power.

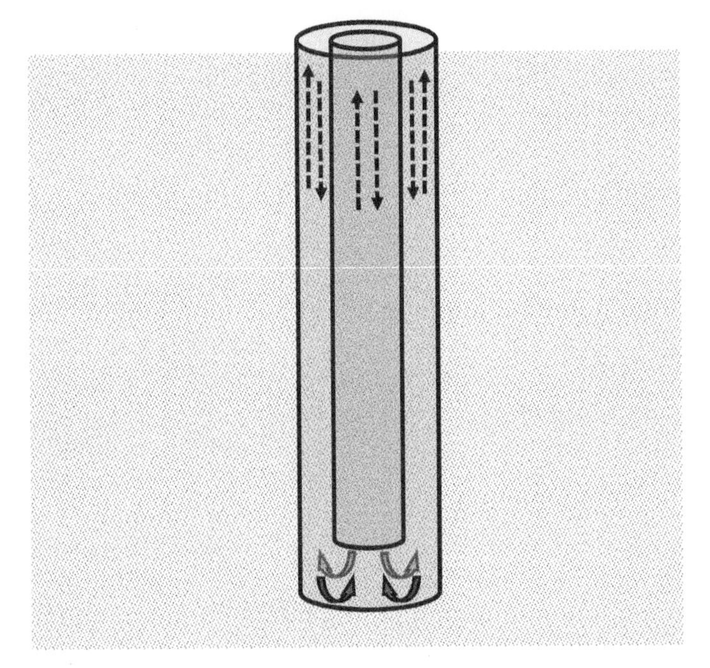

Figure A.6 A schematic of a coaxial heat exchanger. The flow can enter and move downward via the inner pipe and returns through the annulus, or, it can enter via the annulus and return through the inner pipe. Source: D. Ting.

Design a borehole heat exchanger for 0.7 kW of heating or cooling capacity to supplement the need of a residential building. You may start with a simple design by assuming a fixed ground temperature of 10°C and water as the working fluid. The pressure loss (per unit length) can be estimated from that of a fully developed flow in a smooth and straight pipe of borehole diameter. Recall from Chapter 3 that, for an incompressible fluid flowing in a pipe with no losses, as illustrated in Figure A.7, conservation of energy gives

$$P_1 + \tfrac{1}{2}\rho U_1^2 + z_1 \rho g = P_2 + \tfrac{1}{2}\rho U_2^2 + z_2 \rho g \tag{A.8}$$

At the beginning of a representative section of the pipe, the pressure is P_1, the (average) velocity is U_1, the elevation is z_1, and subscript 2 is used to designate these parameters at the end of the representative section. The density is denoted by ρ and the gravity g. Introducing the head loss, h_L, which is in units of height, and dividing Eq (A.8) by the specific weight, ρg, of the moving fluid, we have

$$P_1/\rho g + \tfrac{1}{2}U_1^2/g + z_1 = P_2/\rho g + \tfrac{1}{2}U_2^2/g + z_2 + h_L \tag{A.9}$$

The portion of head loss due to pipe friction, commonly referred to as major head loss, can be deduced from a straight horizontal pipe section, i.e.,

$$h_L = (P_1 - P_2)/\rho g. \tag{A.10}$$

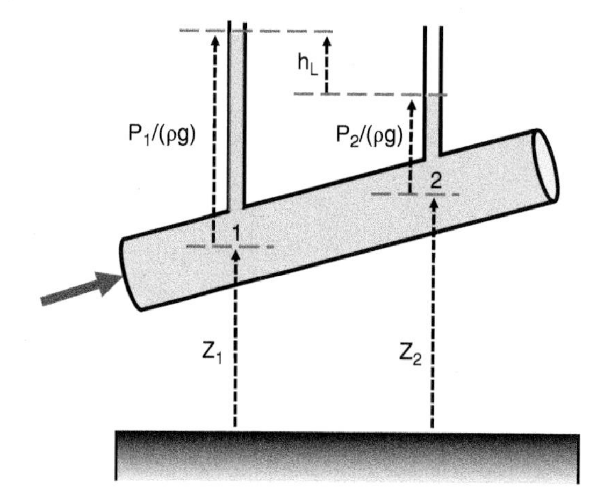

Figure A.7 Pressure head along a pipe where the flowing fluid is a liquid. Source: D. Ting.

For a fully developed flow in a horizontal pipe, we have

$$h_L = \Delta P/\rho g = f\,(L_{pipe}/D)\,(\tfrac{1}{2}U^2/g) \tag{A.11}$$

where the friction factor, f, can be found from the Moody diagram. Other than the major head loss ascribed to pipe friction, h_L, there are other losses such as at the entrance, exit, and end of the inner pipe. These are called minor head loss and they may be neglected for the time being. As a first estimate, you can assume the same major head loss for both annulus and inner pipe flows. This and other assumptions can be relaxed once you have obtained the approximate solution. Also, assume a constant convection heat transfer coefficient throughout the entire length of the borehole, the same heat transfer coefficient for the inner wall of the outer pipe and for the outer and inner walls of the inner pipe, the outer wall of the outer pipe is at the ground temperature, and 100% efficiency of heat exchanged between the working fluid and the indoor air of the building. Optimize the borehole heat exchanger performance. Namely, you want to satisfy the cooling/heating demand with the lowest operating and initial costs. To do so, you want to minimize the pressure loss and, hence, pumping cost, the depth (length) of the borehole, and capital cost. As it happens, reducing pressure loss and borehole length tend to go hand-in-hand. Therefore, focus on the optimal inner-outer diameter ratio and the flow rate.

A.6 Passive Cooling of a Photovoltaic Panel for Efficiency

The energy conversion efficiency of photovoltaic (PV) such as crystalline silicon solar cells, decreases by 0.08% per degree Celsius temperature rise (Radziemska, 2003). This decrease is in absolute efficiency, i.e., a typical commercial PV panel having an energy conversion efficiency of 15% at 20°C will only be harnessing solar energy at 13% when the cell temperature reaches 45°C. Because of this, much research has been conducted on cooling the PV panel. Passive cooling via functioning, low-cost and low-maintenance devices makes good engineering sense. Notably, simple and efficacious physical structures such as winglets (Wu et al., 2017; 2018) and flexible strips (Yang et al., 2019; 2020) can be methodically designed and placed around the edges and on the no-cell space of PV panels to promote natural wind-induced convection. Figure A.8 depicts some of these turbulent vortex generators. The intricacy of the fluid mechanics convoluted by the turbulent vortex generator is literally impregnable. Subtle adjustments in some of the familiar features can lead to sizeable changes in the outcome. The renowned prodigy Blaise Pascal impeccably put it,

> For after all what is man in nature? A nothing in relation to infinity, all in relation to nothing, a central point between nothing and all and infinitely far from understanding either. The ends of things and their beginnings are impregnably concealed from him in an impenetrable secret. He is equally incapable of seeing the nothingness out of which he was drawn and the infinite in which he is engulfed.

To that end, we are not striving to close the chapter, but to further the progress. Judging from recent literature, it is clear that there is still plenty of room for improvement. And who knows, if you are lucky, you may even have a little breakthrough. Thomas Jefferson has advice on boosting luck, "I'm a great believer in luck, and I find the harder I work the more I have of it."

While the vast majority of the convection heat transfer enhancement work has been conducted under the internal flow condition, the exposed upper surface of a solar PV panel is an open atmospheric flow case. What is the best design and best way to place the best turbulent vortex generator to maximize the cooling of a PV panel? Key design parameters include aspect ratio, attack angle, spacing, pattern (row, stagger, tandem), size (and, thus, also Reynolds number, boundary layer). Be careful when transferring (extrapolating) internal flow results to open flow over a flat plate. The idea is to effectively scoop away the heat from the PV panel. To recapitulate, the upper surface of an array of PV panels is fully open to the atmosphere and, hence, it is principally open flow (wind) over a flat surface. On the other side of the panel, the bottom surface is largely confined by the roof or mounting surface. Fittingly, we can transfer the details from forced convection inside a conduit. Focus on the upper surface first. To effectively enhance heat convection, the device should not significantly slow down the incoming wind, as wind is the underlying muscle for spawning vortices and turbulence. For that reason, the design should aim at transforming the incoming wind into an effective and long-lasting heat remover. Consider a typical 1 m by 2 m PV panel with a rated energy conversion efficiency of 17%. For a typical summer day in a temperate location, the solar radiation is $1000\ \mathrm{W/m^2}$, the surface (cell) temperature is 45°C, and the prevailing wind speed is 6 m/s horizontally over the surface of the solar panel. How much energy (power) can you harness? Strategically station the appropriate turbulent vortex generators on the solar panel to maximize passive cooling and avoid blocking the sun and, thus, losing out on the solar energy. How much extra solar power can be reaped with the optimized passive cooling?

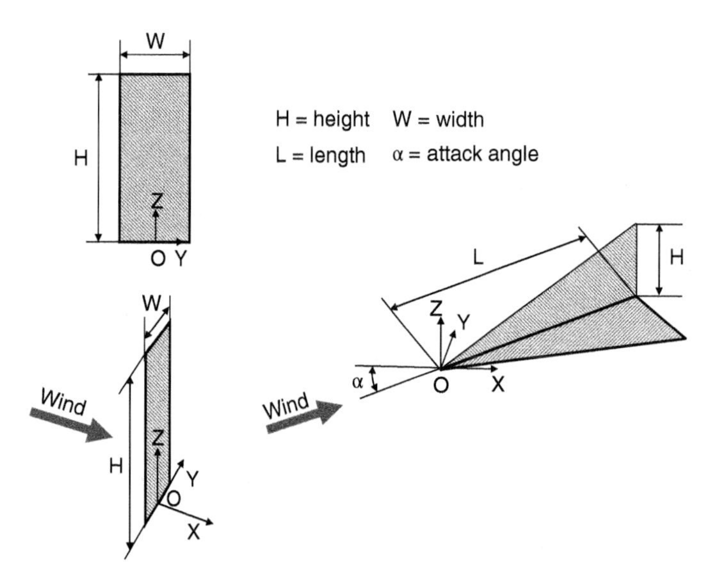

Figure A.8 Some generic turbulent vortex generators for augmenting heat convection. Source: Y. Yang.

A.7 Desert Expedition

Envisage Dr. Jas of the Turbulence and Energy Lab finally coming through with a time dilation chariot, Starship T&E. Incidentally, your class is with her during the premier test flight, which ended up crashing into the middle of the hot and arid part of the Wilderness of Paran in 1440 BCE. To survive, you have to go and get help from Dr. Brown, who just happened to be time traveling in Jerusalem. The erudition from Engineering Design and Optimization of Thermofluid Systems prompted you to anthologize a few pieces of useful parts from Starship T&E together into a deft desert-treading vehicle. The objective (function) is to arrive in Jerusalem in one piece. The constraints include the vastness and aridness of the desert, along with the clumsy pace across the hot sand. For yourself, you need plenty of water to make it, but the water needs to be stored in a compact manner to not further slow down the treading. The design must include features to prevent you from dying of heatstroke in the day and hypothermia at night. All of a sudden, because of your love for animals, a camel, Figure A.9, dawned vividly in a vision. That is it; build your system just like a camel, an intelligent design optimized for desert transportation. "The fundamental claim of intelligent design is straightforward and easily intelligible: namely, there are natural systems that cannot be adequately explained in terms of undirected natural forces and that exhibit features which in any other circumstance we would attribute to intelligence." (William A. Dembski).

Figure A.9 An intelligently designed and optimized desert jalopy, a camel. Source: S. Akhand.

The following highlight but a few unique features of the "ships of the desert." Unlike typical warm-blooded creatures, camels regulate their body temperature over a relatively wide range, enabling them to adapt to the swinging desert temperature (Schmidt-Nielsen et al., 1956; Bouâouda et al., 2014). This is especially true when they are dehydrated. Their feet are not only ingeniously designed not to get blisters, their big toes, making up the large flat feet, are brilliantly created to tread on desert sand, up and down the sand dunes, in an unwavering manner. The hump has been credited for storing water as fat, but there is no real evidence substantiating the conversion of water into fat. It is more likely the energy-rich fat comes from the food they eat, and this stored energy keeps the camel going for days without food. Fat can be broken down into water and a lot of carbon dioxide, while releasing energy. This is why Meerman and Brown (2014) suggest a lot of breathing, to bring in oxygen to break fat into carbon dioxide, for human weight control. A separate design and optimization problem can be formulated to determine the best way to save water and energy. According to Candlish (1981), ethanol is superior for storing both water and energy. The hump also helps in regulating the body temperature with respect to water economy and ambient conditions (Schmidt-Nielsen et al., 1956, Bouâouda et al., 2014). It insulates the body from the scorching sun during the day and supplies heat to warm the rest of the body during the night. And, yes, if you want to last in the desert, pee less. To survive sandstorms, wear a pair of bushy eyebrows along with two rows of camel-hair eyelashes, instead of a pair of goggles which will be clogged by your sweat. In addition, equip yourself with a mask, unless you have camel-like nostrils which can close on demand.

To solidify the illustration above into a computable engineering design and optimization problem, we assume Starship T&E landed 200 km east of Jerusalem. The harsh journey will take 5 days for an average camel treading on a level and straight sand path at 4 km/h for 10 hours a day. Check out the specific terrain and optimize the passage to take. Namely, you need to compare going along the straight passage between your location and Jerusalem and making necessary detours around hills and valleys. Other than the sections of pathway that are dangerously steep, which you need to avoid, account only for the change in speed when ascending and descending and the additional vertical distance that you tread. Another factor you may wish to take into consideration is the ambient temperature. You can use the current temperature or simply assume that the temperature varies between 12°C at midnight and 44°C at midday. Also, to ensure proper functioning of your brain, you need to consume at least 2 liters of water per day. We have assumed that you need to carry a relatively heavy, say, 50 kg, part to Dr. Brown. This is the main reason for designing a "mechanical camel." If the objective is for you to get hold of Dr. Brown and come back in his time machine, then you are better off going by foot. In this case, it is a "camel suit" that you need to design and optimize. The "camel suit" includes a "camel hump backpack" for water, food, heat shield, and thermal storage, "camel footgear" tailor-made for unwavering treading on the hot sand, and the "camel face mask" for protecting your eyes and nose, in particular. Apart from designing the camel suit or the mechanical camel, optimize the route to expedite the journey.

A.8 Fire- and Heat-Resilient Designs

Let us look at another illustration of creativity in nature. Fierce forest fires may kill trees, but not their progeny. How is that possible? The seeds of many trees are encapsulated and, hence, impregnable except under intense heat. By the time a forest fire consumes everything on the ground level and makes its way up a tree, the fire on the ground has smothered. Just when we worry that it is the end of the offspring, such as those seeds inside a pine cone,[1] the capsules protecting them are cracked open by the heat and the heirs of the forest dive into the cool soil, ready to rejuvenate the thicket in no time.

What about living creatures? Apparently, the incredible mound-building termites know a lot about design and optimization. If their meters-size mounds do not wow you, take a closer look at their design details (Figure A.10). Ocko et al. (2019) state that termites collectively consider environmental factors such as heat flow and air exchange into the appropriate construction of their homes. A particular mound is shaped by the response of its internal environment with respect to outside temperature fluctuations. The coupled environmental physics and organism behavior can counsel more sustainable man-made architectures. They even have the foresight of weathering through an apocalypse, that is, they design their mound to survive even a clear-cutting forest fire.

Design a 21-story high-rise block with a circular floor space of 50 m in diameter and the elevator shaft right in the middle. Design the building for maximum natural convection during the warm season. Put measures into the HVAC design to also optimize the heating air circulation during the cold season. What provisions can you devise to mitigate building fires? Given that people who are trapped on the higher levels are probably in the most perilous situation, design an extra means to rescue them in case of fire.

1 Pine cones are often gathered by campers for heartening the campfire. They are thrown into the fire to make the cracking sound.

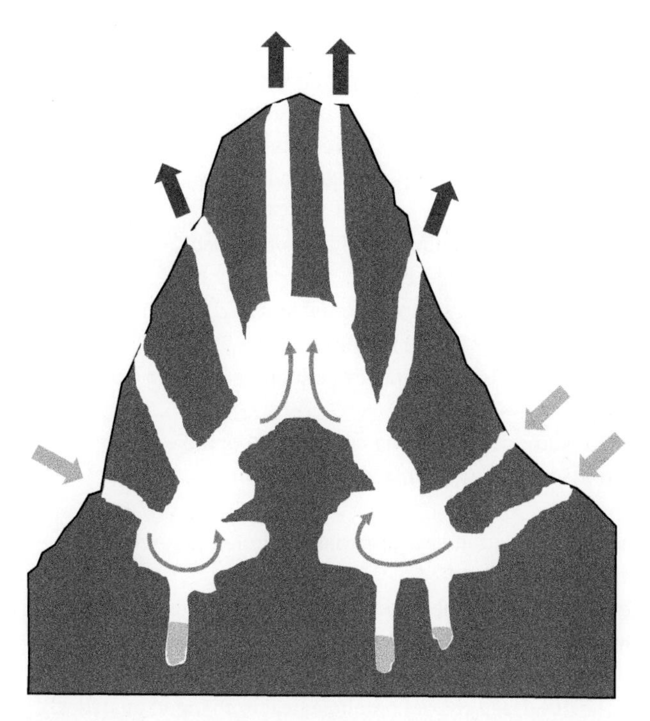

Figure A.10 An intelligently designed and optimized termite mound which can survive an intense forest fire. Source: Y. Yang.

References

Bouâouda, H. Achâaban, M.R., Ouassat, M., et al. (2014). Daily regulation of body temperature rhythm in the camel (*camelus dromedarius*) exposed to experimental desert conditions. *Physiological Reports* 2(9), e12151.

Candlish, J. (1981). Metabolic water and the camel's hump: a textbook survey. *Biochemical Education* 9(3): 96–97.

Carriveau, R., Ebrahimi, M., Ting, D.S-K., and McGillis, A. (2019). Transient thermodynamic modeling of an underwater compressed air energy storage plant: conventional versus advanced exergy analysis. *Sustainable Energy Technologies and Assessments* 31: 146–154.

Cazorla-Marín, A., Montagud-Montalvá, C., Tinti, F., and Corberán, J.M. (2020). A novel TRNSYS type of a coaxial borehole heat exchanger for both short and mid term simulations: B2G model. *Applied Thermal Engineering* 164, 114500.

Cheung, B.C., Carriveau, R., and Ting, D.S-K. (2014a). Multi-objective optimization of an underwater compressed air energy storage system using a genetic algorithm. *Energy* 74: 396–404.

Cheung, B.C., Carriveau, R., and Ting, D.S-K. (2014b). Parameters affecting scalable underwater compressed air energy storage. *Applied Energy* 134: 239–247.

Crotogino, F., Mohmeyer, K-U., and Scharf, R. (2001). Huntorf CAES: more than 20 years of successful operation. Paper presented at Solution Mining Research Institute meeting, Orlando, FL, 23–25 April.

Dai, C., Li, J., Shi, Y., et al. (2019). *An experiment on heat extraction from a deep geothermal well using a downhole coaxial open loop design. Applied Energy* 252: 113447.

Davis, L. and Schainker, R. (2006). Compressed air energy storage (CAES): Alabama Electric Cooperative Mcintosh plant – overview and operational history. *Paper presented at Electricity Storage Association Meeting: Energy Storage in Action*, Knoxville, TN.

Ebrahimi, M., Carriveau, R., Ting, D.S-K., and McGillis, A. (2019). Conventional and advanced exergy analysis of a grid connected under water compressed air energy storage facility. *Applied Energy* 242(15): 1198–1208.

Ebrahimi, M., Ting, D.S-K., Carriveau, R., et al. (2020). Optimization of a cavern-based CAES facility with an efficient adaptive genetic algorithm. Energy Storage, e205.

Gordon, D., Bolisetti, T., Ting, D.S-K., and Reitsma, S. (2018). Experimental and analytical investigation on pipe sizes for a coaxial borehole heat exchanger. *Renewable Energy* 115: 946–953.

Holmberg, H., Acuña, J., Naess, E., and Sønju, O.K. (2016). Thermal evaluation of coaxial deep borehole heat exchangers. *Renewable Energy* 97: 65–76.

Hu, X., Banks, J., Wu, L., and Liu, W.V. (2020). Numerical modeling of a coaxial borehole heat exchanger to exploit geothermal energy from abandoned petroleum wells in Hinton, Alberta. *Renewable Energy* 148: 1110–1123.

Iry, S. and Rafee, R. (2019). Transient numerical simulation of the coaxial borehole heat exchanger with the different diameters ratio. *Geothermics* 77: 158–165.

Liu, J., Wang, F., Cai, W., et al. (2019). Numerical study on the effects of design parameters on the heat transfer performance of coaxial deep borehole heat exchanger. *International Journal of Energy Research* 43(12): 6337–6352.

Luo, Y., Guo, H., Meggers, F., and Zhang, L. (2019). Deep coaxial borehole heat exchanger: analytical modeling and thermal analysis. *Energy* 185: 1298–1313.

Meerman, R. and Brown, A.J. (2014). *When somebody loses weight, where does the fat go? The British Medical Journal 349, g7257*, 16 December.

Ocko, S.A., Heyde, A., and Mahadevan, L. (2019). Morphogenesis of termite mounds. *Proceedings of the National Academy of Sciences of the U.S.A.* 116(9): 3379–3384.

Pan, A., Lu, L., Cui, P., and Jia, L. (2019). A new analytical heat transfer model for deep borehole heat exchangers with coaxial tubes. *International Journal of Heat and Mass Transfer* 141: 1056–1065.

Radziemska, E. (2003). The effect of temperature on the power drop in crystalline silicon solar cells. *Renewable Energy* 28(1): 1–12.

Schmidt-Nielsen, K., Schmidt-Neilsen, B., Jarnum, S.A., and Houpt, T.R. (1956). Body temperature of the camel and its relation to water economy. *American Journal of Physiology* 188(1): 103–112.

Wang, Z., Xiong, W., Carriveau, R., et al. (2019). Energy, exergy and sensitivity analyses of underwater compressed air energy storage in an island energy system. *International Journal of Energy Research* 43(3): 2241–2260.

Wang, Z., Xiong, W., Ting, D.S-K., et al. (2016). Conventional and advanced exergy analyses of an underwater compressed air energy storage system. *Applied Energy* 180: 810–822.

Wu, H., Ting, D.S-K., and Ray, S. (2017).An experimental study of turbulent flow behind a delta winglet. Experimental Thermal and Fluid Science 88: 46–54.

Wu, H., Ting, D.S-K., and Ray, S. (2018). Flow over a flat surface behind delta winglets of varying aspect ratios. *Experimental Thermal and Fluid Science* 94: 99–108.

Yang, Y., Ting, D.S-K., and Ray, S. (2019). Convective heat transfer enhancement downstream of a flexible strip normal to the freestream. International Journal of Thermal Sciences 145: 106059-1–11.

Yang, Y., Ting, D.S-K., and Ray, S. (2020). On flexible rectangular strip height on flat plate heat convection. *International Heat and Mass Transfer* 150: 119269.

Zhang, Y., Yu, C., Li, G., et al. (2019). *Performance analysis of a downhole coaxial heat exchanger geothermal system with various working fluids. Applied Thermal Engineering* 163: 114317.

Index

Page numbers followed by *f* and *t* refer to figures and tables, respectively.

d

e

i

n

o

p

U